Gravity Sanitary Sewer Design and Construction

Second Edition

Prepared by a Joint Task Force of the
Environmental and Water Resources Institute and the
Pipeline Division Committee on Pipeline Planning of the
American Society of Civil Engineers
and the
Collection Systems Subcommittee of the
Technical Practice Committee of the
Water Environment Federation

Edited by Paul Bizier

Library of Congress Cataloging-in-Publication Data

Gravity sanitary sewer design and construction : ASCE manuals and reports on engineering practice no. 60 wef manual of practice no. fd-5/ Prepared by the Joint Task Force on Sanitary Sewers of the American Society of Civil Engineers and the Water Environment Federation

p. cm — (Wef manual ; no. 60)

Includes bibliographical references and index.

ISBN 13: 978-0-7844-0900-8

ISBN 10: 0-7844-0900-5

1. Sewerage—Design and construction. I. Bizier, Paul. II. American Society of Civil Engineers. III. Water Environment Federation.

TD678.G715 2007

628′.2—dc22

2007015717

Published by American Society of Civil Engineers
1801 Alexander Bell Drive
Reston, Virginia 20191

www.pubs.asce.org

ISBN 978-0-7844-0900-8

ISBN 978-1-57278-240-2

Manufactured in the United States of America.

American Society of Civil Engineers/ Environmental and Water Resources Institute

Founded in 1852, the American Society of Civil Engineers (ASCE) represents more than 140,000 members of the civil engineering profession worldwide, and is America's oldest national engineering society. Created in 1999, the Environmental and Water Resources Institute (EWRI) is an Institute of ASCE. EWRI services are designed to complement ASCE's traditional civil engineering base and to attract new categories of members (non-civil engineer allied professionals) who seek to enhance their professional and technical development.

For information on membership, publications, and conferences, contact

ASCE/EWRI
1801 Alexander Bell Drive
Reston, VA 20191-4400
(703) 295-6000
http://www.asce.org

Water Environment Federation

Formed in 1928, the Water Environment Federation (WEF) is a not-for-profit technical and educational organization with 32,000 individual members and 80 affiliated Member Associations representing an additional 50,000 water quality professionals throughout the world. WEF and its member associations proudly work to achieve our mission of preserving and enhancing the global water environment.

For information on membership, publications, and conferences, contact

Water Environment Federation
601 Wythe Street
Alexandria, VA 22314-1994 USA
(703) 684-2400
http://www.wef.org

MANUALS AND REPORTS ON ENGINEERING PRACTICE

(As developed by the ASCE Technical Procedures Committee, July 1930, and revised March 1935, February 1962, and April 1982)

A manual or report in this series consists of an orderly presentation of facts on a particular subject, supplemented by an analysis of limitations and applications of these facts. It contains information useful to the average engineer in his or her everyday work, rather than findings that may be useful only occasionally or rarely. It is not in any sense a "standard," however; nor is it so elementary or so conclusive as to provide a "rule of thumb" for nonengineers.

Furthermore, material in this series, in distinction from a paper (which expresses only one person's observations or opinions), is the work of a committee or group selected to assemble and express information on a specific topic. As often as practicable, the committee is under the direction of one or more of the Technical Divisions and Councils, and the product evolved has been subjected to review by the Executive Committee of the Division or Council. As a step in the process of this review, proposed manuscripts are often brought before the members of the Technical Divisions and Councils for comment, which may serve as the basis for improvement. When published, each work shows the names of the committees by which it was compiled and indicates clearly the several processes through which it has passed in review, in order that its merit may be definitely understood.

In February 1962 (and revised in April 1982) the Board of Direction voted to establish a series entitled "Manuals and Reports on Engineering Practice," to include the Manuals published and authorized to date, future Manuals of Professional Practice, and Reports on Engineering Practice. All such Manual or Report material of the Society would have been refereed in a manner approved by the Board Committee on Publications and would be bound, with applicable discussion, in books similar to past Manuals. Numbering would be consecutive and would be a continuation of present Manual numbers. In some cases of reports of joint committees, bypassing of Journal publications may be authorized.

MANUALS AND REPORTS ON ENGINEERING PRACTICE

No.	Title
13	Filtering Materials for Sewage Treatment Plants
14	Accommodation of Utility Plant Within the Rights-of-Way of Urban Streets and Highways
35	A List of Translations of Foreign Literature on Hydraulics
40	Ground Water Management
41	Plastic Design in Steel: A Guide and Commentary
45	How to Work Effectively with Consulting Engineers
46	Pipeline Route Selection for Rural and Cross-Country Pipelines
47	Selected Abstracts on Structural Applications of Plastics
49	Urban Planning Guide
50	Planning and Design Guidelines for Small Craft Harbors
51	Survey of Current Structural Research
52	Guide for the Design of Steel Transmission Towers
53	Criteria for Maintenance of Multilane Highways
54	Sedimentation Engineering
55	Guide to Employment Conditions for Civil Engineers
57	Management, Operation and Maintenance of Irrigation and Drainage Systems
59	Computer Pricing Practices
60	Gravity Sanitary Sewer Design and Construction (Second Edition)
62	Existing Sewer Evaluation and Rehabilitation
63	Structural Plastics Design Manual
64	Manual on Engineering Surveying
65	Construction Cost Control
66	Structural Plastics Selection Manual
67	Wind Tunnel Studies of Buildings and Structures
68	Aeration: A Wastewater Treatment Process
69	Sulfide in Wastewater Collection and Treatment Systems
70	Evapotranspiration and Irrigation Water Requirements
71	Agricultural Salinity Assessment and Management
72	Design of Steel Transmission Pole Structures
73	Quality in the Constructed Project: A Guide for Owners, Designers, and Constructors
74	Guidelines for Electrical Transmission Line Structural Loading
76	Design of Municipal Wastewater Treatment Plants
77	Design and Construction of Urban Stormwater Management Systems
78	Structural Fire Protection
79	Steel Penstocks
80	Ship Channel Design
81	Guidelines for Cloud Seeding to Augment Precipitation
82	Odor Control in Wastewater Treatment Plants
83	Environmental Site Investigation
84	Mechanical Connections in Wood Structures
85	Quality of Ground Water
86	Operation and Maintenance of Ground Water Facilities
87	Urban Runoff Quality Manual
88	Management of Water Treatment Plant Residuals
89	Pipeline Crossings
90	Guide to Structural Optimization
91	Design of Guyed Electrical Transmission Structures
92	Manhole Inspection and Rehabilitation
93	Crane Safety on Construction Sites
94	Inland Navigation: Locks, Dams, and Channels
95	Urban Subsurface Drainage
96	Guide to Improved Earthquake Performance of Electric Power Systems
97	Hydraulic Modeling: Concepts and Practice
98	Conveyance of Residuals from Water and Wastewater Treatment
99	Environmental Site Characterization and Remediation Design Guidance
100	Groundwater Contamination by Organic Pollutants: Analysis and Remediation
101	Underwater Investigations
102	Design Guide for FRP Composite Connections
103	Guide to Hiring and Retaining Great Civil Engineers
104	Recommended Practice for Fiber-Reinforced Polymer Products for Overhead Utility Line Structures
105	Animal Waste Containment in Lagoons
106	Horizontal Auger Boring Projects
107	Ship Channel Design
108	Pipeline Design for Installation by Horizontal Directional Drilling
109	Biological Nutrient Removal (BNR) Operation in Wastewater Treatment Plants
110	Sedimentaion Engineering: Processes, Measurments, Modeling, and Practice
111	Reliability-Based Design of Utility Pole Structures
112	Pipe Bursting Projects
113	Substation Structure Design Guide

Manuals of Practice of the Water Environment Federation

The WEF Technical Practice Committee (formerly the Committee on Sewage and Industrial Wastes Practice of the Federation of Sewage and Industrial Wastes Associations) was created by the Federation Board of Control on October 11, 1941. The primary function of the Committee is to originate and produce, through appropriate subcommittees, special publications dealing with technical aspects of the broad interests of the Federation. These publications are intended to provide background information through a review of technical practices and detailed procedures that research and experience have shown to be functional and practical.

Water Environment Federation Technical Practice Committee Control Group

ABSTRACT

This Manual provides both theoretical and practical guidelines for the design and construction of gravity sanitary sewers.

The initial chapter introduces the organization and administrative phases of the sanitary sewer project. Subsequent chapters are presented in a sequence detailing the parameters necessary to establish the design criteria, complete the design, and award a construction contract. The Manual concludes with a discussion of the commonly used trenchless and conventional methods of sanitary sewer construction.

This Manual is intended to be of practical use to the designer of a gravity sanitary sewer system and is based upon the experience of engineers in the field of sanitary sewer structural and hydraulic design. Charts, illustrations, and example problem solutions are used liberally throughout to reinforce the text.

Joint Task Force on Sanitary Sewers

Richard Thomasson, Chairman, Chapter 6
Dennis Doherty, Vice-Chair, Chapter 12*
Paul Bizier, EWRI Liason, Chapters 6, 9
Marsha Slaughter, Secretary
Matt Cassel, Chapter 8*
Rao Chitikela
John Christopher, Chapter 8*
Jacques Delleur, Chapter 5
John Duffy
Angie Essner, Chapter 7*
Mike Glasgow
James Joyce, Chapter 4*
Karen Karvazy, Chapter 10*
Lisa Lassi, Chapter 10*
LaVere Merritt, Chapters 3, 5*
Terry Moy
Mike Murphy
Mohammad Najafi, Chapter 12*
Aaron Nelson, Chapters 2, 7*
Paul Passaro, Chapter 1*
Morris Sade
Sam Samandi, Chapter 9
Howard Selznick, Chapter 3*
Mark M. Smith, Chapter 4
Carl Sutter, Chapter 10*
John Trypus, Chapter 1*
Michael VanDine

In addition to the Task Force, Blue Ribbon Panel reviewers included:

Wayne Dillard, Burns & McDonnell
Terry Walsh, Greeley and Hanson
Heidi Dexheimer, G. C. Wallace, Inc.

Staff assistance was provided by Lorna Ernst for the Water Environment Federation and Suzanne Coladonato for the American Society of Civil Engineers.

*principal contributing author.

CONTENTS

FOREWORD

In 1960, a joint committee of the Water Pollution Control Federation (WPCF) and the American Society of Civil Engineers (ASCE) published the Manual of Practice on the Design and Construction of Sanitary and Storm Sewers. In 1964, a second joint committee was formed to revise and expand the Manual; in 1969, the revised edition was published. In subsequent reprintings, the 1969 edition of the Manual was continuously revised to provide information on improved and more current practices.

In 1978, the WPCF authorized preparation of this Manual of Practice devoted to gravity sanitary sewers. In 1979, ASCE entered into an agreement with WPCF to continue their joint publication relationship. Since that time, the Water Environment Federation (WEF, formerly the WPCF) and the Environmental and Water Resources Institute (EWRI) of ASCE have continued to work together on joint publications. As a result, a joint committee of the Water Pollution Control Committee of EWRI, the Pipeline Division of ASCE, and the Collection Systems Subcommittee of WEF's Technical Practice Committee was formed in 2004 to update this Manual.

This Manual should be considered by the practicing engineer as an aid and a checklist of items to be considered in a gravity sanitary sewer project, as represented by acceptable current procedures. It is not intended to be a substitute for engineering experience and judgment, or a treatise replacing standard texts and reference material.

In common with other manuals prepared on special phases of engineering, this Manual recognizes that this field of engineering is constantly progressing with new ideas, materials, and methods coming into use. Other alternatives available to the designer of sanitary sewers include vacuum, pressure, vacuum-pressure, and small-diameter gravity sewers. It is hoped that users will present any suggestions for improvement to the Technical Practice Committee of WEF, to EWRI, and to the Pipeline Division of ASCE for possible inclusion in future revisions to keep this Manual current.

The members of the Committee thank the reviewers of this Manual for their assistance in submitting their suggestions for improvement.

CHAPTER 1

ORGANIZATION AND ADMINISTRATION OF SANITARY SEWER PROJECTS

1.1. INTRODUCTION

Wastewater treatment and collection systems are a major expenditure of public funds, but the wastewater system's function is rarely acknowledged and sewers are seldom seen by the public. Sanitary systems are essential to protecting the public health and welfare in all areas of concentrated population and development. Every community produces wastewater of domestic, commercial, and industrial origin. Sanitary sewers perform the vitally needed functions of collecting these wastewaters and conveying them to points of treatment and disposal.

The various stages of design and construction of sanitary sewer projects require an understanding of the objectives of each stage of the project and of the responsibilities and interests of the parties involved.

Separate sanitary and storm sewers are highly desirable and are used, with few exceptions, in new systems. The major advantages of separate systems, including wastewater treatment plants, are the protection of watercourses from pollution and the exclusion of stormwater from the treatment system with a consequent saving in treatment plant construction and operating cost. Combined sewers are frequently encountered in older communities where it may be extremely difficult or costly to provide separate systems. Separation is desired, where economically feasible, to reduce the magnitude of facilities and energy demand of treatment works.

Municipal sewer systems have been a major topic of regulatory interest since the late 1990s with the introduction of the Sanitary Sewer Overflow Rule and changes in financial accounting and reporting standards for state and local governments. Under the provisions of the Clean Water Act (CWA) of 1972 as amended, owners and operators of municipal wastewater sewer systems are prohibited to release non-permitted discharges into receiving waters of the United States. A National Pollutant Discharge Elimination System (NPDES) permit must be issued to include the allowable discharge points for treated wastewater, combined sewer overflows, and emergency relief overflows. However, Section 301(a) of the CWA does not include sufficient information to indicate if and when a sanitary sewer overflow (SSO) might be permitted. The Environmental Protection Agency (EPA) believes that the number of SSOs can be significantly reduced through improved sewer system capacity, management, operation, and maintenance (CMOM). To address this important issue, the EPA issued a Notice of Proposed Rulemaking on January 4, 2001 to introduce the SSO Rule and CMOM Regulations. The proposed SSO Rule and CMOM regulations would expand the NPDES permit requirements for municipal sanitary sewer collection systems and clarify sanitary sewer overflow prohibitions.

The proposed SSO Rule is intended to clarify the prohibition on SSO discharges into the waters of the United States outlined in Section 301(a) of the Clean Water Act. In June, 1999, the Government Accounting Standards Board (GASB) issued Statement 34 (GASB 34) which changed the financial accounting and reporting standards for state and local governments. Of important relevance, the new standard requires that municipalities record and report depreciation on their wastewater infrastructure assets. Municipalities are required to use an asset management system that maintains a current inventory of their wastewater and associated sewer system infrastructure.

1.2. DEFINITION OF TERMS AND CLASSIFICATION OF SANITARY SEWERS

The following terms as used in this Manual are defined in the *Glossary—Water and Wastewater Control Engineering* (APHA 1981) as follows:

Building Drain: In plumbing, that part of the lowest horizontal piping within a building that conducts water, wastewater, or stormwater to a building sewer.

Building Sewer: In plumbing, the extension from the building drain to the public sewer or other place of disposal. Also called **House Connection.**

Combined Sewer: A sewer intended to receive both wastewater and storm or surface water.

Force Main: Liquid conveyance under pressure to raise liquid to the desired elevation.

Intercepting Sewer: A sewer that receives dry-weather flow from a number of transverse sewers or outlets and, frequently, additional predetermined quantities of stormwater (if from a combined system), and conducts such waters to a point for treatment or disposal.

Lateral Sewer: A sewer that discharges into a branch or other sewer and has no other common sewer tributary to it.

Main Sewer: (1) In larger systems, the principal sewer to which branch sewers and submains are tributary; also called **Trunk Sewer.** In small systems, a sewer to which one or more branch sewers are tributary. (2) In plumbing, the public sewer to which the house or building sewer is connected.

Relief Sewer: (1) A sewer built to carry flows in excess of the capacity of an existing sewer. (2) A sewer intended to carry a portion of the flow from a district that has insufficient sewer capacity to another sewer with available capacity to prevent overloading and possible overflows. Relief sewers may also be used to eliminate overflows to combined sewer outfalls when they are constructed as part of a treatment plant and expansion.

Sanitary Sewer: A sewer that carries liquid and waterborne wastes from residences, commercial buildings, industrial plants, and institutions, together with minor quantities of ground, storm, and surface waters that are not admitted intentionally. See also **Wastewater.**

Separate Sewer: A sewer intended to receive only wastewater, stormwater, or surface water. See also **Combined Sewer, Sanitary Sewer,** and **Storm Sewer.**

Separate Sewer System: A sewer system carrying sanitary wastewater and other waterborne wastes from residences, commercial buildings, industrial plants, and institutions, together with minor quantities of ground, storm, and surface waters that are not intentionally admitted. See also **Wastewater, Combined Sewer.**

Storm Sewer: A sewer that carries stormwater and surface water, street wash and other wash waters, or drainage, but excludes domestic wastewater and industrial wastes. Also called **Storm Drain.**

Outfall: (1) The point, location, or structure where wastewater or drainage discharges from a sewer, drain, or other conduit. (2) The conduit leading to the ultimate disposal area.

Outfall Sewer: A sewer that receives wastewater from a collecting system or from a treatment plant and carries it to a point of final discharge.

Trunk Sewer: A sewer that receives many tributary branches and serves a large territory.

Waste Water: In a legal sense, water that is not needed or that has been used and is permitted to escape, or which unavoidably escapes from ditches, canals or other conduits, or reservoirs of the lawful owners of such structures. See also **Wastewater.**

Wastewater: The spent or used water of a community or industry, which contains dissolved and suspended matter.

1.3. PHASES OF PROJECT DEVELOPMENT

Conception and development of typical sanitary sewer projects comprise the following phases:

1.3.1. Preliminary or Investigative Phase

The objective of this phase is to establish the broad technical and economic bases for environmental assessments, policy decisions, and final designs. The importance of this phase cannot be overemphasized. Inadequate preliminary work will be detrimental to all succeeding phases and may endanger the successful completion of the project or cause the owner to undertake planning which may not produce the most economical or efficient result. This phase usually culminates in an engineering report, which includes items such as:

- Statement of the problem and review of existing conditions.
- Capacities and conditions required to provide service for design period.
- Method of achieving the required service—if more than one method is available, an evaluation of each alternative method.
- General layouts of the proposed system with indication of stages of development to meet the ultimate condition when the project warrants stage development.
- Establishment of applicable engineering criteria and preliminary sizing and design that will permit preparation of construction and operating cost estimates of sufficient accuracy to provide a firm basis for feasibility determination, financial planning, and consideration of alternative methods of solution.
- Various available methods of financing and their applicability to the project.
- An assessment of the anticipated environmental impacts of construction and review of the long-term effects on the environment. The environmental assessment, when required, must be sufficiently thorough and objective to enumerate the environmental consequences of the project. Measures to mitigate any negative impacts should be set forth. Projects that have community-wide consequences and that are supported by federal funds may be subjected to detailed evaluation through Federal Environmental Impact Statement procedures.
- Operation and maintenance factors related to confined space entry requirements, as regulated by the Occupational Safety and Health Administration (OSHA), should be considered during the preliminary engineering phase. Design consideration should be given by

the engineer to facilitate safe ingress and egress into manholes and sewers, platforms, safety ladders, etc. Manholes should be accessible for scheduled maintenance. The accessibility of back line systems is critical and easements are recommended for maintenance purposes.

- The selection of conveyance materials is critical. The corrosion induced by microbial action, low pH levels, and hydrogen sulfide gas must be considered when materials and linings are evaluated. Some materials will require protection from stray currents and the debilitating effect they will have on conveyance systems. Polyethylene wrapping, polyvinyl chloride (PVC) lining, and cathodic protection are standard responses to these issues.
- The construction of a sanitary sewer system will inevitably require that streets and arterial highways be excavated or trenchless methods be utilized (see Chapter 12). Further, wetlands may be affected and the system may require approvals from many jurisdictions, municipalities, counties, and state and federal agencies. This coordination and permitting in itself may weigh heavily on the construction methods utilized—the cost of traffic interruption, loss of revenue in commercial districts, and environmentally sensitive land disturbance are critical elements in the planning process and may require that the engineer and owner evaluate trenchless construction methods for new construction during this phase. The trenchless methods may include, among others, microtunneling, horizontal directional drilling, and pipe bursting.
- Analysis of risk associated with infrastructure failure and well-reasoned discussions on system component locations and materials selected.

It must be recognized that the preliminary engineering report is not a detailed working design or plan from which a sanitary sewer project can be constructed. Indeed, such detail is not necessary to meet the objectives of the preliminary or investigative phase or the environmental assessment. Nonetheless, proper preliminary engineering is the fundamental initial step of final planning. (For additional information concerning the Survey and Investigation phases of sanitary sewer project development, see Chapter 2.)

1.3.2. Design Phase

The design phase of a sanitary sewer project comprises the preparation of construction plans and specifications. These documents form the basis for bidding and performance of the work; they must be clear and concise. Design therefore consists of the elaboration of the preliminary plan to include all details necessary to construct the project.

1.3.3. Construction Phase

This phase involves the actual building of the project according to the plans and specifications previously prepared.

1.3.4. Operation

Although this Manual is devoted to matters of design and construction, the efficient operation of a sanitary sewer system is an important element to consider during the development of such projects.

1.4. INTERRELATIONS OF PROJECT DEVELOPMENT PHASES

Since all phases of sanitary sewer projects are interrelated, the following points are applicable:

1. The capacity, arrangement, and details of a sanitary sewer system will not be satisfactory unless the preliminary or investigative phase is current and properly completed.
2. Adequate preliminary engineering and estimating are essential to sound financial planning, without which subsequent phases of the project may be placed in jeopardy.
3. Environmental assessment documentation is intended to provide a single source of comparison and evaluation of all development, construction, and operation phases of the project.
4. Inadequate design or improperly prepared plans and specifications can lead to confusion in construction, higher costs, failure of the project to meet intended functions, or actual structural or hydraulic failure of component parts.
5. Proper execution of the construction phase is necessary to produce the quality and features intended by adequate design. Moreover, the value of the design can be lost by incompetent or careless handling of the construction phase.
6. All sanitary sewer projects have certain features requiring operation and maintenance. Unless they are anticipated and provided for, the usefulness of the project will be impaired.
7. The cost estimates for these sometimes expensive capital projects require updating based on elapsed time and modifications to the project. Often, changes are made that may have a significant impact on the overall final cost. The updating process will ensure that all parties are aware of the financial impact of the project as the design process evolves from the initial preliminary design to the final biddable documents.
8. The impact of the combination of limited space and major renovation to existing systems will add cost for, among others, bypass pumping and inherent delays due to the need for sequential construction. Cost

estimates must factor in these hidden charges to the project. The soft costs of contract bonds and insurance may also be significant as the project construction costs increase.

The disruption to infrastructure, streets, utilities, and sediment and erosion control requires the coordination of the project with other impacted utilities and regulators having jurisdiction within the public right of way (ROW). This analysis and coordination may have a significant impact on route selection.

1.5. PARTIES INVOLVED IN DESIGN AND CONSTRUCTION OF SANITARY SEWER PROJECTS

Engineering projects, including sanitary sewers, are the result of the combined efforts of several interested parties. The owner, engineer, and contractor are the principal participants. Legal counsel, financial consultant, various regulatory agencies, and other specialists also are involved to varying degrees. Responsibilities of these individuals or organizations are summarized as follows.

1.5.1. Owner

The owner's needs initiate the project and he provides the necessary funds. The owner is party to all contracts for services and construction and may act directly or through any duly authorized agent. The owner most often is the collective citizens of a governmental unit whose affairs may be handled by various legislative and administrative bodies. The owner may also be a private group.

When the owner is a governmental unit, its business may be conducted by one of the following, depending on the organization of the unit and the laws controlling its operations:

- City councils or similar bodies, carrying out sanitary sewer projects as only one of many duties for the given unit.
- A special commission or board of a governmental unit dealing with more limited areas of interest than are usually are handled by a city council. Such boards or commissions may be concerned with sanitary sewer projects alone or with a governmental unit's general utility system. The geographical limits of responsibility of such boards or commissions usually coincide with those of the parent governmental unit.
- A legislatively established district, agency, or authority with unique geographical limits and whose affairs are administered by a separate and distinct administrative board or commission. Such units commonly are referred to as districts or authorities—for example,

the County Sanitation Districts of Los Angeles County, which includes 74 separate cities and the unincorporated county. Often, the responsibilities of such sewering districts are limited to main trunk sanitary sewers, intercepting sanitary sewers, treatment facilities, and outfalls, leaving lateral sewers as the responsibility of the individual governmental units within the area served by the larger district.

The fund-raising powers of the first two bodies are usually regulated by the same laws, which apply to financing by the parent governmental unit. Fund-raising powers for a specially constituted district may show considerable variation due to the many differences in legislative provisions for special district formation and financing.

Temporary private ownership of sanitary sewer projects is sometimes encountered in new developments. The developer may construct the sanitary sewer system and later transfer title to the appropriate governmental unit in accordance with local regulations. In some instances, wastewater systems and treatment works are under the permanent ownership of private utility companies. The design and construction of these private systems require that the agency that will or may assume authority maintain a presence during design and construction process.

1.5.2. Engineer

The engineer has the responsibility of supplying the owner with the basic information needed to make project implementation policy decisions, detailed plans and specifications necessary to bid and construct the project, consultation, general and resident inspection during construction, and services necessary for the owner to establish satisfactory operation and maintenance procedures. The engineer's responsibilities are all of a professional character and must be discharged in accordance with ethical standards by qualified engineering personnel.

Engineering for sanitary sewer projects may be performed either by engineering departments which are a part of a governmental unit, or by private engineering firms retained by the owner for specific projects. In many instances, sanitary sewer projects are a joint effort by both types of organizations.

1.5.3. Contractor

The contractor performs the actual construction work under the terms of the contract documents prepared by the engineer. The construction agreement is between the contractor and the owner. One or more contractors may perform the work on a single project.

The functions of the contractor may be carried out by an owner's employees especially organized for construction purposes, but such practice for sanitary sewer projects of any magnitude is not common.

The complexity of many sanitary projects may require that the general contractor have overall responsibility for associated subcontractors, and an independent construction management firm to represent the owner.

1.5.4. Other Parties

Many other parties may be involved at various stages of a project. Some of these are:

- *Legal Counsel.* All public works projects are subject to local and state laws; competent legal advice is required to ensure compliance with these laws and the avoidance of setbacks because of legal defects in the project. Special legal counsel may also be required in connection with financing the project, particularly where a bond issue is involved.
- *Financial Consultant.* Advisory services with respect to project financing are often required and may be provided as a separate and specialized service. Such services are occasionally provided as part of a general financing agreement with a financing agency.
- *Regulatory Agencies.* The most frequently encountered regulatory body is the state Health Department, Department of Environmental Quality, or, in some states, a specially designated water pollution control agency which usually adopts minimum standards pertaining to features of design, plans, and specifications for sanitary sewer projects. Other regulatory bodies having jurisdiction may include agencies such as municipality or sanitary sewer districts; local, regional, or state planning commissions; federal agencies concerned with water pollution control; and federal or state agencies having functional control of navigable waters.
- *Pipe Manufacturers.* These manufacturers produce the specified pipe materials and often will embark on focused developmental research to prove or improve a specific pipe product.

1.6. ROLE OF PARTIES IN EACH PHASE

The roles of the owner, engineer, and contractor with respect to each other in the different phases of the project are distinct. Unauthorized assumption of roles and duties of one of the parties may result in delays, failures, and/or contractual controversies. The legal counsel of the owner usually will be responsible for all legal issues pertaining to funding and compliance with the local purchasing statutes.

1.6.1. Preliminary or Investigative Phase

The owner and the engineer are the principal parties involved in the preliminary phase of sanitary sewer projects. It must be recognized that all policy decisions relating to the project, arranging for financing, etc., rest solely in the hands of the owner.

1.6.2. Design Phase

The design phase, up to the time of soliciting and receiving construction bids, involves both the owner and the engineer. Designs prepared by the engineer are normally subject to the approval of the owner. The engineer may recognize preferences of the owner and be guided by these preferences when they are consistent with good engineering practice. The engineer must recognize and conform to the legal, procedural, and regulatory requirements governing each project.

1.6.3. Construction Phase

In his relationship with the contractor, the engineer must exercise authority on behalf of the owner. The engineer determines whether the work is substantially in accordance with the requirements of the contract documents. He must avoid direct supervision of the contractor's construction operations. If he does control or direct the acts of the contractor, he may become involved as a third party in any legal action brought against the contractor.

1.6.4. Operation Phase

Full information on the intended functioning of all parts of a project should be furnished to the owner by the engineer. The owner's staff must assume final responsibility for operation at the time the project or any part of it is completed and accepted by him. In some cases, the engineer, by special agreement with the owner, may provide advisory services in connection with operation and maintenance procedures for a period of time after initial operation.

1.7. CONTROL OF SANITARY SEWER SYSTEM USE

Of all public utilities, sanitary sewer systems are probably the most abused through misuse. This situation results from a misconception that a sanitary sewer can be used to carry away any unwanted substance or object that can be put into it. The absence of adequate regulations setting forth proper uses and limitations of the system, and the lack of enforcement of existing regulations by those responsible for operation of the system, tend to foster such a misconception. Abuse of the sanitary sewer system can result in extensive damage and can compound the problems of wastewater treatment. Without proper maintenance and control, a sanitary sewer system may become a hazard to public safety and may increase operating costs unnecessarily.

The following are common consequences of sanitary sewer system misuse:

- Explosion and fire hazards resulting from discharge of explosive or flammable substances into the sanitary sewer.
- Sanitary sewer clogging by accumulations of grease, bed load, and miscellaneous debris.
- Physical damage to sanitary sewer systems resulting from discharge of corrosive or abrasive wastes.
- Surface and groundwater overload resulting from improper connections to sanitary sewers.
- Watercourse pollution resulting from discharge of wastewater to storm sewers.
- Interference with wastewater treatment resulting from extreme wet-weather flows or from wastes not amenable to normal treatment processes.
- The overloading effect of rainfall-induced inflow and infiltration must be evaluated and action taken to reduce or eliminate these elements to ensure the overall effective operation of a wastewater collection and treatment system. The presence of high rates of infiltration and the negative impact of same to the treatment process are not trivial. The cost of treating this non-wastewater flow, in metered communities, presents a tangible annual cost that can be used for the determination of current dollars that can be expended to correct these problems.
- The proposed SSO Rule and CMOM regulations, if enacted, will have a significant impact on the legal responsibility of the owner of the large or small sanitary sewer collection system to ensure that their systems function efficiently. In addition, the fixed asset reporting requirements of GASB 34 will also require that these critical assets are valued and enhanced on a continuing basis. Financial institutions will review the value of depreciated infrastructure to assess the bond rating of a borrowing agency or utility.

In the organization of sanitary sewer projects, as well as in the management of completed systems, provision must be made for the controlled use of the sanitary sewers and enforcement of appropriate regulations. A comprehensive report on sanitary sewer ordinances can be obtained from the Water Pollution Control Federation (WPCF 1975). Their publications may be helpful in determining the adequacy of existing regulations or preparing new ones.

1.8. FEDERAL AND STATE PLANNING AND FUNDING ASSISTANCE

1.8.1. Federal Assistance

Federal programs to provide financial assistance for the construction of publicly owned wastewater treatment and transport systems began in the middle 1950s and have grown to major significance. In many states, federal financial assistance programs for qualified wastewater systems are supplemented by state grant and loan programs. The engineer is frequently responsible for locating such sources of financial assistance and preparing the necessary applications for the owner.

The Water Pollution Control Act of 1972, amended as the Clean Water Act of 1977 and 1981, was the most significant source of federal assistance. This law initially required state water pollution control agencies and certain sub-state agencies to conduct planning for water quality management. Currently, funding for wastewater infrastructure is in the form of low-interest loans issued by the states. These loans are repaid and create a revolving account from which other projects can be funded. Availability of financial assistance for sanitary sewers varies from time to time and from state to state, since it is subject to the policies of the federal government and of individual states.

The second most significant source of federal assistance is the Farmers Home Administration (FmHA). Grants and loans for such systems are available primarily to small, rural communities. Other federal grants and loans are available from the Department of Housing and Urban Development (HUD), the Economic Development Administration (EDA), and a variety of regional agencies such as the Appalachian Regional Commission (ARC) and the Coastal Plains Regional Commission (CPRC). Contact should be made with state and federal elected officials to ascertain the availability and sources of funding for wastewater treatment projects.

Federal assistance programs have provided significant aid to local wastewater agencies, but they are often limited by federal appropriations and generally have substantial requirements and administrative procedures which must be followed. A review of current funding allocations and regulations is necessary to determine availability of federal funds at any given time.

1.8.2. State Assistance

A majority of the 50 states operate programs to assist local governments in the planning and financing of wastewater projects, including sanitary sewers. Many states have area planning and development commissions which prepare or assist in the preparation of planning documents for sanitary sewers for communities within their jurisdiction. States operate grant and loan programs ranging from the very modest up to 100% of the cost of wastewater projects when supplemented by various federal assistance programs. Such programs are usually administered by the state water pollution control agencies or health departments. The current status of state and local programs must be reviewed as a part of project design activities. In order to distribute scarce resources more efficiently, some jurisdictions issue low-interest loans to fund many sewer system projects and to provide a revolving funding source.

1.9. LOCAL FUNDING

Local funding methods include the following:

1.9.1. General Obligation Bonds

These bonds are backed by the full faith and credit of the issuer and often are paid for by the levy of general property taxes. In addition to or instead of taxes, revenues from service charges or other sources may be used to meet bond payments. Advantages of these bonds over other bonds include lower interest rates because of the substantial security provided and the ease with which they may be sold. However, in most states the issuance of general obligation bonds by local government is subject to constitutional and statutory limitations and to voter approval.

1.9.2. Special Assessment Bonds

Such bonds are payable from the receipts of special benefits assessments against certain properties or recipients of benefit. Assessments that are not paid become a lien on the property. Assessments may be based on front footage, area of parcels of property, or other bases. In most cases, the prior consent of landowners representing some statutory percentage of the total property to be assessed is required to implement special assessments.

1.9.3. Revenue Bonds

These bonds are payable from charges made for services provided. Such bonds have advantages when agencies lack other means of raising capital; they can be used to finance projects which extend beyond normal agency boundaries. The success of revenue bonds depends upon economic justification for the project, reputation of the agency, methods of billing and collection, rate structures, provisions for rate increases, financial management policy, reserve funds, and forecast of net revenues.

1.9.4. Pay-As-You-Go Financing

This method entails gathering sufficient funding prior to and during construction. Funds may be gathered through a system of increased user charges and/or connection fees. Advantages of this method are elimination of interest cost and voter authorization. The principal disadvantage is significantly higher charges during the period when funds are being collected.

1.9.5. Revenue Programs and Rate Setting

Various methods are in use to obtain the revenue needed to operate, maintain, replace, extend, and enlarge wastewater systems. The principal methods include the levy of ad valorem taxes, service charges, connection charges, and combinations of these. Each agency's revenue program and rate settings reflect the needs of an individual community and its local policy. No single revenue-generating program can be considered ideal for every situation.

Alternatives should be examined in terms of equity among user groups, promotion of water conservation, and implementation and updating requirements. Because the aim is to achieve fairness to all users and beneficiaries—domestic, commercial, and industrial—it becomes necessary to consider the question of fairness as a whole and not simply from the standpoint of any one class. The procedures for allocating user costs are easier to understand in specific examples than in the general terms considered here. Examples are contained in various publications prepared by ASCE, WEF, and the American Water Works Association (AWWA).

1.9.5.1. Ad Valorem Taxes

Taxes on real estate may be used as a primary revenue source for wastewater service in the event the agency has not accepted a federal grant and, therefore, is not subject to restrictions under the Clean Water Act or by taxing limitations. Real estate values and real estate taxes often reflect ability to pay, which may be a primary criterion in reducing hardship for some citizens. On the other hand, the value of real estate may have little relationship to the cost of service which is provided to individual users. There is often a preference for payment of service as a tax rather than as a service bill because the tax may be deductible from federal and state income taxes, whereas a service charge is not.

The strongest objection to ad valorem taxation as a method of cost recovery comes from those who believe that benefits are strictly proportional to quantity of wastewater or a combination of quantity and waste characteristics.

The simplicity of an ad valorem tax system is a marked advantage. It requires no wastewater measurement or sampling program for charge purposes. The required accounting and billing work is minimal. Clearly this procedure will require minimal overhead cost, but the lack of individual user accountability may not encourage conservation.

1.9.5.2. Service Charges

Service charges provide financial support of wastewater systems based on some measure of actual physical use of the system. Measures of use include volume of wastewater; volume of wastewater plus quantity of pollutant matter; number or size of sewer connections; type of property, such as residential, commercial, or manufacturing; number and type of plumbing fixtures, water-using devices, or rooms; uniform rates per connection; and percent of water charge. A recent development is that, in some jurisdictions, services changes are being applied by governing bodies to account for infrastructure improvements to mitigate environmental impacts.

1.9.5.3. Connection Charges

A one-time charge at the time a user connects to a wastewater system is used to generate revenue. Charges vary from a nominal inspection fee to a full, prorated share of the cost of the entire wastewater system. In some cases connection fees reflect costs to provide new capacity in the wastewater system. Connection fees are not generally used for operation and maintenance, but are used for purposes such as payment of debt service and financing wastewater system expansion.

1.9.5.4. Combination Taxes and Service Charges

Combination systems have often resulted from an original ad valorem tax system when additional sources of revenue were required and funds raised from user revenues were unpopular or were politically precluded from use to fund operation and maintenance costs. The Clean Water Act requires that grantees implement a system of service charges based on usage to pay for operation and maintenance costs.

The combination system can result in payment for operation of a wastewater treatment system by nonusers who do not benefit from these facilities. One can argue, however, that a wastewater collection system creates a safe environment for all residents and may be a charge on a *pro rata* basis to all residents.

This option should be evaluated carefully prior to implementation. Issues of current public policy in a region or geographic area relative to financing of a service which benefits a segment of a jurisdiction by general taxation is a slippery slope indeed.

1.10. SAFETY

A goal in sanitary sewer projects is to eliminate unsafe conditions and unsafe acts. To be effective, desire and enthusiasm for safety must be encouraged and supported at all levels of employment in the owner organization. Among the numerous laws, rules, and regulations which govern safety is the Occupational Safety and Health Act of 1970.

1.10.1. Investigations

- Investigations or surveys, such as an EPA Sanitary Sewer Evaluation Survey (SSES), should be conducted with safety as a paramount consideration in accordance with regulations for equipment, training, and size of sewer entry crew.
- The engineer must provide a work environment for his workers that recognizes hazards likely to cause death or serious physical harm. This can be accomplished with the utilization of proper safety equipment and procedures.

1.10.2. Design

- Safety factors are to be considered by the engineer for reducing the ultimate strength of a material to a working strength. The factor of safety will vary, depending on the type of material and its use. (A comprehensive discussion of design considerations is presented in Chapter 9.)
- The engineer must be familiar with safety practices and regulations such as OSHA standards and how their requirements may affect the design of the sanitary projects. Several of these design standards include safety landings in manholes and confined space procedures.
- Sanitary sewers should be separated from gas and water mains and other buried utilities. Many states have regulations that set minimum horizontal and vertical separation requirements.
- Ventilation should provide for air to enter the sanitary sewer system and provide for the escape of gases. Currently, the trend is to utilize

solid or watertight manhole covers in streets and water courses. With the decrease in ventilation through holes in manhole covers, inconspicuous vents or forced draft may be needed to provide adequate ventilation.

1.10.3. Construction

- The construction contract documents should require the contractor to adhere to all laws and regulations which bear on the project construction and to be responsible for safety at the construction site.
- The OSHA standards require the employer (contractor) to provide employees with a safe and healthful place of employment. Sections of the OSHA standards state that it shall be the responsibility of the employer to initiate and maintain such programs as may be necessary to provide safe conditions, and that frequent and regular inspections be made by competent persons designated by the employers.
- The engineer should specifically instruct his own field personnel to follow safety precautions while visiting the construction site. Field personnel should be issued hardhats and eye goggles or other appropriate protective equipment for use while visiting the site.
- The terms of the contract documents entered into for a construction project play an important role in determining whether or not an engineer has any duty in regard to the safety of the contractor's employees. Litigation by injured employees of contractors naming the engineer as a party to a suit often occurs. Unless the engineer has specifically accepted responsibility for construction site safety, the contract documents must provide that the contractor is responsible for construction site safety. The engineer must nevertheless take appropriate action if a potential safety hazard is observed.

1.10.4. Operation and Maintenance

- A safety program is necessary for the operation and maintenance of a sanitary sewer system. Safety is the responsibility of every individual, not only for personal protection but also for the protection of fellow employees.
- Safety equipment such as traffic control devices, safety harnesses, tools for manhole cover removal, gas detectors and blowers with duct discharge for positive displacement of manhole atmosphere, and rubberized cloth gloves must be available for protection where needed.

1.11. NATIONAL ENVIRONMENTAL POLICY ACT OF 1969

The National Environmental Policy Act of 1969 (NEPA) established that Congress is interested in restoring and maintaining environmental quality. It directs all federal agencies to identify and develop methods and procedures to ensure that environmental factors be given appropriate consideration in decision making along with economic and technical considerations. NEPA applies also to any project implemented by federal funds. A number of states have adopted legislation that parallels NEPA for state and local activities.

If the nature of the proposed project and its impact does not obviously reflect the need for the preparation of an Environmental Impact Statement (EIS), an initial Environmental Assessment (EA) reviews the anticipated effects of the proposed action. A conclusion is drawn as to whether the proposed action would significantly and adversely affect the environment and therefore require the preparation of an EIS. If there is no significant adverse impact on the environment, then the assessment normally serves as the basis for a negative declaration, or Finding of No Significant Impact (FONSI). Wastewater projects can affect the types and intensities of land use by providing facilities to accommodate new development. The location of new or expanded facilities affects the location of new development. With limited public funds, a decision to finance a specific facility limits funds for other facilities locally or in other areas and for similar or other purposes. A compatible relationship needs to exist between the potential growth-accommodating effects of facilities and the ability of a region to support additional growth.

The assessment should address relative impacts of population growth on natural resources such as air and water, and on visual conditions, community characteristics, archaeology, and other cultural values. It should clearly illustrate the range of impacts resulting from various growth levels, ranging from "no-growth" or "no project" to build-out.

Socioeconomic effects, employment, availability of housing, and the costs of providing projected populations with other public services and utilities should be assessed. Those long-term environmental goals which conflict with economic needs must also be identified in assessment of secondary impacts.

An EA is an environmental report prepared expressly to determine whether a proposed action requires the preparation of an environmental impact statement. The EA may be based on procedures such as checklists, matrices, networks, overlays, and specific studies. The environmental assessment should:

- Describe the proposed action.
- Describe the environment to be affected.

- Identify all relevant environmental impact areas.
- Evaluate the potential environmental impacts.
- Identify adverse impacts that cannot be avoided should the action be implemented.
- Identify irreversible and irretrievable commitments of resources.
- Discuss the relationship between local short-term uses of man's environment and long-term productivity.
- Identify conflicts with state, regional, or local plans and programs.
- Evaluate alternatives to the proposed action.
- Discuss any existing controversy regarding the action.

It is important to quantify impacts, where possible, to permit a clear picture of the issues for discussion and decision-making purposes. For projects involving minimal impacts or improvements to existing infrastructure, the governing body may determine that a categorical exclusion is appropriate, which has significantly fewer permitting requirements.

1.12. CAPACITY, MANAGEMENT, OPERATIONS, AND MAINTENANCE (CMOM)

The EPA, under the provisions of the Clean Water Act, is proposing to clarify and expand existing permit requirements for owners of municipal sanitary sewer collection systems. These changes, if enacted, will apply to virtually all municipal sanitary sewer collection systems and their satellite system contributors.

The objective is to promote good management, operation, and maintenance procedures that will result in reduced sanitary sewer overflows which can lead to environmental and public health risks, such as beach closings and contamination of ecological systems in the waterways.

The number of sanitary sewer overflows is estimated at approximately 40,000 per year. It is anticipated that by increasing the accountability for municipal sewer system owners and providing direction as proposed in the CMOM programs, these numbers will be reduced significantly to the betterment of the environment and public health. This particular rule addresses SSOs, not combined sewer overflows (CSOs). A brief description of the program elements follows:

- *Capacity Assurance, Management, Operation, and Maintenance Programs.* These programs will help communities ensure they have adequate wastewater collection and treatment capacity and incorporate many standard operation and maintenance activities for good system performance. When implemented, these programs will provide for efficient operation of sanitary sewer collection systems.

- *Notifying the Public and Health Authorities.* Municipalities and other local interests will establish a locally tailored program that notifies the public of overflows according to the risk associated with specific overflow events. The EPA is also proposing that annual summaries of sewer overflows be made available to the public. The proposal also clarifies existing recordkeeping requirements and requirements to report to the state.
- *Prohibition of Overflows.* The existing Clean Water Act prohibition of sanitary sewer overflows that discharge to surface waters is clarified to provide communities with limited protection from enforcement in cases where overflows are caused by factors beyond their reasonable control or severe natural conditions, provided there are no feasible alternatives.
- *Expanding Permit Coverage to Satellite Systems.* Satellite municipal collection systems are those collection systems where the owner or operator is different from the owner or operator of the treatment facility. Some 4,800 satellite collection systems will be required to obtain NPDES permit coverage to include the requirements under this proposal.

Additional information is obtainable on the EPA's web site, www.epa.gov.

1.12.1. Government Accounting Standards Board Statement 34 (GASB 34)

GASB 34 fundamentally requires each permitee to value its infrastructure assets. The objective of introducing this process is to ensure that the infrastructure of each satellite system is enhanced and maintained to ensure proper operation and to maintain its value on the balance sheet.

Clearly, the straight-line depreciation method will reduce its value to zero following the end of the period of useful life. This period may vary from jurisdiction to jurisdiction, based on the provisions of the local bond law. The terminal zero value at the conclusion of the depreciation period may not be applicable in the case of buried pipelines; these systems will always have some functional value. The methods used to determine that value include condition, serviceability, capital investment, and rehabilitation.

At some point, the annual expenditure required to respond to faulty collection systems will equal or exceed the debt service incurred to correct these deficiencies. The added benefit to the owner is the enhanced value of its sewer collection system and reduction or redirection of the operating budget allocated for this purpose.

1.13. MEASUREMENT UNITS

In the United States, sanitary engineering technology has been based on measurements expressed in the foot-pound-second system previously prevalent in most English-speaking countries (WPCF 1976). Metrication throughout the world is proceeding at varied paces; however, standardized measuring units of Le Système International d'Unités (SI) are becoming

TABLE 1-1. Applicable SI Base Units

Quantity	SI Unit	Symbol	Definition
Base Units:[a]			
Length	meter	m	The meter is the length equal to to 1,650,763.73 wavelengths in vacuum of the radiation corresponding to the transition between the levels $2p_{10}$ and $5d_5$ of the krypton-86 atom.
Mass	kilogram	kg	The kilogram is equal to the mass of the international prototype of the kilogram.
Time	second	s	The second is the duration of 9,192,631,770 periods of the radiation corresponding to the transition between the two hyperfine levels of the ground.
Supplementary Units:[b]			
Plane Angle	radian	rad	The radian is the unit of measure of a plane angle with its vertex at the center of a circle and subtended by an arc equal in length to the radius.
Solid Angle	steradian	sr	The steradian is the unit of measure of a solid angle with its vertex at the center of a sphere and enclosing an area of the spherical surface equal to that of a square with sides equal in length to the radius.

[a]The radian and steradian may be used advantageously in expressions for derived units to distinguish between quantities of a different nature but of the same dimension.
[b]In practice, the symbols rad and sr are used where appropriate, but the derived unit "1" is generally omitted.
Source: www.physics.nist.gov, accessed March 7, 2007.

TABLE 1-2. Applicable SI-Derived[a] Units Expressed in Terms of Base Units

Quantity	SI Unit	SI Symbol
Acceleration	Meters per second squared	m/s^2
Area	Square meters	m^2
Density	Kilograms per cubic meter	kg/m^3
Specific volume	Cubic meters per kilogram	m^3/kg
Velocity	Meters per second	m/s
Viscosity, kinematic	Square meters per second	m^2/s
Volume	Cubic meters	m^3

[a]Even though the selection of the meter-kilogram-second (mks) system yields coherent units more easily than those used, for example, in the centimeter-gram-second (cgs) system, not all of the units are convenient for all applications. Consequently, provision is made for multiples and submultiples of the base unit.

Source: www.physics.nist.gov, accessed March 7, 2007.

TABLE 1-3. Applicable SI-Derived Units with Special Names

Quantity	SI Unit	Symbol	Expression in Terms of Other Units	Expression in Terms of SI Base Units	Definition
Force	newton	N	—	$m \cdot kg \cdot s^{-2}$	The newton is that force which, when applied to a body having a mass of 1 kilogram, gives it an acceleration of 1 meter per second per second.
Pressure	pascal	Pa	N/m^2	$m^{-1} \cdot kg \cdot s^{-2}$	The pascal is the pressure or stress of 1 newton per square meter.

Source: www.physics.nist.gov, accessed March 7, 2007.

the rule rather than the exception. This Manual, where practical, presents both metric and conventional systems.

The SI makes use of only seven base units, which are divided into three classes:

1. *Base Units.* These are seven well-defined units that are dimensionally independent: meter (m), kilogram (kg), second (s), ampere (A), candela (cd), Kelvin (K), and mole (mol). Base unit definitions applicable to sewer design are given in Table 1-1.
2. *Derived Units.* These are expressed algebraically in terms of base units. Several derived units have been given special names and symbols that may themselves be used to express other derived units in a simpler way than in terms of base units. Some of the derived units applicable to sewer design are shown in Tables 1-2 and 1-3.
3. *Supplementary Units.* These are two units that are defined neither as base units nor as derived units. The units, radian (rad) and steradian (sr), are shown in Table 1-1.

REFERENCES

"Glossary." Water and wastewater engineering. (1981). APHA, AWWA, WPCF, ASCE, New York, N.Y.

Financing and charges for wastewater systems. (1981). A Joint Committee Report of APWA, ASCE, and WPCF, New York, N.Y.

Water Pollution Control Federation (WPCF). (1976). Units of expression for wastewater treatment. Manual of Practice No. 6, WPCF, Washington, D.C.

WPCF. (1975). Regulation of sewer use. Manual of Practice No. 3. WPCF, Washington, D.C.

CHAPTER 2
SURVEYS AND INVESTIGATIONS

2.1. INTRODUCTION

Surveys and investigations produce the basic data needed for the successful conception and/or development of a sanitary sewer design project. The fundamental importance of any survey or investigation requires that it be carried out competently and thoroughly if an effective project is to result.

The term "survey," as used in this Manual, refers to the process of collecting and compiling information necessary to develop any given phase of a project. In one sense, it may include observations relating to general conditions affecting a project, such as historical, political, physical, environmental, and fiscal matters. In another sense, a survey may comprise the precise instrument measurements necessary for the engineering design.

The term "investigation" is often used interchangeably with "survey." Its use in this Manual, however, usually refers to the assimilation and analysis of the data produced by surveys to arrive at policy and engineering decisions.

Surveys and investigations for the preliminary phase of a project are broad in nature, with emphasis on covering all factors relating to a project and determining the relative importance of each. The preliminary phase of a project may likely originate from a second or third party as a result of an SSES following the guidelines set forth in the ASCE Manual and Report on Engineering Practice No. 62 (ASCE 1994).

Surveys and investigations for design and construction phases are more precise and detailed, usually being limited by the scope of the project. However, the level of accuracy and the nature of survey and investigation work also varies according to the type of construction being incorporated and the general location. For example, there would be a significant difference between a project that is designed to rehabilitate an existing sewer system, one being designed to augment an existing system, and one that is being designed as a totally new system. Furthermore, there is a significant difference between systems located in urban and rural environments, so an engineer must keep all factors in mind when developing the scope of the survey and investigation phases.

Methods of conducting a survey vary widely, depending on the phase of development under consideration and the objectives. Proper surveys require broad knowledge of the particular field and an understanding of the problems to be solved in the phase of the project for which the survey is being conducted. Knowledge of the various aspects of sanitary sewer design set forth in other chapters of this Manual will lead to recognition of the specific information needed from a survey for any given project. The objectives of the survey for the several project phases and the type of information required for each phase are discussed in this chapter.

2.2. TYPES OF INFORMATION REQUIRED

Several different kinds of information, applicable in varying degrees to the different project phases, may be collected during the course of surveys for a typical project. These include:

2.2.1. Physical

- Topography, surface and subsurface conditions, details of paving and other surfaces to be disturbed, pertinent above- and underground utilities and structures, soil characteristics, water table elevations, and traffic control needs.
- Locations of streets, alleys, easements, and obstructions; required rights-of-way; and all similar data necessary to define the physical features of a proposed sanitary sewer project, including preliminary horizontal and vertical alignment.
- Details of the existing sanitary sewer system to which a proposed sanitary sewer may connect.
- Pertinent information relative to possible future extension of the proposed project by annexation or service agreements with adjacent communities or areas.
- Locations of historical and archaeological sites, and of people, plant, and animal communities and any other environmentally sensitive areas.

- Approvals (right-of-entry agreements) required to enter existing (system owner) easements or rights-of-way through or across property(s) owned by others.
- Access needs for possible construction equipment and future maintenance personnel and vehicles.

2.2.2. Developmental

- Existing populations and future population trends and density in the area to be served.
- Type of development (i.e., residential, commercial, or industrial).
- Historical and experience data relating to existing facilities which may affect proposed sanitary sewers.
- Comprehensive regional master plans of other agencies, especially the Ocean Water Act (as amended) Section 208, Areawide Plans, and Section 201, Facilities Plans.
- Location of future roads, airports, parks, industrial areas, etc., which may affect the routing and location of sewers.

2.2.3. Political

- Present political boundaries and probability of annexation of adjacent areas.
- Possible service agreements with adjacent communities and satellite systems; feasibility of multi-municipal or regional system.
- Existence and enforcement of industrial waste ordinances regarding pretreatment or limitations on the concentrations of damaging substances.
- Requirements for new waste ordinances to achieve desired results.
- Effectiveness and adequacy of present political subdivision to undertake the project; desirability of a new organization to sponsor the same.

2.2.4. Sanitary

- Quantity and strength of municipal wastewater to be transported.
- Water use data and flow gagings, where appropriate, to establish dry- and wet-weather flow rates from existing similar areas.
- Capacity and condition of existing sanitary sewer system.
- Other pertinent data necessary to establish the required design criteria and capacity for the given project, including infiltration/inflow requirements and ordinances.

2.2.5. Financial

- Information relative to existing authorization or policies, obligations, and commitments bearing on financing of the proposed sanitary sewer.
- Amounts, retirement schedule, and refinancing penalty of outstanding bonds and unobligated bonding capacity available for the proposed project.
- Availability of federal or state assistance through grants or loans.
- Taxable valuation, existing tax levies, and any limits affecting the proposed project.
- Schedule and methods of developing existing sanitary sewer service rates and revenues generated.
- Property plats as required for sanitary sewer assessments and special methods of assigning assessments.
- History of local construction factors and operating costs and conditions affecting cost.
- All similar data necessary to establish a feasible financing program for the proposed project.

2.3. SOURCES OF INFORMATION

As with all sources of information, the engineer should evaluate the relative accuracy of the information gathered during the surveys to determine its suitability to the design needs of the project. The sources of information, whether they be digital or hard copy, should be referenced for future records and conveyed to the system owner upon project completion. It is also important to note that resolved instances of conflicting data sets should be duly recorded and also reported to the system owner such that its records can be updated to reflect the true data and source. Possible sources of information sought by surveys for sanitary sewer projects include:

2.3.1. Physical

- Existing maps and sewer system plans, including United States Geological Survey (USGS) topographic maps, city plats and topographic maps, state highway plans and maps, tax maps, and local utility records and plans. Depending upon the system owner, much of the needed information may be contained in a Geographic Information System (GIS). However, the engineer should confirm the accuracy of the content.

- Aerial photographs.
- Instrument surveys, including approximate surveys by such devices as hand level, total station, and Global Positioning System (GPS), all of which may be useful for preliminary work.
- Photographs and/or videos of complex surface detail to supplement instrument surveys, and photographs and/or videos to show detail of existing sewer systems.
- Borings and test pits, either by hand or by machine, for determining subsurface soil and water conditions.
- Underground structure locations should be investigated to confirm the utility records. The engineer should follow CI/ASCE Standard 38-02 (ASCE 2002).
- Local 208 agency, historical preservation office, archaeological society, department of natural resources, soil conservation service, or local universities to identify any environmentally sensitive areas.

2.3.2. Developmental

- Census reports.
- Planning and zoning reports and maps.
- General field and/or aerial photo examination to note type, degree, and density of development.
- Criteria of regulatory agencies having jurisdiction over the project.
- Engineering reports or studies of related projects in the area.

2.3.3. Political

- Enabling or authorizing legislation.
- Municipal and state laws.
- Conferences with owner and other officials.
- Comprehensive plans established by planning agencies.
- Local and area meeting reports and minutes.

2.3.4. Sanitary

- Canvass of significant industries to determine type and amount of waste.
- Flow gagings and sampling in existing sanitary sewers to establish dry- and wet-weather flow characteristics from similar areas.
- Records of water pumpage and water sales.
- Design basis and operational characteristics of existing sanitary sewers from system records.

- Federal, state, and local requirements and ordinances for infiltration/inflow.
- Records of existing treatment plant influent characteristics.

2.3.5. Financial

- Pertinent records of the owner's fiscal officer.
- Auditor's or treasurer's records relating to tax levies.
- Operating statements and reports of income and expenses for the sanitary sewer, water, and other utility departments.
- Ordinances or laws and bond indenture governing outstanding bonds, and procedures for financing and contracting the proposed project.
- Assessment plats and schedules for prior projects to show methods in use in the locality.
- Tax maps showing subdivision and ownership of property to be affected by special assessments.

2.4. SURVEYS FOR DIFFERENT PROJECT PHASES

The objectives of the typical and differing phases of project development must be understood to ensure that the survey to be conducted is meaningful. Typical phases of project development are discussed in Chapter 1. A discussion of the objectives and the nature of surveys for each phase follows.

2.4.1. Preliminary Surveys

This phase of development is concerned with the broad aspects of the project, including required capacity, basic arrangement and size, probable cost, environmental assessment, and methods of financing. Accordingly, information is required in sufficient detail to show general physical features affecting the layout and general design. The scope as well as the feasibility and environmental aspects of the project development must be considered at this time.

When the project involves rehabilitation of an existing sewer system, the preliminary survey must contain horizontal and vertical locations of pipe inverts and manholes, date and material of construction, dry- and wet-weather flow rates, and other condition-related data. Much of this information may be collected during an SSES and/or may be located in the owner's asset management system or GIS. This should also include a Quality Level D or C utility location as defined by CI/ASCE Standard 38-02 (ASCE 2002).

When a connection to an existing sanitary sewer system is proposed, the preliminary survey must contain flow data for use in establishing the design capacity and sanitary sewer layout.

Extreme precision and detail are neither necessary nor desirable in this phase, but all data obtained must be reliable. The type and extent of the information needed for any given project will usually become apparent as the work progresses, and this may vary widely depending on the size and complexity of the project. Since preliminary surveys are used to develop the information upon which the estimates of the engineering report are based, these must be thoroughly and competently prosecuted. Sufficient allowance must be made for items affecting the total cost of the project, such as trends in construction costs, pavement removal and replacement practices, backfill methods, sheeting, dewatering, and unusual quantities of difficult excavation. Otherwise, costs will be underestimated.

Occasionally, photogrammetric methods may be advantageous in obtaining a portion of the data needed in making a preliminary survey.

2.4.2. Design Surveys

Surveys for this phase form the basis for engineering design as well as for the preparation of plans and specifications. Design surveys are concerned primarily with obtaining physical and sanitary wastewater flow data, rather than developmental, environmental, or financial information. In contrast with the preliminary surveys, design surveys must contain all the detail and precision the design engineer needs to correlate his design and the resulting construction plans with actual field conditions. Design surveys involve the use of surveying instruments in establishing the accurate location of pertinent topographic features. Photogrammetric methods also may be used in obtaining this information.

When the project involves rehabilitation of an existing sewer system, the preliminary survey shall follow the guidelines set forth in the ASCE Manual and Report on Engineering Practice No. 62 (WEP/ASCE 1994). During this phase it is important to quantify the number and location of system defects to be repaired, as well as to identify suitable forms of repair. This should be followed by a comparative hydraulic analysis of the system capacity demands with those of the rehabilitated system to confirm that adequate capacity is available after rehabilitation and future buildout (if applicable).

Presumably, the preliminary phase will have established the general extent of the project so that the area to be covered by the design survey can be defined. However, further definition of location within the general area may be required during the course of the design survey. The design survey also may extend to some degree beyond the proposed construction limits in order that possible future expansion may be facilitated.

It is obvious that accurate surveys are required to produce accurate designs and plans. Vertical control usually is established by setting benchmarks throughout the project, the elevations of which have been confirmed by level circuits to within 0.01 ft (3 mm). Although property and street lines are often used for horizontal control, control traverses or coordinate systems may be desirable. A special problem frequently encountered in design surveys is the nature and extent of existing underground utilities and structures which must be cleared or displaced by the new sanitary sewer. Such information, insofar as practical, must be obtained during the design survey to establish right-of-ways, minimize utility relocation costs, obtain lower construction bids, and prevent changing or realignment after the commencement of the project.

Where the accurate location of important substructures cannot be ascertained by other means, and conflict is possible, excavation to determine location, elevation, and detail at the point of crossing is warranted. Such substructure information may be determined by the owner, engineer, or contractor prior to sewer construction, with alignment or invert elevations revised accordingly. However, it should be noted that design projects set in an urban environment will be more likely to encounter various underground intangibles, such as abandoned or unrecorded utilities; old building or structure foundations; left-in-place sheeting and support piles; unusual fill or backfill material; multiple layers of road surfaces; abandoned railroad beds, tracks, or ties; and so on. For this reason it is highly recommended that Quality Level A utility location is obtained for these scenarios where space is limited and complicated by multiple utilities, and the engineer should follow CI/ASCE Standard 38-02 (ASCE 2002).

2.4.3. Construction Surveys

Surveys for this phase are concerned almost exclusively with physical aspects. Construction surveys are required to establish control for line and grade, to check conformity of construction, and to establish settlement levels on existing structures immediately adjacent to the construction where necessary.

2.5. INVESTIGATIONS

Investigations may take many forms but always are directed toward determining the most feasible, practical, and economical methods of achieving a desired result. On small sanitary sewer projects, these may involve no more than an on-the-spot decision to use conventional minimum standards for a simple gravity flow extension to an existing sewer system. Larger projects, on the other hand, may have several alternatives—all of which must be considered. Projects involving relief of existing sewer systems, for example, usually require extensive studies before the design flow capacity and the method of correction can be ascertained. In addition, many of the studies conducted as part of the relief alternatives analyses, or independently as an SSES, result in identification of system features that need to be corrected and can be corrected via system rehabilitation. In those instances, some recommendations are simple whereas others may have multiple options that are ultimately decided by owner preference.

The following questions are typical of those to be resolved by investigation:

- What is the extent of area to be served and what is the pattern of present and future land use? Has an area zoning plan been adopted? How does the area relate to a regional sanitary sewer plan?
- Are there known problem areas in the system (SSOs or recurring maintenance locations)? Is there a significant system response to wet weather? Are there areas that cannot be maintained due to limited access, such as unmaintained rights-of-way? Are permanent access roads or pathways an option in those areas?
- What general arrangement of the system will best fill the need? What easements or rights-of-way are required for this arrangement?
- Does or will the system parallel and or cross a waterway (creek, stream, river) or affect tidal or nontidal wetlands? If so, a stream migration analyses may be beneficial to determine needed protective features necessary for future stream migration scenarios.
- What part of the wastewater flow will be intercepted for treatment from an existing sewer system?
- Are there combined sewers in the system? How will flow from combined sewers be handled?
- If a relief system is being constructed, does the owner want to maintain control of the operation between the existing and relief lines, or minimize future bypass pumping needs? If so, flow control devices and methods should be evaluated and discussed with the system owner.
- What are the estimated present and future wastewater flows?
- Shall sanitary sewers all discharge to one point for treatment or will treatment be provided at more than one location?
- How will requirements of other agencies (e.g., state and county highway departments, railroads, other utility organizations) dictate specific locations for crossing, rights-of-way, installation, and details of materials or construction?
- If a regional sewer system is anticipated, what are the long-term environmental effects of exporting wastes from a given groundwater basin to another?

REFERENCES

American Society of Civil Engineers (ASCE). (2002). "Standard guideline for the collection and depiction of existing subsurface utility data." CI/ASCE Standard 38-02, ASCE, Reston, Va.

Joint Task Force of the Water Environment Federation and the American Society of Civil Engineers (WEP/ASCE). (1994). Existing sewer evaluation and rehabilitation, second ed., ASCE Manuals and Reports on Engineering Practice No. 62, ASCE, Reston, Va.

CHAPTER 3
QUANTITY OF WASTEWATER

3.1. INTRODUCTION

Sanitary sewers are constructed to transport the wastewater of a community to a point of treatment and disposal. Estimating quantities of wastewater is the first step in designing new sewer pipelines for (1) new future development or redevelopment, or (2) to develop a capital improvement plan (CIP) for rehabilitating or replacing existing sewers. A CIP would also include physical inspection and other testing of the sewers for conditions analysis.

There are three general categories of wastewater flow in sanitary sewers: residential, nonresidential, and infiltration/inflow (I/I). Residential wastewater is from dwellings such as homes, apartments, and condominiums, as well as from transient residences or group quarters such as prisons, dormitories, and hotels. Nonresidential wastewater is from offices, retail stores, shopping malls, warehouses, factories, schools, prisons, hospitals, churches, mosques, synagogues, and community centers; it is often termed "commercial" and/or "industrial." I/I includes both infiltration and inflow, which are defined as follows (American Public Health Association 1981). Infiltration is groundwater entering a sanitary sewer system through joints, porous walls, and cracks. Inflow is extraneous flow that enters a sanitary sewer from sources other than infiltration, such as connections from roof leaders, basement drains, land drains, and manhole covers. Inflow typically results directly from rainfall or irrigation runoff. Each category of wastewater flow has a multistep forecasting procedure as summarized in Table 3-1.

For new or rehabilitated sewer design, the tractive force approach to self-cleansing evaluation (see Chapter 5) requires that the information

TABLE 3-1. Summary of Forecasting Procedures[a]

	Residential Wastewater	Nonresidential Wastewater	Other Wastewater (Section 3.7)		Infiltration/ Inflow (Section 3.8)
Step 1	Determine planning horizon or design period (Section 3.2).	Determine planning horizon or design period (Section 3.2).	Large industries—case-by-case basis	Flow monitoring	Determine area affected or length and diameter of pipeline.
Step 2	Forecast population or number of dwelling units, including transients and group quarters (Section 3.3).	Forecast land uses and/or number of employees (Section 3.4).			Determine/ estimate infiltration/ inflow allowance.
Step 3	Determine average unit flows based on population or dwelling units (Section 3.5).	Determine average unit flows based on land use and/or number of employees (Section 3.5).			
Step 4	Develop average flows (Section 3.6).	Develop average flows (Section 3.6)			Determine average unit flows.

[a]Develop design and peak and minimum flows (see Section 3.9). If necessary, develop alternative forecasts for sensitivity analysis (see Section 3.10).

from step 4 in Table 3-1 be used to establish four flow rates for each sewer reach:

- The average daily flow, Q_{avg1}, for the low flow period in the pipe, usually when it is first placed into service. Derivation of Q_{avg1} is presented in Sections 3.3 through 3.6.
- The design minimum flow, Q_{min}, which is the largest hourly flow during the low-flow week in the life of the system. This flow is used for self-cleansing design using the tractive force model with the objective of transporting the sediment during low-flow periods. By using Q_{min} for tractive force analysis to set minimum slopes in sewer

design, the potential for sulfide-related corrosion, as discussed in Chapter 4, can be minimized. The most direct way of generating Q_{min} is with the equation $Q_{min} = Q_{avg1} \times P_1$, where P_1 = a peaking factor discussed in Section 3.7.
- The average daily flow, Q_{avg2}, for the design day for the system. Derivation of Q_{avg2} is also presented in Sections 3.3 through 3.6.
- The capacity design flow, Q_{max}, which is the largest 1-hour flow on the design day for the system. This is normally generated by using an appropriate peaking factor P_2 with Q_{avg2} in the equation $Q_{max} = Q_{avg2} \times P_2$. Q_{max} is estimated for the *end* of the design period (Section 3.2) to determine the hydraulic capacities of sanitary sewers and downstream facilities such as pumps and the treatment plant. The peaking factor P_2 is also discussed in Section 3.7. Downstream from pumping stations, Q_{min} and Q_{max} are determined by pump capacity and, indirectly, by Q_{avg1} or Q_{avg2} from upstream of the pumping station. In other words, Q_{max} in a gravity sewer downstream of a pumping station occurs when the pump station is operating at planned full capacity at the end of the design life. Q_{min} occurs when the pumping station is operating at planned largest pumping rate at the beginning of service at the pumping station.

Previous versions of this Manual and similar references have emphasized use of historical per capita wastewater flows and population trends. However, per capita wastewater flows for a given area may not accurately reflect wastewater flow in that area because of nonresidential land uses—which are related to employment or specialized activities, not population—and trends in lower indoor water consumption from water conservation activities. Furthermore, I/I quantities are not related to population. In other words, extrapolating historical per capita wastewater flow is not the best practice.

Engineers may prefer to be on the conservative (high) side in forecasting wastewater quantities for sewer capacity design. The rationale is that conservative estimating provides a contingency or safety factor to handle wastewater from unanticipated population growth and land development. Some consequences of forecasting too low and undersizing sewers are as follows:

- If growth exceeds the forecast, capacity of the pipeline might be reached prematurely, resulting in (1) overflows and backups, especially if there are significant quantities of I/I; and (2) costly construction of a parallel (relief) sewer or increasing the sewer diameter by pipe bursting.
- Development may be constrained by the lack of capacity, although some communities may use this mechanism to limit development.

On the other hand, forecasting too high often leads to oversizing of sewers, some consequences of which are the following:

- Construction costs will be higher without a compensating benefit. However, the size of the pipe itself is not the major cost factor in sewer construction; excavation and backfill are the major cost components and do not vary significantly between commercially available pipe sizes, especially for the smaller diameters. For example, construction cost of a 12-inch (300-mm) pipe will be about 10% to 25% greater than a 10-inch (250-mm) pipe, whereas the hydraulic capacity of the 12-inch pipe is 64% greater than the 10-inch pipe (for the same slope and friction factor).
- Self-cleansing power for a given flow rate and pipe slope is better in a smaller-diameter pipe than in a larger one. For low flows in larger pipes laid on flat slopes, the reduced self-cleansing power may not be sufficient to transport solids, resulting in solids accumulation and odor problems. This problem is compounded when the larger-diameter pipe is laid at a flatter slope than the smaller-diameter pipe would have been laid.
- Excess capacity in the pipeline may encourage or justify population growth and land use development that are not desired by elected officials.

The consequence of an *accurate* forecast is that the sewer will accommodate existing and future wastewater flow from officially sanctioned development (Boland, 1998).

There is a great deal of literature on forecasting water demand based on variables such as socioeconomic characteristics (e.g., household income, housing type, household size, employment), seasonal variations, economic conditions, climate, and the price of water. Such forecasts can be disaggregated into indoor water use, which affects wastewater quantities, and outdoor water use, which does not. Detailed discussion of such models is beyond the scope of this Manual, but water demand forecasts by the water supply authority may be useful to the engineer in wastewater forecasting (Jones et al., 1984; Baumann et al., 1998; Mays, 2004).

3.2. DESIGN PERIOD

The length of time considered in forecasting flows and setting capacities of sanitary sewers is referred to as the design period. Proper engineering practice requires that the design period must be established prior to the design of the sanitary sewer. The design period is related to the planning horizon developed by the planning department having jurisdiction over the area and the expected useful life of the sewer pipe.

Residential, commercial, and industrial development in most political subdivisions are governed by a general plan or zoning ordinance. There may also be an accompanying master plan for installation and/or rehabilitation of sanitary sewers within its jurisdiction. These plans permit orderly expansion of facilities based on sound engineering without resort to costly "crisis" action. The planning horizon for the general plan or the sewer master plan can be as little as 10 years and as much as 50 years.

In some cases, no specific planning horizon is indicated in the general plan. Instead, planners use the so-called saturation or build-out population or land use. This is the maximum population and commercial or industrial development that could occur from the existing general plan or zoning ordinance. Engineers can forecast flows based on such populations or land use without regard to time frames.

It is also possible that growth and development may be phased over the long term. In cases where sanitary trunk sewers would initially serve relatively undeveloped lands adjacent to metropolitan areas, it may not be feasible to design and construct initial facilities for more than a limited future development period. Nevertheless, easements and rights-of-way for future facilities can be secured in advance of development. This situation may need to be tempered to avoid imposing large current costs on a relatively small current population by building reserve capacity far beyond what is likely needed during a reasonable design period.

Engineers have typically used 50 years for the expected useful life of the pipe, based on historical estimates of how long a pipeline can carry wastewater before the pipe deteriorates or collapses. In the case where the planning horizon is greater than the expected useful life of the pipe, the planning horizon can be used as the design period, with the understanding that some sewers may need replacement or rehabilitation before the end of the design period. In the case where the planning horizon is less than the expected useful life of the pipe, the engineer may consider basing the sewer design on saturation or build-out figures or on phased development.

3.3. POPULATION OR DWELLING UNIT FORECAST

Once the design period or planning horizon has been established, the next step in calculating *residential* wastewater flow is to forecast the population or number of dwelling units and persons per dwelling unit, both for the time of start of service and at the end of the design period.

Previous versions of this Manual and other historical references give several mathematical and graphical methods for population forecasting.

Instead of using past trends to extrapolate future population, a better way is to use data on population and residential land use for the area in question from documents held by the jurisdiction's planning department, such as the general plan land use element, zoning ordinance, or development plans for the planning horizon or design period. These data may include population or residential dwelling unit densities in terms of dwelling units (DU)/acre (1 acre = 0.4 hectares). Typically, there may be ranges of DU/acre and each range may have different persons/DU. For example:

Low density	≤4 DU/acre	3.0 persons/DU
Medium density	4–10 DU/acre	2.8 persons/DU
High density	>10 DU/acre (multifamily units)	2.5 persons/DU

Engineers should consult with planners or demographers to determine which figures should be used. Professional planners should also be able to inform the engineer as to which parts of the community will be developed before others. Engineers should keep in mind that general plans or zoning ordinances might be periodically updated; the most current versions should be used.

Population forecasts may also be adjusted by a housing vacancy rate. Obviously, forecasts would be larger if this rate was excluded (e.g., 0% vacancy), but would be more accurate if this was included.

Population forecasts may include transients in hotel and motel rooms; group quarters such as school dormitories, prisons, jails, long-term health care facilities (nursing homes); and seasonal-use facilities such as campgrounds and resorts. Transient or seasonal population estimates, such as for hotels and motels, can be converted to equivalent full-time residents by using multipliers of the population fraction that is considered transient or seasonal. Alternatively, these facilities can be included in the land use forecast in Table 3-2 for institutions or commercial developments because they may have kitchens and/or laundries.

These population or dwelling unit forecasts should be done for the beginning and end of the design period to obtain the *residential* portion of Q_{avg1} and Q_{avg2}. For accurate hydraulic design of sewers, these calculations are applied at many key points in the existing or proposed sewer system. The area to be served should be divided into subareas defined by topography, land uses, and pipe layout (existing or proposed). The most downstream point of the subarea is the key point, which is usually a manhole or pump station where the flow rate is larger than at upstream key points. There are a variety of computer applications that can perform such calculations rapidly. An example of a population and dwelling unit forecast for a subarea at the *end* of the planning period is shown in

TABLE 3-2. Example of a Residential Dwelling Unit (DU) or Population Forecast for a Subarea at the Beginning or End of the Design Period

Residential Land Use (1)	Area, acres (2)	DU/ Acre (3)	No. of DUs (4) = (2) × (3)	Persons/ DU (5)	% Developed at End of Planning Horizon or Design Period (6)	Total Population (7) = (4) × (5) × (6)
Low Density/ Rural	100	3	300	3.0	80	720
Medium Density	250	8	2,000	3.2	70	4,480
High Density	150	16	2,400	2.5	60	3,600
Resort Hotel	—	—	300 (rooms)	1.0 (reflects seasonal or transient population)	70 (occupancy rate)	210
Total						9,010

Note: acre × 0.4 = hectare.

Table 3-2. There would be another version of Table 3-2 for the *beginning* of the planning period.

3.4. LAND USE AND/OR EMPLOYEE FORECASTS

Once the design period or planning horizon has been established, the next step in calculating *nonresidential* wastewater flow is to forecast the land use and/or number of employees. Nonresidential wastewater is generated by employees in a manner similar to residential (i.e., toilets, sinks, showers, and so forth). In each case, the business does not consume water in products made or services rendered (i.e., services as opposed to manufacturing). Wastewater from manufacturing is considered separately in Section 3.7.1.

As with residential population or dwelling unit forecasts, the best source of data on nonresidential land use and number of employees is the community's general plan land use element, zoning ordinance, or development plans for the planning horizon or design period, or for build-out or saturation with no specific date. This forecast may include both net land area and gross floor area. The number of employees depends on the type of business and size of the building, which can be determined by floor-to-area ratios (FARs) and floor area per employee.

As with residential land use, there may be many types of commercial/industrial land use. Typical categories include:

- Office
- Commercial retail (neighborhood and regional)
- Commercial services (neighborhood and regional)
- Institutional
- Manufacturing and raw material processing
- Transportation
- Communication and utilities
- Industrial and commercial complexes
- Mixed urban or developed land
- Public assembly, recreational, cultural, and entertainment
- Resource extraction and exploitation.

Engineers should consult with planners or demographers to determine which categories are relevant and the applicable FARs. Professional planners may also tell the engineer which parts of the community will be developed before others. Engineers should further keep in mind that general plans or zoning ordinances might be periodically updated.

These land use or employee forecasts should be done for the beginning and end of the design period to obtain the *nonresidential* portion of Q_{avg1} and Q_{avg2}. For accurate hydraulic design of sewers, these calculations are applied at many key points in the existing or proposed sewer system. The area to be served should be divided into subareas defined by topography, land uses, and pipe layout (existing or proposed). The most downstream point of the subarea is the key point, which is usually a manhole or pump station where the flow rate is larger than at the upstream key points. There are a variety of computer applications that can perform such calculations rapidly. An example of a land use or employee forecast for a subarea at the *end* of the planning period is shown in Table 3-3. There would be another version of Table 3-3 for the *beginning* of the planning period.

3.5. AVERAGE UNIT FLOWS

Once the population or dwelling unit (Section 3.3) and land use or employees (Section 3.4) have been forecast, the third step is to apply unit flows to obtain Q_{avg1} and Q_{avg2}. Unit flows are typically based on volume per unit time per person, dwelling unit, floor area, land area, or employee. The best sources of unit flows are previous or current wastewater flow

Table 3-3. Example of a Land Use or Employee Forecast for a Subarea at the Beginning or End of the Design Period

Land Use (1)	Area, acres (2)	Floor to-Area Ratio (FAR) (3)	Floor Area, ft^2 (4) = (2) × (3) × 43,560	Average Floor Area/ Employee/ft^2 (5)	% Developed at End of Planning Horizon or Design Period (6)	Total No. of Employees (7) = (4) ÷ (5) × [100 − (6)] (rounded)
Office	20	0.7	609,840	300	60	1,220
Retail	15	0.5	326,700	1,000	70	229
Industrial	10	0.3	130,680	2,000	60	39
Total						1,488

Note: acre × 0.4 = hectare; $ft^2 \times 0.093 = m^2$

measurements for similar land uses. Unfortunately, wastewater flow from specific parcels or geographical areas is not routinely measured or metered. Wastewater at a treatment plant, however, is usually monitored for a variety of reasons, such as allocating capital and operating and maintenance (O&M) costs among agencies using the plant, designing upgrades to the plant, and sizing new equipment. However, such flows may represent many different land uses and unit flows may not reflect future development.

Water consumption is usually metered and can provide a source of information for unit wastewater flows. Such data must be for *indoor* water use, such as water that enters the sewer system from household, office, and commercial or retail appliances and fixtures (e.g., toilets, showers, sinks, dishwasher, and laundry). The data record should represent time when there is little or no outdoor water use, such as landscape irrigation or car washing. Long-term records are preferable because short-term data may include and periods when water use is not typical, such as droughts, which could unrealistically skew the data. In some parts of the United States this is winter water consumption because there is little outdoor water use when it rains or snows. If only annual average daily water use data are available, it may be possible to determine indoor water use with an estimated percentage of total annual demand. Such percentages are a function of climate and area irrigated (lot size and building footprint area). In general, the percent indoor water use is lower for arid regions, as shown in Table 3-4.

In some cases, water agencies have worksheets that can be adapted to calculate per capita indoor water use. For such worksheets, the engineer

TABLE 3-4. Residential Indoor Water Use as a Percentage of Total Annual Water Use[a]

Location	% Indoor Use
Santa Barbara, Calif.	52
San Diego, Calif.	37
Goleta, Calif.	50
Riverside, Calif.	50
Palm Springs, Calif.	20–25
Albuquerque, N.M.	60
Brisbane, Australia	47
Northern Territory, Australia	40
Florida (statewide average)	50
Boulder, Colo. (citywide average)	57
Colorado Springs, Colo.	63
Denver, Colo.	40
Scottsdale, Ariz.	33
California (statewide average)	56
Pennsylvania (statewide average)	93

Sources: Quoted on City of Boulder, Colo. web site based on publications of cities listed or data logger studies, http://bcn.boulder.co.us/basin/local/residential.html, accessed October 27, 2005.

[a]Statewide averages for California and Pennsylvania are from the EPA web site: www.epa.gov/watrhome/you/chapt1.html, accessed October 27, 2005.

must obtain data (usually from the water purveyor) or make reasonable assumptions about such factors as:

- Number of showers per day and length of each shower (time in minutes × 2 to 5 gal/min).
- Number of toilet flushes per day (typically three to five) and gallons per flush (typically 1.6 to 5.0).
- Use of faucets (washing, shaving, toothbrushing, food preparation, dishwashing) and time used daily.
- Frequency of use of dishwasher and clothes washer, and gallons per use.

The City of Tampa's website has an example (http://www.tampagov.net/dept_water/conservation_education/Customers/Water_use_calculator

.asp, accessed October 27, 2005). Although this is for residential use, it can be adapted to nonresidential use with number of employees instead of residents.

Many engineers use unit flows from textbooks or figures from other jurisdictions. These data may be secondary sources quoting from older references and should be used with caution. These older data may not reflect (1) recent water conservation measures and installation of fixtures and appliances that use less water; and (2) lifestyle changes, such as a decreasing number of stay-at-home parents. In both cases, residential unit factors are reduced.

To illustrate how water conservation measures affect water demand forecasts, in 1984 the Texas Water Development Board predicted that the most likely water demand in 2000 would be 25 million acre-feet (1 acre-foot = 1,234 m^3). Six years later, assuming the adoption of major conservation measures, the Board predicted water usage of about 14 million acre-feet. In fact, in 2000, Texans used 16.4 million acre-feet, slightly above the 1990 prediction but some 9 million acre-feet below the 1984 most-likely scenario (Texas Water Development Board, undated). In a study for Seattle, Washington, after installation of various water conservation devices in a sample of 37 single-family detached homes, average per capita indoor residential use dropped from 63.6 gal/capita/day to 39.9 gal/capita/day—a 37% decrease. This decrease was consistent for six months after the initial data readings were taken (Deoreo et al., 2001).

Some jurisdictions may require the engineer to use allowances from its own regulations. Such allowances may be quite conservative (high) and lead to overdesign of sewers. If better data exist, those data should be used.

Tables 3-5 and 3-6 are summaries of unit factors from recent studies.

3.6. AVERAGE FLOWS

Average wastewater flows (Q_{avg1} and Q_{avg2} from Section 3.1) are the product of the residential and employment data (Sections 3.3 and 3.4) and unit water factors (Section 3.5). To obtain flows for hydraulic design of sewer pipelines:

1. Q_{avg1} and Q_{avg2} must be converted to design minimum, Q_{min}, and design maximum Q_{max} (Section 3.9).
2. Other wastewater flows from infiltration/inflow (Section 3.8) and industrial uses (Section 3.7) must be added.

An example of Q_{avg1} or Q_{avg2} calculation is shown in Table 3-7.

TABLE 3-5. Residential Wastewater Average Daily Flows

Location and Land Use	Unit Flow, gal/ capita/day	Remarks
Waterloo/Cambridge, Ontario	71	Indoor water use, spring and fall, 1996; ~1,000 homes each location (AWWARF 1999).
Seattle, Wash.	57	
Tampa, Fla.	66	
Lompoc, Calif.	66	
Eugene, Ore.	84	
Boulder, Colo.	65	
San Diego, Calif.	58	
Denver, Colo.	69	
Phoenix, Ariz.	78	
Scottsdale/Tempe, Ariz.	81	
Walnut Valley Water District, Calif.	68	
Las Virgenes Municipal Water District, Calif.	70	
East Bay Municipal Water District, Calif.	64	1994 survey of indoor water use of 600 homes (Darmody et al. 1996).
Rural Wisconsin (specific location not identified)	43	Indoor water use monitored from 11 rural homes in for 14–77 days each in early 1970s. Flows may be low because of use of outhouses (EPA 1978).
Nationwide	66 (range = 57.3–73.0)	Indoor water use for 210 residences (Brown and Caldwell 1984).
Phoenix, Ariz.	71 (range = 65.9–76.6)	Indoor water use for 90 residences over a three-month period in 1989 (Anderson and Siegrist 1989).
Tampa, Fla.	51 (range = 26.1–85.2)	Indoor water use for 25 residences over a three-month period in 1993 (Anderson et al., 1993).
Milwaukee Metropolitan Sewerage District, Wisc.	64	Monitoring from six strictly residential areas in July and August 1976 (Milwaukee Metropolitan Sewerage District 2005).
East Bay Municipal District, Calif. (Oakland and vicinity)	60	Indoor water use for service area, calendar year 2004; excludes 5 gal/capita/day for leaks (EMBUD 2005).
Los Angeles, Calif.	90	Estimates based on water consumption (City of Los Angeles 1992).
Stamford, Conn.	80	Wastewater flow based on billing records (Jeanette Brown, City of Stamford, Conn. Water Pollution Control Authority, personal communication, 2005).
Winnipeg, Manitoba	56 average 43–77 range	Based on indoor residential water use, excluding leaks, 1992. Range reflects the number of people per household: higher figure is fourlowefigure is for one (Griffin and Morgan, undated)
Morgan Hill, Calif.	79–114	Low-medium density residential wastewater flow meter data for 679 homes and assumed three persons/DU 1990–1991. Range of flows is probably due to infiltration/inflow.

Note: gal/day × 3.8 = L/day.

TABLE 3-6. Nonresidential Wastewater Average Daily Flows

Location and Land Use	Unit Flow, gal/acre/day (gad) or gal/employee/day (ged)	Remarks
Los Angeles, Calif.	30 ged	Estimates based on water consumption (City of Los Angeles 1992).
East Bay Municipal District, Calif. (Oakland and vicinity)	30 ged	Based on indoor water consumption for faucets and toilets from residential data.
Stamford, Conn.	6–32 ged	Wastewater flow based on billing records. Range reflects commercial establishments without cafeterias to those with cafeterias, showers, and gyms (Jeanette Brown, City of Stamford, Conn. Water Pollution Control Authority, personal communication, 2005).
Morgan Hill, Calif. (mixed use)	2,450 gad	Flow monitoring in manhole for two weeks in April, 1992. Tributary area was 96 devel oped acres: 24 retail, 31 hospital, 41 industrial, and 25 homes.

Note: gal/day × 3.8 = L/day; acre × 0.4 = hectare.

TABLE 3-7. Example of Average Wastewater Flow Computation (Q_{avg1} or Q_{avg2})

	Demographic Factor (Table 3-2 or 3-3)	Unit Flow Factor	Average Flow, gal/day
Residential	9,010 population	60 gal/person/day	540,600
Nonresidential	1,488 employees	30 gal/employee/day	44,640
Total			585,240

Note: gal/day × 3.8 = L/day.

3.7. OTHER

3.7.1. Industrial

Industrial wastewater typically is from a firm that uses water in manufacture of products. For manufacturing industries, especially large factories, water use and wastewater generation should be developed on a case-by-case basis. Industrial wastewater quantities may vary from little more than the normal domestic rates to many times those rates. The type of industry to be served, the size of the industry, operational techniques, and the method of on-site treatment of wastewater are important factors in estimating wastewater quantities. Furthermore, peak discharges may be the result of flows contributed over a short time frame, such as a 10-hour working period. Peak discharge to the sanitary sewer sometimes can be reduced by the use of detention tanks or basins arranged to discharge at smaller rates over longer periods or to discharge only during hours when wastewater flows are small.

Water audits can be performed on individual industrial plants over a period of time. That time period should reflect seasonal and other variations in water use by that industry. Water supply can be monitored and tracked through the industrial process to determine how water is consumed and disposed. Such audits are typically conducted by the water supply agency to help industries use water wisely, reduce demand, and thereby conserve water supply. The results of such audits may be useful for wastewater forecasting.

3.7.2. Flow Monitoring

To obtain estimates of industrial or any other type of wastewater flow, a program of flow monitoring may be warranted. This can be a permanent or temporary installation in a manhole or other structure. Individual customer water meters are also a type of flow monitoring that can be used to estimate wastewater flows (Sections 3.1 and 3.5). The following general discussion refers to flow monitoring related to wastewater only.

Flow monitoring can occur at:

- Key locations in the sewer system.
- Individual industrial plants (as part of water audits above).
- Individual developments—residential or commercial.
- Wastewater treatment plants.

Results of flow monitoring are often used for cost and revenue allocations so that certain areas of the jurisdiction are charged at a rate based on flow. In a multi-jurisdictional setting, flow monitoring results can be used to properly allocate the costs of the sewer system and treatment plant among the jurisdictions. For sewer design, flow monitoring results can help in deriving unit flow factors as described in Section 3.5. For example, a flow monitor may be installed in a manhole downstream of a 200-unit residential development. The results would give flow per dwelling unit for that type of residence that could be used elsewhere in the jurisdiction for a similar housing type.

The choice of a location for a flow monitor for flow forecasting purposes is based on several factors:

- *Characteristics of land use upstream of the monitor location*. Ideally, the land use should be homogeneous, such as single-family residential, multi-family residential, retail, office, hospital, school, etc.
- *Site sewer and/or manhole characteristics*, such as hydraulic features (sags), grease buildup, cracks or other damage, as well as the possibility of variances between recorded pipe sizes and slopes and actual field conditions. For best results, flow direction should not change abruptly going through the manhole and the manhole should not have debris, brick, or any other objects that might disrupt the flow.
- *I/I contribution*. If there are I/I issues upstream of the monitor location, the I/I contribution would need to be separated from the base sanitary flow. Meter site selection can be accomplished after reviewing the collection system maps and preliminary field inspection of any sanitary sewer overflow (SSO) locations. Each monitoring site should be selected so that the footage of the collection system upstream of the meter can be isolated for the purposes of determining extraneous I/I. Installation of rainfall meters across the study area to measure rainfall intensity and duration throughout the monitoring period can assist in establishing wet-weather capacity for SSO analysis.

For flow forecasting, a monitoring period can be as little as a few weeks. To perform a detailed analysis of I/I and pipe capacity, the period could be up to six months to obtain data during both dry and wet weather. In such cases, the objective is to monitor multiple rain events of varying intensities to accurately assess the inflow response for each event.

Information obtained during the monitoring period can be used to determine the following:

- Average daily flow—dry weather.
- Peak flow—dry weather.
- Average daily flow—wet weather.
- Peak flow—wet weather.
- Peak inflow rates.
- Total I/I volume.

A detailed discussion of flow meter technologies is beyond the scope of this Manual. A general overview follows. Flow meters may be categorized as direct-discharge and flow-velocity. The direct discharge method is based on computing the flow using easily measured or sensed variables, such as flow depth or pressure. Some examples are flumes, weirs, magnetic meters, and ultrasonic devices. Sensors that read flow depth or pressure are of two basic types: wetted or submersible sensors, where the level/velocity sensor is mounted in the flow stream and the sensor is secured to a mounting band that fits snugly in the pipeline; and non-contact sensors that are mounted in a level position above the flow stream so that the signal is aimed at the flow and does not hit the pipe or invert walls. Flow-velocity devices are based on measurement of velocity applied to the cross-sectional area of the flow. Some examples are propeller meters and dye tracers (Metcalf & Eddy, 1981). Regardless of method used, results from flow monitoring should be verified using existing pipe slopes and diameters to check whether there are anomalies in the results.

3.8. INFILTRATION/INFLOW

Ideally, sanitary sewers should convey sanitary wastewater only. Sewer pipe, regardless of material, is not designed, manufactured, or installed to allow extraneous water to enter the pipeline. As a practical matter, however, sanitary sewer design capacity must include an allowance for the extraneous water components of I/I that inevitably become a part of the total flow.

3.8.1. Infiltration

Groundwater can enter sanitary sewers as infiltration through pipe joints, broken pipe, cracks, openings in manholes, and similar defects in sanitary sewer structures. Defective service connections also can contribute appreciable quantities of infiltration. The presence of infiltration is based on a groundwater table at or above the pipeline invert for at least a portion of the year. Areas with deep groundwater below the pipe do not generally show infiltration problems; however, this does not preclude dealing with inflow issues, discussed in Section 3.8.2.

Before the use of compression-type gasket pipe joints, the bulk of infiltration in structurally sound pipe entered at pipe joints in older sanitary sewers built with either cement-mortar, hot-poured bituminous, or cold-installed bituminous materials. These joints were seldom satisfactory because of the initial difficulty in constructing a watertight pipe joint and normal deterioration of the joint material with time. Often, joint material would slough to the pipe invert, resulting in a leaky joint at the crown.

Other sources of infiltration into new sanitary sewer systems can be traced to defects in pipe soil foundation or pipe strength, faulty installation practices, or service connections. An engineer evaluating capacity of older sewers should consider infiltration as a source of flow in that pipeline if groundwater is present at or above the pipe invert at least part of the year.

Use of compression-type gasket pipe joints made it possible to reduce groundwater infiltration and leakage significantly. A detailed discussion of pipe joints and joint materials is found in Chapter 8.

The best way to estimate infiltration quantities is with flow metering as described in Section 3.7. In the absence of flow metering data, design of extensions to existing systems should consider past practices and trends in infiltration, with allowances made where necessary.

The selection of a capacity allowance to provide for infiltration should be based on the physical characteristics of the tributary area, the type of sewer pipe and pipe joints to be used, and sewer pipes in the existing sanitary sewers. In general, the design infiltration allowance is added to the peak rate of flow of wastewater and other components to determine the actual design peak rate of flow for the sanitary sewer. A survey of infiltration allowances, which typically include inflow discussed in the next section, is summarized in Table 3-8.

3.8.2. Inflow

Inflow is extraneous flow that enters a sanitary sewer from sources other than infiltration, such as connections from roof drains, basement drains, land drains, and manhole covers. Inflow typically results directly from rainfall or irrigation runoff.

Historical tests made on manhole covers submerged in only 1 inch (25 mm) of water indicate that the leakage rate per manhole may be from 20 to 75 gpm (1.3 to 4.7 L/sec), depending on the number and size of holes in the cover (Rawn, 1937). Newer manholes with solid covers and without pick holes would contribute less inflow. Illegal roof drain connections also can overload smaller sanitary sewers. Rainfall of 1 inch/hr (25 mm/hr) on 100 m^2 (1,080 sq ft) of roof area, for example, would contribute inflow of up to 11 gpm (0.7 L/sec).

3.9. PEAK AND MINIMUM FLOWS

Design peak and minimum flows can be calculated by applying peak factors to the average flows developed in previous sections or can be estimated directly using the fixture unit method. The following procedure

TABLE 3-8. Infiltration/Inflow Allowances for Various Sewer Agencies

Agency Location	Infiltration/Inflow Allowance	Comments[a]
Edmonton, Alberta	0.28 L/sec/ha for manholes not located in sag locations; 0.4 L/sec/ha for man holes in sag locations.	www.edmonton.ca/portal/server.pt/gateway/PTARGS_0_2_284_220_0_43/http%3B/CMSServer/COEWeb/infrastructure+planning+and+building/water+and+sewers/sewer+guidelines.htm, accessed October 27, 2005.
State of Oregon	<2,000 gal/day/acre	www.deq.state.or.uswq/wqrules/340Div52ApxA.pdf, accessed October 27, 2005.
Lethbridge, Alberta	2.25 m^3/day/ha	www.lethbridge.ca/NR/rdonlyres/82DC4DE1-C54A-4981-AFDF-1925E7699054/1077/Section5SANITARY SEWER.pdf, accessed October 25, 2005.
Augusta, Ga.	25 gpd/inch-diameter/mile	
London, Ontario	8,640 L/ha/day (0.100 L/sec/ha)	Environmental and Engineering Services Department, The Corporation of the City of London. October, 2003.
Kingston, Ontario	0.14 L/ha/sec from residential land, 28 m^3/ha/day from industrial, commercial, and institutional lands.	
Union County, N.C.	300 gpd/inch-diameter/mile	
State of New Hampshire	100 gpd/inch-diameter/mile of pipe for sizes to 48 inches. 200 gpd/inch-diameter/mile for sizes more than 48 inches.	www.gencourt.state.nh.us/rules/envws700.html, accessed October 27, 2005.
Municipality of The County of Antigonish, Nova Scotia	500 Imperial gpd/inch-diameter/mile of pipe	www.antigonishcounty.ns.ca/MunicipalServiceSpecs.pdf, accessed October 27, 2005.
Maryland Department of the Environment	400 gal/day/acre	www.mde.state.md.us/assets/document/water/WastewaterCapacityMgmtGuidance.pdf, accessed October 27, 2005.

Note: gal × 3.8 = L; acre × 0.4 = hectare; in. × 25.4 = mm.
[a]Web search on "infiltration allowance" + sewer.

could be used for the peaking method—separately for the beginning and the end of the design period:

1. Tabulate the average flows for each subarea; this would include all of the subarea flows—including residential, employee, industrial, and I/I flows. Each resulting flow is the estimate of average daily flow coming from the subarea.

2. Accumulate the subarea average flows down through the entire system. The resulting accumulated flows are the estimates of average daily flows point-by-point (typically manhole-by-manhole) down through the entire sewer system.
3. Apply peaking factors to each point (manhole) down through the system, thus generating the peak 1-hour flows for all sewer reaches in the system. The peak flows from Q_{avq1} are the Q_{min} values needed for self-cleansing design; those from Q_{avq2} are the Q_{max} values needed for capacity design.

3.9.1. Peak Factors

Peak factors discussed in this section (P_1 and P_2 from Section 3.1) combined with the average flows from Section 3.6 give peak and minimum flows that form the basis of hydraulic design for sewer pipelines. Historically, often only the peak flows were determined, but tractive force, self-cleansing design requires that minimum flows be established as well. Again, these minimum flows are not the smallest flows but, rather, the largest 1-hour flow that occurs during the low-flow week of the design life of the sewer reach being evaluated.

The flow of wastewater (exclusive of groundwater infiltration and inflow) will vary continuously throughout any one day, with the lowest flows usually occurring between 2 A.M. and 6 A.M. and the highest flows occurring during the daylight hours. The I/I component usually remains reasonably constant throughout the day except during and following periods of rainfall.

The best sources for peak factors are recent flow measurements for similar land uses. However, records of existing wastewater flows or water demands are rarely complete enough to permit estimates of wastewater minimum or peak flow.

There are some available data on peaking factors. A summary of measured peaking factors for 24 sewer agencies in the United States is presented in Table 3-9. These factors include I/I, not just sanitary dry-weather flow. As such, relatively wet areas (such as the northwest) had higher peaking factors, probably because of the influence of I/I. Relatively arid areas (such as the southwest) had the lowest peaking factors.

The data in Table 3-9 show that, in general, peak factors for maximum wastewater flow vary inversely with population served (i.e., average flows for entire cities). However, for individual sewers, peaks may be much higher than shown because average flows will likely be smaller than the data in this table [the smallest average flow was 6.0 mgd (22.8 million L/day)].

In general, peak factors decrease with increasing average flow. The farther downstream in a sewer, the greater the average flow because the

TABLE 3-9. Overall Peak Factors for Various Sewer Agencies[a]

Agency Location	Peak Factor	Comments
Southeast	2.0	
Central	2.4	
Southwest	1.8	
Northwest	3.8	
Northeast	2.3	
Agency Size		
Large	2.1	Population served >500,000
Medium	2.6	Population served 100,000–500,000
Small	3.0	Population served <100,000

Source: American Society of Civil Engineers. (1999). "Optimization of collection system maintenance frequencies and system performance." ASCE/EPA Cooperative Agreement No. CX824902-01-0, ASCE, Reston, Va.

[a]For measured values only. Data reported as estimates are not included. Figures are rounded to nearest 0.1.

population of area served is greater. Consequently, peak factors typically decrease for downstream reaches. Higher peak factors for smaller areas occur because small flows are sensitive to changes. For example, a slight change in water use could mean a relatively large increase or decrease in total wastewater flow for the area served. On the other hand, slight changes in water use in larger areas would result in relatively small increases or decreases in total flow for the area served.

As with rainfall-runoff calculations, the "time of concentration" in sanitary flows determines when peak flows from a given area reach the sewer. Although time of concentration, as applied in rainfall-runoff calculations, is not used in sewer flow calculations, the principle is similar in that peak flows from consecutive areas do not reach a given point in a sewer at the same time and, as a consequence, do not add. The accumulated peak flow tends to attenuate in downstream sewer reaches where there are larger areas and populations served and the average flows are greater. Most empirical peak flow equations are based on the general form $PF = K \div Q^{-n}$ or $PF = K \times Q^{-n}$, where PF = peak factor, K = a coefficient, Q = average flow, and n = an exponent < 0.5 (see, for example, Metcalf & Eddy 1981). Q in this format can also be population or some other demographic factor.

Figures 3-1 through 3-4 are examples of the variations in rates of flow for situations in which dry-weather wastewater flows are expected to govern. Figure 3-1 shows the ratios of peak and minimum flows to average daily wastewater flow recommended for use in design by various

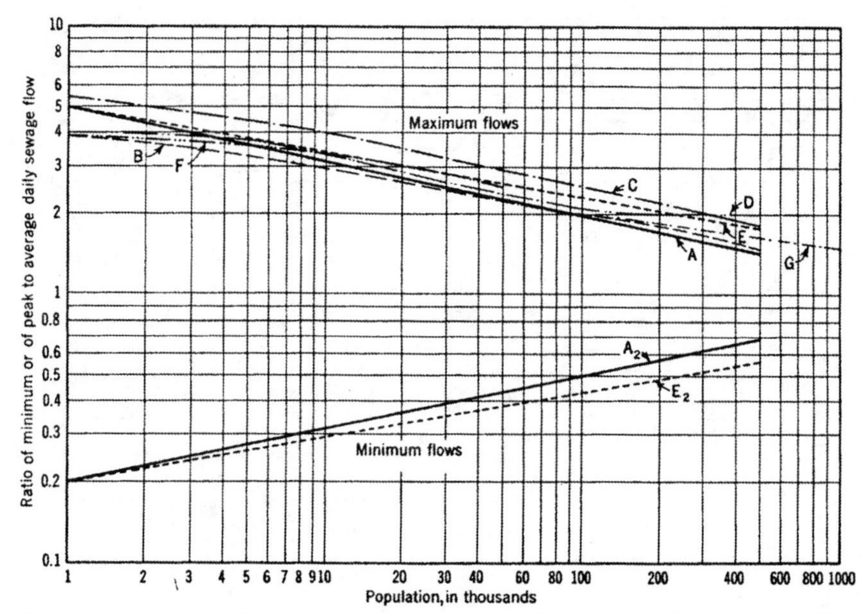

*Curve A source: Babbit, H. E., "Sewage and Sewage Treatment." 7th Ed., John Wiley & Sons, Inc., New York (1953).

Curve A_2 source: Babbit, H. E., and Baumann, E. R., "Sewage and Sewage Treatment." 8th Ed., John Wiley & Sons, Inc., New York (1958).

Curve B source: Harman, W. G., "Forecasting Sewage at Toledo under Dry-Weather Conditions." *Eng. News-Rec.* **80**, 1233 (1918).

Curve C source: Youngstown, Ohio, report.

Curve D source: Maryland State Department of Health curve prepared in 1914. In "Handbook of Applied Hydraulics." 2nd Ed., McGraw-Hill Book, Co., New York (1952).

Curve E source" Gifft, H. M., "Estimating Variations in Domestic Sewage Flows." *Waterworks and Sewarage*, **92**, 175 (1945).

Curve F source: "Manual of Military Construction." Corps of Engineers, United States Army, Washington, D.C.

Curve G source: Fair, G. M., and Geyer, J. C., "Water Supply and Waste-Water Disposal." 1st Ed., John Wiley & Sons, Inc., New York (1954).

Curve A_2, B, and G were constructed as follows:

Curve A_2, $\frac{5}{P^{0.107}}$

Curve B, $\frac{14}{4 + \sqrt{P}}$

Curve G, $\frac{18 + \sqrt{P}}{4 + \sqrt{P}}$

FIGURE 3-1. Ratio of extreme flows to average daily flow compiled from various sources.

authorities. Figure 3-2, based on dry-weather maximums, is the modification of a chart originally prepared for the design of sanitary sewers for a group of 18 cities and towns in the Merrimack River Valley in Massachusetts. The ratios given are approximately correct for a number of other municipalities in the same general area. Figure 3-3 was developed by the

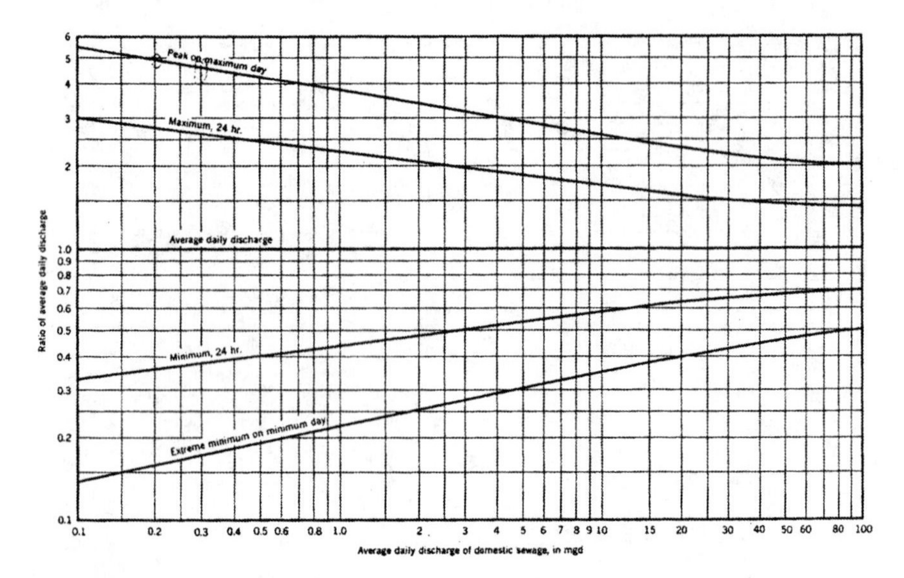

FIGURE 3-2. Ratio of extreme flows to average daily flow in New England. (Mgd × 3,785 m³/day).

Bureau of Engineering, City of Los Angeles, and has been in use since 1962; for this curve, $PF = 2.64 \times Q^{-0.095}$. The Los Angeles curve has been checked with actual flow measurements and found to be generally accurate (McKibben et al. 1994). Figure 3-4 shows peak residential wastewater flow for the City of Toronto.

Engineers should use caution with curves or parameters developed historically, since changing lifestyles (such as more women in the workforce instead of staying at home; water conservation; and increasing use of reclaimed water for residential irrigation) may result in attenuated peaks from residential flow.

3.9.2. Fixture Units

In the 1940s, Dr. Roy Hunter of the National Bureau of Standards developed and published a methodology for determining necessary pipe sizing by estimating maximum demand on the delivery and drainage systems (Hunter 1940a, 1940b). This was developed because one of the major code concerns was then, and is now, pipe sizing for supply and drainage piping in a building. For supply and drainage, each type of fixture or appliance (such as shower, toilet, sink, dishwasher, and laundry) is assigned a "fixture unit value."

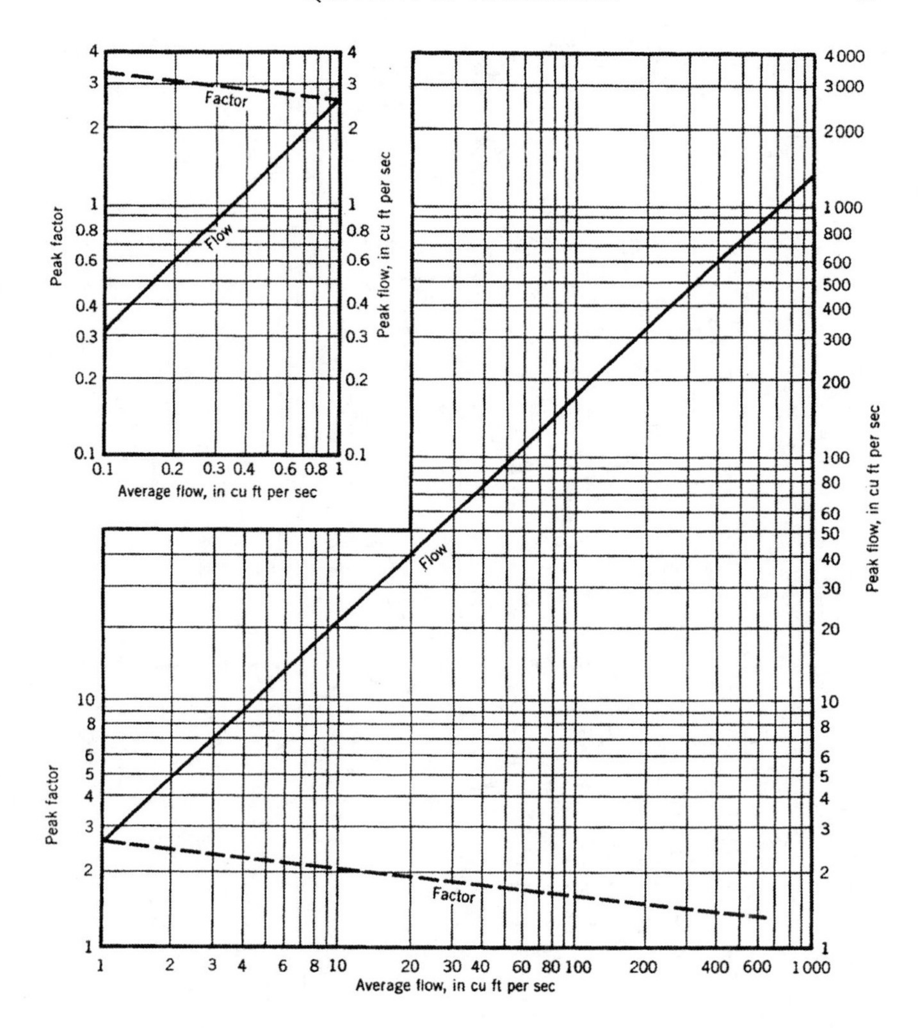

FIGURE 3-3. Ratio of peak flow to average daily flow in Los Angeles. (cfs × 1.7 = m^3/min.)

It would be possible to calculate pipe sizes based on simultaneous use of all fixtures in a building. However, the likelihood of this scenario is very small. Actual peak flow would be a function of the probability of fixture use at any time. Dr. Hunter used probability theory to assess the problem of plumbing design loads and assumed that the operation of the principal fixtures in a plumbing system could be considered as random events. His goal was to quantify, on the basis of probability, a means of sizing waste conduits based on usage demand at a probable "not to exceed" fixture rate (Breese 2001).

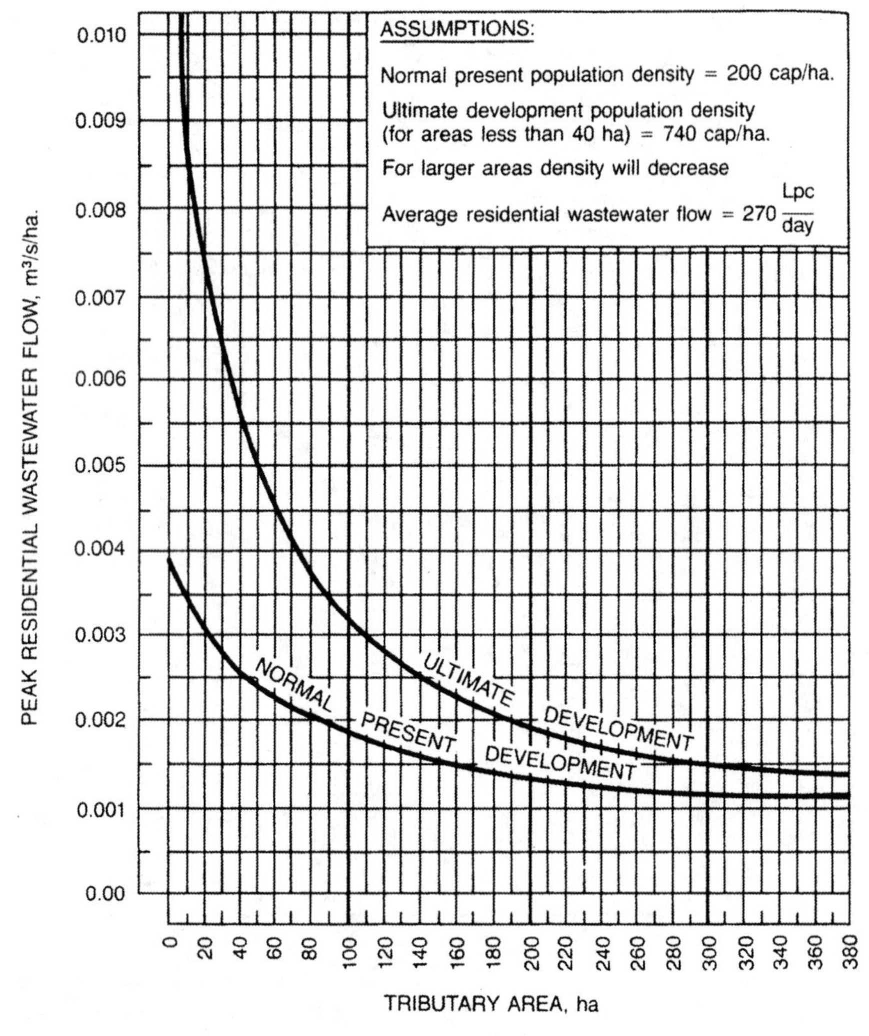

FIGURE 3-4. Peak residential wastewater flow for the City of Toronto, Ontario, Canada (m^3/sec/ha × 14.5 = cfs/acre; L/cap/day × 0.26 = gal/cap/day; ha × 2.5 = acre). Source: Department of Public Works, Toronto, Canada, 1980.

Although modified over the years, Hunter's basic work is still used as the basis for pipe sizing in a plumbing system and it appears in many plumbing codes. However, the original intent of Hunter's work was for design of building interior water supply and drainage piping and appurtenances, not sanitary sewers. Furthermore, Hunter cautioned that judgment must be exercised regarding which values to use in his probability function to obtain satisfactory results. His own assumptions were for nearly continuous use of toilets, bathtubs, showers, and faucets. This may be true in large office buildings, hotels, and multi-family dwellings, but may not be warranted for lower-density residential or commercial areas (AWWA 2004).

Use of the probability function is based on assumed values for certain parameters for water supply:

- Average duration of flow for a given type of fixture for one use (e.g., flushing a toilet).
- Average time between successive operations of a given fixture (e.g., time between flushing toilets).
- Flow from a given fixture for each use (e.g., gal/flush for a toilet).

These values were based on experimental observations in the late 1930s for flush toilets and bathtubs only. For toilets, flow ≅ 27 gpm (1.7 L/sec) which gives ~4 gal/flush (15 L/flush) based on 9 seconds to fill. Today's toilets use much less water, as low as 1.6 gal/flush (6.1 L/flush). For other fixtures controlled by faucets, such as lavatories and sinks, where flows and times are a matter of personal habits or preferences, figures are derived based on reasoning that flows from these fixtures would be much less than the peaks from toilets and/or bathtubs. In Hunter's original 1940 publication (Hunter 1940a), there is no discussion of showers, dishwashers, and clothes washers, probably because such fixtures and appliances were not in widespread use at that time.

According to Hunter, the same probability functions and fixture units can be generally be applied to estimates of wastewater loads. Consequently, it is possible to use a fixture unit count in a group of buildings to estimate the peak wastewater flow rate. The general procedure is as follows for building *water supply*, which is a conservative estimate of wastewater flow:

1. Estimate the number of fixture units according to tables in the Uniform Plumbing Code (UPC) or the applicable plumbing code for the jurisdiction.
2. Use the curves in AWWA Manual of Practice M22 (AWWA 2004) to estimate peak demands for indoor water use. This AWWA manual also has a section on irrigation water use, which should not be used for wastewater forecasts.

For drainage, the procedure is not based on deriving peak flow rates but is an indirect derivation of pipe size as follows:

1. Estimate the number of drainage fixture units according to tables in the UPC or applicable plumbing code for the jurisdiction.

2. Use tables in the UPC or applicable plumbing code to determine the pipe size based on the maximum number of drainage fixture units that are allowed on a horizontal pipe. In the UPC, the tables are based on a pipe slope of 0.25 inch/ft (20.9 mm/m or ~0.021) and should be adjusted for actual pipe slopes.

It is suggested that the fixture unit method be used to check results of the previous forecasting methods and *not* be used as a primary forecasting tool.

3.9.3. Summary

The following procedure could be used for the peaking method—separately for the beginning and the end of the design period:

1. Tabulate the average flows for each subarea; this would include all of the subarea flows—including residential, nonresidential, and I/I flows. Each resulting flow is the estimate of average daily flow coming from the subarea.
2. Accumulate the subarea average flows down through the entire system. The resulting accumulated flows are the estimates of average daily flows point-by-point (typically manhole-by-manhole) down through the entire sewer system.
3. Apply peaking factors to each point (manhole) down through the system, thus generating the peak 1-hour flows for all sewer reaches in the system. The peak flows from Q_{avq1} are the Q_{min} values needed for self-cleansing design; those from Q_{avq2} are the Q_{max} values needed for capacity design.

3.10. UNCERTAINTY IN FORECASTS

Forecasts of population, housing, land use, and/or employment described in Sections 3.3 and 3.4 may be based on the best available information, but are not statements of fact. They are considered opinions about the future of a community that cannot be verified until the forecast dates arrive. Consequently, there is a level of uncertainty in forecasting. Engineers can deal with this uncertainty in a logical, rational manner.

- *Two or more planning scenarios.* Often, planners will have multiple planning scenarios that result in multiple forecasts of population, housing, land use, and/or employment. There may also be several unit water factors from Section 3.5 and/or peaking factors from Section 3.9.1. Engineers may then develop two or more wastewater forecasts, typically a maximum and a minimum, to see how those forecasts affect the choice of pipe size.

- *Sensitivity analysis.* With judicious use of computer software, multiple planning scenarios can be analyzed quickly and efficiently. By varying the any of the components of forecasts, such as population, employment, unit factors, etc., the effect on pipeline size(s) can be determined. This type of analysis can be used to answer questions such as what happens to sewer capacity if, for example:
  a. A certain development is approved (or not approved) in 10 years.
  b. The jurisdiction annexes an adjacent area.
  c. There is redevelopment in a certain area that significantly increases water use.

Sensitivity analysis based on optimized designs (lowest cost for the given flow rates and constraints) will provide valuable information to decision makers. Optimization can give cost differential information based on the specific system under design rather than on generalized rules of thumb, and thus allow for better decisions as to capacity and other variables.

REFERENCES

American Society of Civil Engineers (ASCE). (1999). Optimization of collection system maintenance frequencies and system performance. ASCE/EPA Cooperative Agreement No. CX824902-01-0, ASCE, Reston, Va.

American Public Health Association, et al. (APHA) (1981). Glossary, Water and wastewater control engineering, third edition, APHA, Washington, D.C.

American Water Works Association (AWWA). (2004). "Sizing water service lines and meters." *Manual of Water Supply Practices M22*, AWWA, Denver, Colo.

American Water Works Association Research Foundation (AWWARF). (1999). *Residential end uses of water.* AWWARF, Denver, Colo.

Anderson, D. L., and Siegrist, R. L. (1989). "The performance of ultra-low-volume flush toilets in Phoenix." *J. Am. Water Works Assoc., 81*(3), 52–57.

Anderson, D. L., Mulville-Friel, D. M., and Nero, W. L. (1993). "The impact of water conserving plumbing fixtures on residential water use characteristics in Tampa, Florida." *Proc. Conserv93 Conf.*, December 12–16, Las Vegas, Nev., ASCE, AWRA, and AWWA.

Baumann, D. D., Boland, J. J., and Hanemann, W. M. (1998). *Urban water demand management planning*. McGraw-Hill, New York, N.Y.

Boland, J. J. (1998). "Forecasting urban water use: Theory and principles." *Urban water demand management planning*, D. D. Baumann, J. J. Boland, and W. M. Hanemann, eds., McGraw-Hill, New York, N.Y., 77–94.

Breese, J. (2001). "Solving the mixed system problem." *Plumbing Engr., 29*(3), 39–48.

Brown and Caldwell. (1984). *Residential water conservation projects.* Research Rep. 903. U.S. Department of Housing and Urban Development, Office of Policy Development, Washington, D.C.

City of Los Angeles. (1992). *Sewer design manual, Part F.* Los Angeles Bureau of Engineering, Los Angeles, Calif.

Darmody, J. et al. (1996). "Water use surveys—An essential component of effective demand management." *Proc., 1996 Ann. Conf. American Water Works Association,* Toronto, Ontario, Canada, June.

Deoreo, W. D. et al. (2001). "Retrofit realities." *J. Am. Water Works Assoc., 93*(3), 58–72.

East Bay Municipal Utilities District (EMBUD). (2005). *Draft urban water management plan.* EMBUD, Oakland, Calif.

Griffin, D., and Morgan, D. (Undated). "A new water projection model accounts for water efficiency." Canada Mortgage and Housing Corporation, Ontario, Canada. Available online at http://www.cmhc-schl.gc.ca/en/inpr/su/waco/waar/waar_001.cfm, accessed February 27, 2007.

Hunter, R. B. (1940a). *Methods of estimating loads in plumbing systems.* Rep. BMS65. U.S. Department of Commerce, National Bureau of Standards, Washington, D.C.

Hunter, R. B. (1940b). *Plumbing manual.* Rep. BMS66. U.S. Department of Commerce, National Bureau of Standards, Washington, D.C.

Jones, C. V., et al. (1984). "Municipal water demand." Statistical and management issues. *Studies in water policy and management, No. 4,* Westview Press, Boulder, Colo.

Maryland Department of the Environment. (2006). "Wastewater capacity management plans." 2006 Guidance Document. Available at www.mde.state.md.us/assets/document/water/WastewaterCapacityMgmtGuidance.pdf, accessed May 29, 2007.

Mays, L. W. (2004). *Urban water supply management tools.* McGraw-Hill, New York, N.Y.

McKibben, J. W., Bramwell, D., and Gautsch, J. M. (1994). "Wastewater collection system planning with GIS in a large system." *Proc., Urban and Regional Information Assoc. Annual Conf.,* Milwaukee, Wisc.

Metcalf & Eddy, Inc. (1981). *Wastewater engineering: Collection and pumping of wastewater.* McGraw-Hill, New York, N.Y.

Milwaukee Metropolitan Sewerage District (MMSD). (2005). *Cost recovery procedures manual.* MMDS, Milwaukee, Wisc.

Rawn, A. M. (1937). "What cost leaking manhole?" *Waterworks and Sewage, 84*(12), 459.

Texas Water Development Board. (Undated). "Water Quality in Texas: Supply and Demand Projections." Available at www.texasep.org/html/wqn/wqn_6fut.html, accessed May 29, 2007.

U.S. Environmental Protection Agency (EPA). (1978). Management of small waste flows. *EPA 600/2-78-173,* EPA Municipal Environmental Research Laboratory, Washington, D.C.

CHAPTER 4

CORROSION PROCESSES AND CONTROLS IN MUNICIPAL WASTEWATER COLLECTION SYSTEMS

4.1. INTRODUCTION

Sanitary sewers are considered a corrosive environment where many different forms of corrosion can occur. All corrosion processes found in sewers adversely affect both the structure and function of a wastewater collection system, some more than others.

There are many different ways for wastewater collection systems to be damaged from corrosion. This chapter discusses the various types of corrosion common to typical municipal wastewater collection systems and offers proven mitigation alternatives. This chapter can assist the utility in determining where in the collection system corrosion may be occurring and what can potentially be done to correct or control the problem. The information in this chapter is not intended to eliminate the need for professional corrosion engineering assistance.

4.2. CORROSION

4.2.1. History of Corrosion

Humans have been trying to understand and control corrosion for as long as they have been using metal objects as tools and implements. The first corrosion noted by ancient humans was likely corrosion of early metal tools. With significant effort by early man, these tools were wrought from naturally occurring metal ores abundant in the Earth's crust. Unlike our perplexed early ancestors, we know much more about these early identified corrosion processes and have found a few more in the process.

Sewers, collection structures, and pump stations experience unique corrosion processes which not only affect metals but also can destroy concrete, steel, or the entire structure over time. Most of us know that steel will rust and that rusting is accelerated by moisture. Since sewers convey water, we should anticipate that all surfaces associated with sewers should also be wet and we should thus take precautions with iron-based metals. Sewers are also prone to the production of hydrogen sulfide gas, which can be biologically oxidized to sulfuric acid directly on the walls and surfaces of sewer pipes and structures. The combination of moisture and acidic conditions makes sewers particularly corrosive and warrants special attention and consideration.

There is also evidence that hydrogen sulfide gas concentrations in sewers are increasing due to the reduction of metals in wastewater required by the Clean Water Act. Metal ions in wastewater react with sulfide to form an insoluble precipitate so it cannot be released as hydrogen sulfide gas. Certain heavy metals also exhibit a toxic effect on the anaerobic bacteria that produce sulfide in sewers. Removing these metals has caused a general increase in wastewater sulfide and hydrogen sulfide gas. Increases in sewer hydrogen sulfide gas concentration cause an increase in sulfuric acid production and corrosion.

4.2.2. Sources of Corrosion

Corrosion, as experienced in the municipal wastewater industry, can be defined as:

> Any unintentional chemical, physical, biological or electrical process involving the gradual deterioration, degradation or destruction of collection system components that is due to the performance of the intended function of the system.

Rust is simply metal (usually iron-based) that has been oxidized back to its natural state by reacting with oxygen in the air. The presence of water increases the rate of this reaction. Figure 4-1 illustrates external surface oxidation of a steel sluice gate.

Other corrosion processes affecting metal components can also be electrochemical in nature, having the essential features of a battery. If the conditions of the situation allow a current to flow from an area of higher potential (anode) to an area of lower potential (cathode), significant damage to both the anode and the cathode can occur. If a metal component is exposed to different oxygen concentrations in different areas, electrons will flow and cause corrosion. The part of the metal with the lowest oxygen concentration will act like the anode while the higher oxygen concentration part will act like the cathode. This is called an oxygen concentration cell. Depending

FIGURE 4-1. External surface oxidation of a steel sluice gate.

upon the particular situation, the metal can become pitted, flaky, chalky, or discolored, or might disappear completely.

The flow of electrons can also occur between two different metals if placed into contact with each other. The flow of electrons between two metals in contact with each other is governed by the susceptibility of the particular metals to lose or accept electrons; this is termed galvanic corrosion. This effect can be concentrated locally to form a pit or it can extend across a wide area to produce general wastage. Pitting can also provide sites for fatigue initiation and can allow external corrosive agents to penetrate deeper and further degrade the metal. Figure 4-1 also illustrates the severe pitting of an aluminum handrail bolted to a reinforced concrete walkway (the white element on the left side of the photo).

In this picture, electrons are leaving the aluminum (anode) and traveling through the concrete to the reinforcing steel (cathode). The sacrifice of electrons by the aluminum reduces or stops the oxidation of the reinforcing steel.

The flow of electrons in metals can also be induced by unrelated local sources, such as electrical substations, trolley or subway power feeds, or even high-tension power lines. The presence of external stray currents can cause the flow of electrons in metal sewer pipes, causing part of the sewer to act as the anode and part as the cathode, resulting in increased corrosion.

Some industrial discharges contribute to corrosion either by a physical characteristic of the wastewater or through biochemical reactions and interactions that take place in the sewer. Low pH, high biochemical oxygen demand (BOD), high sulfate, and high temperatures can all contribute to accelerated corrosion in wastewater collection systems.

One form of corrosion experienced in some collection systems is also known as erosion. Although erosion may not appear to fall into the realm of corrosion, it fits the classical definition of corrosion as discussed above. In some circumstances, flowing water and the debris load carried in a sewer can cause physical erosion and removal of the invert of the pipe. Although rare, erosion has been known to cut through the reinforcing steel of concrete pipe, causing structural degradation of the sewer.

Perhaps the most well-known collection system corrosion problem is the destruction of concrete pipe and structures by acid produced from hydrogen sulfide gas. Billions of dollars are spent annually repairing sewer damage caused by just this single corrosion process. The damage is not caused by hydrogen sulfide gas alone; common bacteria present in all sewers have the special ability to consume hydrogen sulfide gas and excrete sulfuric acid. These bacteria live above the waterline in sewers where exposure to moisture, carbon dioxide, and oxygen are present. Since bacteria cause this type of corrosion, it is termed microbiologically induced corrosion (MIC).

In this chapter, corrosion processes have been separated into biological and nonbiological processes for convenience of discussion. MIC unquestionably causes the most damage to collection systems and is the result of complex biological and chemical reactions taking place in the sewer. Due to the importance and complexity of MIC in wastewater collection systems, it is discussed in Section 4.3.3.2 and in more detail in Section 4.4.

4.3. NONBIOLOGICAL CORROSION PROCESSES

Although some of the most dramatic corrosion experienced in collection systems results from biological processes generating hydrogen sulfide gas and sulfuric acid, serious damage and loss of service life also occurs through nonbiological corrosion processes. These processes employ chemical, electrical, and physical methods to corrode both metals and concrete; they are discussed in the following sections.

4.3.1. Oxidation of Metals

All metals exhibit a tendency to be oxidized, some more easily than others. In their natural form as ores, most metals are highly oxidized and are combined with other elements in a relatively stable energy state. Workable metals are extracted from ores through the application of a considerable amount of energy (heat) that allows metal atoms to come together and then cool into a crystalline structure with strong electrical bonds. This is what gives metals their unique strength, but it also changes their energy state to a higher, unstable level. Metals are constantly trying to return to their natural, low-energy state through oxidation, which is what we call rusting in the case of iron-based metals. When these metals are then placed in certain environments they will combine chemically with oxygen and other elements, transfer electrons, and form oxidized compounds in an attempt to return to their more natural state with lower, stable energy levels.

Most metals, particularly iron, will start to corrode on contact with water. The oxygen in the water reacts with the iron to form ferrous and ferric hydroxides. Moisture in the air, along with acids, bases, salts, and other solid and liquid compounds, can also cause or enhance oxidation. Metals will also corrode when exposed to gaseous chemicals such as acid vapors, formaldehyde gas, ammonia gas, and sulfur-containing gases.

Most metals commonly used in wastewater collection systems are divalent cations, meaning they have two extra positive charges in their basic atomic structure. The atoms are constantly seeking negative charges to satisfy their electrical stability. It just so happens that oxygen (from water molecules) and sulfide (both found abundantly in sewers) have two excess negative charges (electrons) to trade. Both oxygen and sulfide form ionic compounds upon contact with metals. Hydrogen sulfide will readily give up its two weakly bonded hydrogen atoms to form a strong ionic bond with sulfide ($S^{=}$) to form a metal sulfide. This is what makes hydrogen sulfide exposure turn the surface of copper and brass to a black or dark blue compound of copper and nickel sulfide. Figure 4-2 illustrates this process.

4.3.1.1. Oxidation Controls

The oxidation of metals occurs at the surface. Protecting the surface of metals against contact with oxidizing compounds will prevent oxidation. This simple solution is more difficult to achieve than it sounds. Some metals oxidize to form a thin coating of oxidation product which protects the underlying bare metal. Aluminum is a prime example of this type of oxidation. The thin patina of aluminum oxide on the surface of aluminum protects it against further oxidation. If the surface of aluminum is scratched or abraded to remove this patina, oxygen in the air will quickly re-form this oxidation layer.

Ferrous metals must be protected with a coating to prevent oxidation in the moist, sulfide-prone environment of a sewer. The coating and protection of ferrous metals is an industry unto itself with many techniques

FIGURE 4-2. Color change of brass due to direct reaction with H_2S.

and products available to the engineer. It should be noted that surface profile and preparation are crucial to the success of any coating process.

One of the best ways to prevent metal oxidation in sewers is to avoid the use of metals, or restrict metals to those that can withstand the corrosive environment of a sewer. The only common metal alloy that has been shown to resist the corrosive sewer environment satisfactorily is Grade 316 stainless steel (316 SS). 316 SS is the standard molybdenum-bearing grade of stainless steel, second in importance to 304 SS among the austenitic stainless steels. Austenitic stainless steels have high ductility, low yield stress, and relatively high ultimate tensile strength when compared to typical carbon steel.

A carbon steel, on cooling, transforms from austenite to a mixture of ferrite and cementite. With austenitic stainless steel, the high chrome and nickel content suppress this transformation, keeping the material fully austenitic upon cooling. (The nickel maintains the austenite phase upon cooling and the chrome slows down the transformation so that a fully austenitic structure can be achieved with only 8% nickel.) Along with nickel and chrome, the addition of molybdenum gives 316 SS better overall corrosion resistance properties than 304 SS, particularly higher resistance to low-concentration sulfuric acid environments such as sewers.

Figure 4-3 shows a submersible pump station with 304 SS pump guide rails that have corroded due to the presence of hydrogen sulfide gas.

Stainless steel has a very thin and stable oxide film rich in chrome. If damaged, this film re-forms rapidly by reaction with the atmosphere. The surface can also be chemically passivated to enhance corrosion resistance, if desired (passivation reduces the anodic reaction involved in the corrosion process).

4.3.2. Electrochemical Corrosion

The following paragraphs summarize the basics of electrical corrosion as it applies to the municipal wastewater industry. The discussion is not intended to be a complete treatment of the subject of electrical corrosion, but the reader will gain an appreciation of the causes and implications of corrosion so that appropriate measures can be taken.

The type of corrosion mechanism and its rate of attack depend on the exact nature of the environment (air, soil, water, seawater) in which the corrosion takes place. In today's wastewater collection systems, the waste products of various chemical and manufacturing processes find their way into the air and waterways, and sewers generate many other unfavorable substances. Many of these substances, often present only in minute amounts, act as catalysts, accelerants, or inhibitors of the corrosion

FIGURE 4-3. Corrosion of 304 SS guide rails under moderate H_2S exposure.

process. The corrosion engineer then needs to be on the alert for the effects of these contaminants.

The first step in preventing material corrosion is understanding its specific mechanism. The second and often more difficult step is designing a type of prevention.

4.3.2.1. Oxygen Concentration Cell Corrosion

One of the simplest examples of electrochemical corrosion is oxygen concentration cell corrosion of metal pipe in a collection system. An oxygen concentration cell can develop at any point on a metal surface where there is a difference in oxygen concentration between two points. For instance, if aerobic and anaerobic environments both exist on the surface of the same piece of metal or pipe, an oxygen concentration cell can form. Typical locations of oxygen concentration cells in collection systems are iron or steel gravity sewers or force mains flowing partially full. The area of the pipe in contact with the anaerobic slime will contain no oxygen, whereas the upper pipe is exposed to atmospheric oxygen and dissolved oxygen in the bulk flow. Oxygen cells can also develop under gaskets, wood, rubber, plastic, and other materials in contact with the metal surface where anaerobic conditions can form. Corrosion will occur at the area of low oxygen concentration (anode), which can be anywhere at or below the waterline.

Oxygen concentration cell corrosion can be prevented by sealing the entire surface of the pipe or structure from contact with oxygen with a coating or lining or by using nonferrous materials. Although the conditions for oxygen concentration cell corrosion exist everywhere, they are typically not a significant corrosion concern in collection systems.

4.3.2.2. Galvanic Corrosion

Perhaps the best-known of all corrosion types is galvanic corrosion, which occurs at the point of contact between two metals or alloys with different electrode potentials. An example of this might be aluminum in contact with steel or iron. The aluminum becomes anodic and suffers the effects of oxidation while protecting the iron. Whenever two dissimilar metals are in contact, the potential for the flow of electrons exists. And whenever electrons flow, one of the metals is being oxidized.

A tabulation of the relative strength of this tendency to oxidize is called the galvanic series. Knowledge of a metal's location in the series is an important piece of information to have in making decisions about its potential usefulness in contact with other metals. Table 4-1 is a listing of the galvanic series of common metals used in wastewater engineering.

An obvious area of concern is the use of one type of metal in bolts, screws, and welds to fuse together pieces of another metal. The combination to be desired is the large anode–small cathode combination, rather

TABLE 4-1. Galvanic Series Listing

Alloy	Voltage Range Versus Reference Electrode[a]
Magnesium	−1.60 to −1.63
Zinc	−0.98 to −1.03
Aluminum Alloys	−0.70 to −0.90
Cast Irons	−0.60 to −0.72
Steel	−0.60 to −0.70
Aluminum Bronze	−0.30 to −0.40
Red Brass, Yellow Brass	−0.30 to −0.40
Copper	−0.28 to −0.36
400 Series Stainless Steels	−0.20 to −0.35
Monel	−0.04 to −0.14
300 Series Stainless Steels	−0.00 to −0.15
Hastelloy C-276	+0.10 to +0.04
Graphite	+0.30 to +0.20

[a]A saturated calomel electrode.

than the reverse. Bolts, screws, and so on should be made of the metal less likely to be oxidized so that the bolt or weld is cathodically protected. For example, you can bolt a large aluminum hatch cover down with steel bolts, but do not use aluminum hangers to support a ductile iron pipe.

Similar electrical potentials may also be developed between two areas of a component made of a single metal as a result of small differences in composition, structure, or the conditions to which the metal surface is exposed. Just as in the oxygen cell corrosion discussed above, that part of a metal component which becomes the corroding area is called the anode; that which acts as the other plate of the battery is called the cathode and does not corrode, but it is an essential part of the system. Of great importance is the conductivity of the corroding solution—in this case, wastewater. When large areas of the surface are in contact with wastewater of high conductivity, such as that containing seawater, the attack on the anodic metal may spread far from its contact point with the cathodic metal.

4.3.2.3. *Stray Current Corrosion*

Stray currents, which can cause corrosion in metals, may originate from direct-current distribution lines, substations, or street railway systems, etc. and flow into a pipe system or other steel structure. Alternating currents very rarely cause corrosion. The corrosion resulting from external stray currents is similar to that experienced from galvanic cells (which generate their own current), but different remedial measures may be indicated. Just like the galvanic cell, the corroding metal from stray currents is again the anode from which current leaves to flow to the cathode. Soil and water characteristics also affect the corrosion rate in the same manner as with galvanic-type corrosion.

However, stray current strengths may be much higher than those produced by galvanic cells and, as a consequence, corrosion may be much more rapid. Another difference between galvanic-type currents and stray currents is that the latter are more likely to operate over long distances, since the anode and cathode are more likely to be remotely separated from one another. Seeking the path of least resistance, the stray current from a foreign installation may travel considerable distances along a pipeline, causing severe corrosion only where it leaves the line. Knowing when stray currents are present becomes highly important when remedial measures are undertaken, since a simple sacrificial anode system is likely to be ineffectual in preventing corrosion under such circumstances. Testing and monitoring can be performed to measure the potential threat to metal pipelines from stray current corrosion. If in the vicinity of powerful direct current sources, metal pipelines should be thoroughly protected against stray current corrosion by consulting with a qualified corrosion professional.

4.3.2.4. Soil Corrosion

The response of iron and carbon steel to soil corrosion depends primarily on the nature of the soil and certain other environmental factors, such as the availability of moisture and oxygen. These factors can lead to extreme variations in the rate of corrosion. For example, under the worst conditions buried steel pipe may perforate in less than one year, whereas archeological digs in arid desert regions have uncovered iron tools that are hundreds of years old.

In general, some basic guidelines can be formulated to generally characterize soil corrosion and its severity.

> *Soils with high moisture content, high electrical conductivity, high acidity, and high dissolved salts will be most corrosive.*

One important mechanism in the soil corrosion processes is oxygen concentration cell corrosion (also discussed above), in which the oxygen concentration in the soil varies from place to place. An underground pipe that passes from clay to gravel will have a high oxygen concentration in the gravel region and almost no oxygen in the impermeable clay. The part of the pipe in contact with the clay becomes anodic and suffers damage.

A similar situation is found where a pipe passes under a road. The section under the road (which is more difficult to get to for repair) is oxygen-deprived and will suffer the greatest damage.

4.3.2.5. *Electrochemical Corrosion Controls*

Preventive measures for oxygen cell corrosion involve the use of protective coatings and modification of the moisture and oxygen environment. Coatings are the most common protection used to slow the rate of atmospheric corrosion. Many other materials, such as plastics, ceramics, rubbers, and even electroplated metals, can be used as protective barriers. The corrosion resistance of a metal can be greatly increased by the proper choice of alloys. For example, nickel, chrome, and molybdenum added to steel will increase its corrosion resistance. This is why 316 SS is the preferred metal choice for collection systems.

What can you do to minimize galvanic corrosion? First, always try to eliminate the cathodic metal by making all parts of a structure out of the same material. When this is not possible, use nonmetallic, nonabsorbent washers and insulators between the dissimilar metals to prevent current flow. For example, use plastic or ceramic washers and sleeves to isolate bolts as they pass through a plate of a different alloy instead of fiber and paper washers, which absorb water.

Selection of the proper alloys and connectors is also critical in preventing damage from galvanic corrosion. Knowledge of the position of a metal on the galvanic series is an important factor in the proper selection of metals for collection system use. If dissimilar metals are to be in contact, make the voltage difference between them as small as possible (see Table 4-1, Galvanic Series Chart). When the two metals in a galvanic couple are close together on the series, such as yellow brass and copper, their voltage ranges overlap and either one can be the anode, depending on the exact exposure conditions. When the metals are far apart on the chart, there is a greater tendency for electrons to flow from the anodic metal to the cathodic metal. On rare occasions it may be necessary to use an impressed current to stop galvanic corrosion. Connecting a low-voltage direct current source to the metal to be protected provides a ready source of electrons to the cathode and stops anodic oxidation.

Stray current corrosion can be minimized by identifying sources of stray currents in the vicinity of buried metal infrastructure and insulating or reducing their effect. In the absence of this knowledge, an impressed current with a higher potential than the stray current(s) will protect the structure.

The cure for soil corrosion can be isolation from low-oxygen environments (wrapping pipe in plastic), cathodic protection, or both. Cathodic protection involves the use of a sacrificial anode such as zinc or aluminum. In this situation, the metal to be protected is connected electrically to a buried piece of scrap zinc or aluminum that will take its place as the anode. The anode is destroyed by the corrosion reaction, leaving the cathode intact. This technique is still used extensively to protect underground gas, sewer, and water pipelines from many forms of electrochemical corrosion. Wrapping pipe in plastic sheeting before burial insulates the pipe from varying oxygen concentration soils and groundwater and reduces the flow of electrons from the pipe.

4.3.3. Industrial Discharges

Some industries and other sewer users discharge materials into the collection system that will adversely impact corrosion rates. Some utilities regulate discharges into their collection system for the purpose of protecting infrastructure and preventing upsets to the treatment plant. Many chemicals that are not commonly regulated can have dramatic impact on collection system corrosion.

4.3.3.1. Chemical Industrial Discharges

- *Low pH.* Low-pH discharges are commonly regulated primarily to protect downstream biological treatment processes. Low-pH discharges destroy natural alkalinity in the wastewater which is needed by the treatment plant to offset the impacts of nitrification. If added in sufficient quantity, they can depress the local pH of the wastewater, causing damage to concrete and other cementitious sewer materials. Typical Type II Portland cement concrete can withstand an external pH of 6 nearly indefinitely with only minor surface damage; however, a pH shift to 5 increases the corrosion damage significantly. Regulating discharges to a minimum pH of 6 at the point of discharge is a reasonable and prudent means to prevent concrete damage from low-pH discharges.

  Low-pH discharges also impact metal corrosion by providing a rich source of hydrogen ions (H^+) to react with electrons lost from metals in an oxygenated environment. By consuming electrons, the hydrogen ions are completing the circuit and causing the flow of electrons and accelerated corrosion of the metal. Low-pH waters also enhance the rusting oxidation process by making the oxidized iron more soluble, leading to accelerated corrosion. If the pH of water in contact with aluminum drops below 6, the protective layer of oxide is dissolved, which leads to surface metal loss and potential damaging corrosion.
- *High pH.* High-pH discharges do very little harm to wastewater infrastructure in the normal pH ranges between 7 and 9. Most utilities regulate discharge pH to a maximum of 9 to prevent upset to the

downstream biological treatment facility. When the pH of wastewater exceeds 9, there is greater risk of damage to immersed aluminum since the natural protective covering of aluminum oxide is dissolved in high-pH solutions. Typically, wastewater pH values above 10 would be required to damage immersed aluminum. The use of aluminum should therefore be minimized around caustic storage tanks and inside the spill containment zone. Aluminum flashing should not be used over caustic piping insulation for this same reason. Figure 4-4 shows what happens to aluminum in contact with leaking caustic. Note that the metal is completely dissolved and missing.

- *Reverse Osmosis (RO) Water.* Many industries that intensively use metals make attempts to recover the metals for economic or regulatory reasons. Metal plating, photograph finishing, and microchip

FIGURE 4-4. Aluminum corrosion due to contact with high pH (caustic).

manufacturing are the most common industries recovering metals today. Many times, reverse osmosis equipment is used to capture these metals or other compounds for recovery or recycling. The RO process produces essentially pure water as a reject stream when used in this fashion. Pure water, with little or no buffering by other compounds and elements, can be extremely corrosive to both concrete and metals. When pure water produced by an RO system comprises more than 10% of the flow in the receiving wastewater stream, it can cause serious damage or collapse of a concrete or metal sewer. Concrete sewers have been completely destroyed in less than a year by wastewater containing more than 50% RO-produced water. Utilities should be aware of the serious impacts of pure water RO discharges and should include this discharge on the list to be monitored and controlled. RO water can be rebuffered by allowing it to come into contact with carbonate-containing rocks for a sufficient time before discharge to the sewer. The carbonate rocks will be destroyed in the process and will need replacement.

- *Solvents.* Solvents are commonly regulated by receiving utilities, although there is no standard list of prohibited solvents. Most often, solvent discharges are regulated (based upon flammability and health impacts) to concentrations low enough where municipal wastewater systems are unaffected from a corrosion standpoint. However, many more plastic- and petrochemical-derived materials are being used in wastewater collection systems. Although rare, some of these pipe materials can be damaged or dissolved by high concentrations of some solvents and industrial waste products.
- *High Sulfide.* Wastewater with high sulfide concentrations should be prevented due to the subsequent release of hydrogen sulfide gas and microbiological corrosion. Many utilities regulate sulfide discharges to 0.5 mg/L or less to mitigate corrosion caused by direct sulfide discharges.

4.3.3.2. Microbiologically Induced Corrosion-Enhancing Discharges

Some industrial discharges can accelerate or aggravate MIC. For a detailed discussion of how these industrial discharges accelerate corrosion and how to quantify their effects, see Section 4.4. Some common MIC-enhancing industrial discharges include:

- *High Biochemical Oxygen Demand (BOD).* BOD is a food source for the bacteria involved in sulfide generation and sulfide-related corrosion. Industrial discharges containing high BOD that result in elevated sewer BOD concentrations above 250 mg/L can allow these bacteria to flourish and generate elevated concentrations of sulfide, resulting in elevated corrosion rates. Most utilities either regulate high-BOD discharges or place a surcharge on customers based upon BOD. Typically, the surcharge is based upon the cost of treatment at the plant without regard to the increased odor and corrosion caused by such discharges. The rate of MIC is proportional to wastewater BOD concentration.

- *High Temperature*. High-temperature discharges accelerate biological corrosion by increasing the metabolic rate of the bacteria causing the corrosion. The rate of biological sewer corrosion will double with each 13 °F (7 °C) rise in water temperature. The severity of sewer corrosion will significantly increase downstream of high-temperature discharges. Some utilities regulate the temperature of a discharge without regard to the downstream temperature increase. Many utilities regulate the temperature of discharges to prevent more than a 5 °F (3 °C) rise in the downstream flow. Figure 4-5 presents data that indicate an increase in sulfide production with increasing wastewater temperature.

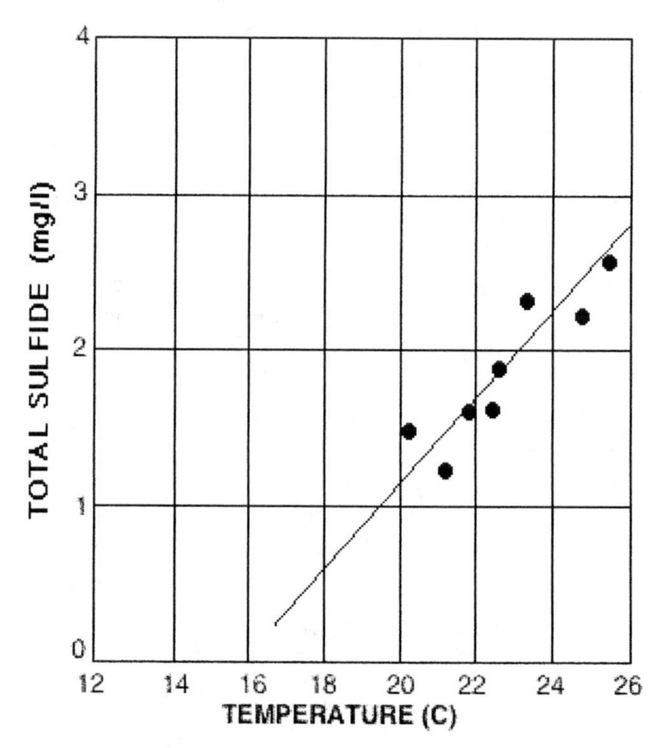

FIGURE 4-5. Water temperature effect on sulfide generation.
Adapted from Pomeroy, R., and F. D. Bowlus. (1946). "Progress Report on Sulfide Control Research," *Sew. Works J.*, 18(4): 597–640.

- *High Sulfate.* The primary source of sulfide in wastewater collection systems is sulfate that has been biologically converted (reduced) to sulfide. Bacteria living in the submerged slime layer in the sewer use sulfate ($SO_4^{=}$) as a source of oxygen and discharge sulfide ($S^{=}$) as a by-product. High-concentration sulfate discharges allow these sulfate-reducing bacteria to flourish and produce elevated concentrations of sulfide, resulting in increased MIC. Figure 4-6 shows data from an experiment to determine the effect of initial sulfate concentration on final sulfide concentration.
- *Seawater.* Although seawater is not an industrial discharge, it does enter wastewater collection systems through infiltration and seawater-flush collection systems. The average seawater sulfate concentration is around 3,000 mg/L, so a little seawater significantly increases the wastewater sulfate concentration. Sewers under the influence of seawater always experience higher rates of sulfide generation, corrosion, and odor complaints than do other sewers. In high-seawater-concentration systems, the increased conductivity of the wastewater also often results in increased electrochemical corrosion processes in metals. The high ionic strength of seawater

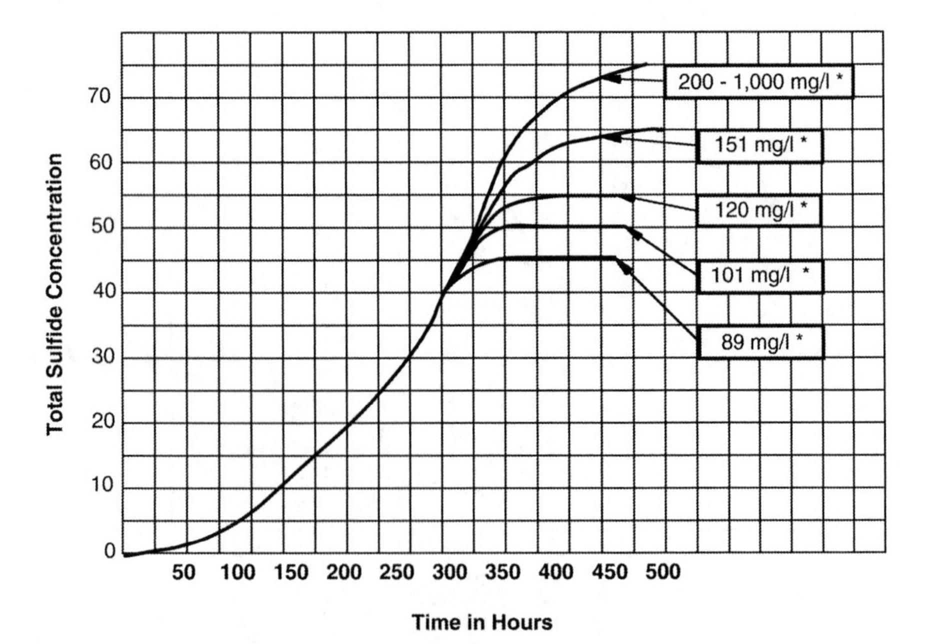

FIGURE 4-6. Sulfate effect on sulfide generation in sewers.
*Initial sulfate concentration.
Adapted from Pomeroy, R., and F. D. Bowlus. (1946). "Progress Report on Sulfide Control Research," *Sew. Works J.*, 18(4): 597–640.

conducts electrons faster than fresh water, allowing more rapid destruction of affected metals.

4.3.3.3. Industrial Discharge Controls

Regulating industrial discharges is a full-time job for most utilities. Many federal, state, and local regulations pertain to the control, pretreatment, and discharge of industrial wastes being discharged to publicly owned treatment works (POTWs). Discharge regulations commonly control pH, temperature, fats, oil and grease, and BOD, and provide measures to monitor these parameters. Some typically unregulated constituents of industrial wastewaters can dramatically impact downstream MIC. As an example, most industries are not monitored for sulfate or sulfite concentrations. Excessive sulfate and sulfite discharges into municipal wastewater provide a rich source of oxidized sulfur, which is used by the slime layer biology and converted to sulfide. The increased sulfide leads to increased hydrogen sulfide generation and increased sulfuric acid corrosion impacts. Figure 4-7 illustrates the change in hydrogen sulfide concentration downstream of a high-sulfate industrial source during a period when the industry was not discharging. As the data suggest, the high-sulfate discharge was responsible for more than 85% of the hydrogen sulfide being generated.

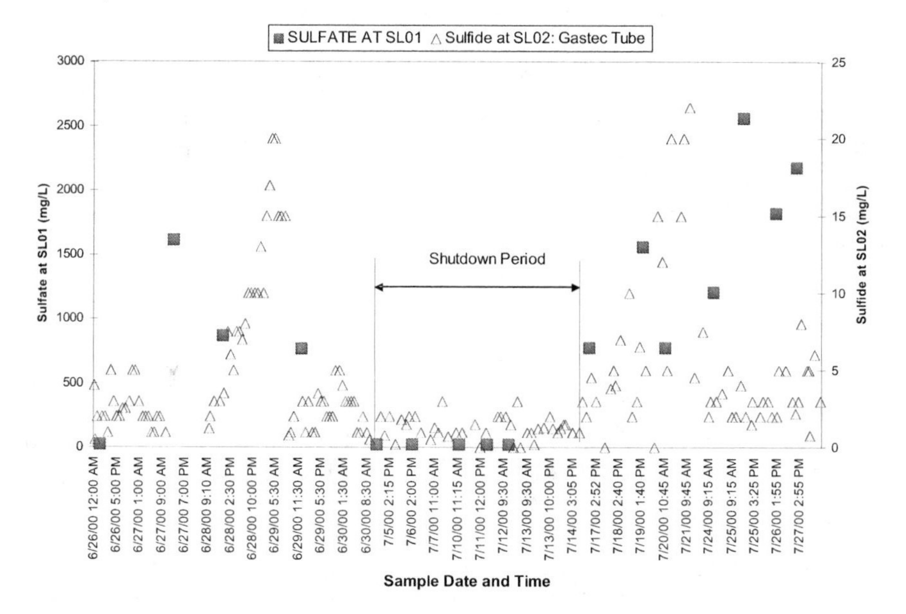

FIGURE 4-7. Effect of industrial sulfate discharge on sewer sulfide generation and hydrogen sulfide gas production.

Regulating industries for excessive sulfate and sulfide has begun in some agencies. Because background wastewater sulfate concentrations vary by location, regulating specific concentrations must be left up to local authorities. Commonly, industrial discharges are allowed to increase the local sewer background sulfate concentration by some agency-established percentage to avoid excessive sulfide-related odor and corrosion impacts.

4.3.4. High-Sulfate Groundwater

High sulfate groundwater concentrations (above 100,000 mg/L), when continuously present, can cause damage over time to exposed concrete and calcium carbonate aggregates. The sulfate reacts with the calcium of concrete, turning it into structurally weak and soluble calcium sulfate (gypsum). Sulfate corrosion is *not* to be confused with MIC, which is caused by biological oxidation of hydrogen sulfide gas to sulfuric acid. Concentrations of sulfate high enough to cause sulfate corrosion of concrete rarely exist in domestic wastewater. Even if such high concentrations were present inside the sewer, they would not support sulfate corrosion due to the protective layer of bacteria covering the concrete surface. However, the high sulfate would cause severe odor and MIC problems.

Damaging concentrations of sulfate can be found in some groundwaters, particularly those under the influence of geothermal or volcanic geology. Scandinavian countries, Hawaii, Japan, and some localized areas of China, Russia, and the United States have groundwater sulfate concentrations high enough to cause severe sulfate corrosion. The use of Type V sulfate-resistant cements can often control sulfate attack from high-sulfate groundwaters. In severe environments (sulfate $> 20{,}000$ mg/L), the outside surfaces of concrete pipes are protected with a coating or liner at the time of burial to protect against sulfate corrosion. Of course, if any of the high-sulfate groundwater would enter the sewer through infiltration it would accelerate MIC, similar to a high-sulfate industrial discharge.

4.3.5. Erosion

4.3.5.1. Water Erosion

A form of corrosion found in municipal wastewater collection systems actually results from physical erosion of concrete and steel by wastewater and its components. Sewer tunnels are in common use today. In order to use these deep, large-diameter systems, the wastewater must be dropped (sometimes considerable distances) into the tunnel. As the water falls downward through the dropshafts, it gains kinetic energy which is dissipated when it strikes the bottom. If the water impinges directly onto concrete or steel, the force of the water and the grit and debris in the water will eventually physically erode the surface. If left to continue, the erosion will eventually breach the pipe and potentially cause structural failure of the pipe. Allowing the water to fall into a plunge pool causes the kinetic energy of the water to be dissipated harmlessly by the water, as illustrated in Figure 4-8.

The plunge pool absorbs the kinetic energy of the falling wastewater without damage to the concrete underneath. Similar erosion can be experienced along very steep-sloped sewers carrying wastewater at supercritical velocity.

4.3.5.2. Debris Erosion

Erosion can also destroy the invert of some collection system trunk sewers and tunnels. All sewers convey a certain amount of grit, gravel, and bottomload. When the conveyed materials are large enough and hard enough, the impacts of tumbling hard debris can take a toll on the concrete invert of a sewer. Invert erosion is more likely in large interceptors that receive flow from older corroding concrete pipe sewers with river gravel (100% silicate) aggregate. Upstream MIC dissolves the concrete matrix and loosens the acid-resistant river gravel aggregate, which then falls into the flow and gets washed downstream. If the volume of the former aggregate is large, the impacts caused by tumbling rocks down the sewer act like micro-jackhammers which, over time, will remove the concrete in the invert of the sewer. In one severe case, erosion actually cut

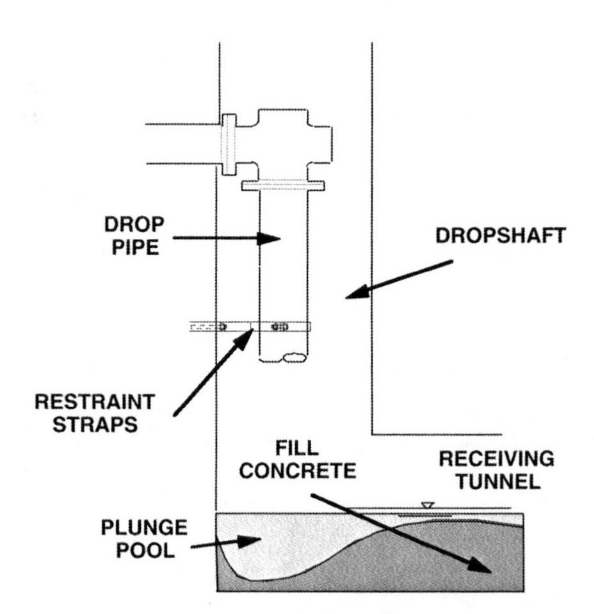

FIGURE 4-8. Typical plunge dropstructure for energy dissipation.

through the concrete and the reinforcing steel in the bottom of the pipe, causing structural compromise. It is possible for erosion to cut completely through a concrete, steel, or plastic pipe. Erosion has been most often associated with major concrete sewers downstream of corroding, unprotected concrete pipe with river gravel or granite aggregate. Most erosion has been noticed in pipes with a diameter range between 42 and 84 inches (1,067 mm and 2,134 mm) and average daily flow velocities greater than 2.5 ft/sec (0.75 m/sec).

4.3.5.3. Erosion Controls

Control of erosion can be accomplished by intercepting the tumbling aggregate in gravel or sand traps constructed in-line on the sewer. Replacing or sliplining the sewer will restore the pipe to pre-erosion structural conditions with a possible loss of pipe carrying capacity; however, the erosion will continue unless the debris is removed or prevented from entering the sewer. It is unlikely that any material that could be reasonably applied to the invert of a sewer would resist erosion for any significant period of time without removal of the debris. Figure 4-9 is a tracing of measured invert erosion in a 72-inch- (1,830-mm)-diameter reinforced concrete pipe (RCP) trunk sewer.

The erosion profile in Figure 4-9 indicates nearly 1.25 inches (32 cm) of concrete and reinforcing steel missing from a steep-sided V-notch only

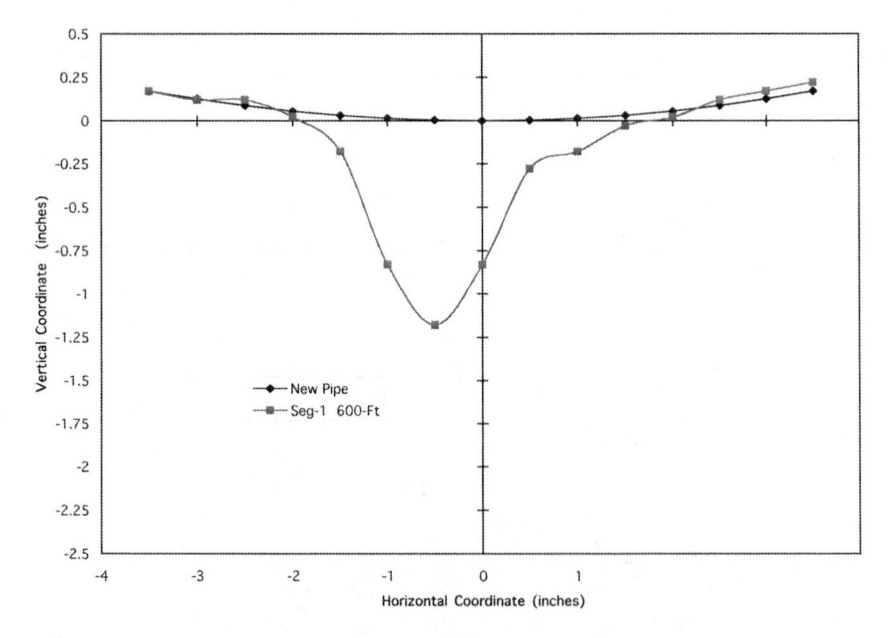

FIGURE 4-9. Measurement of invert erosion compared to original pipe.

3 to 4 inches (75 to 100 cm) wide. Outside this range there was little invert erosion. Sewers with diameters smaller than 84 inches (2,134 mm) are more prone to invert erosion due to the effect of concentrating the impacts along the centerline of the pipe. In a small-diameter pipe, the rock is continually forced back along the centerline of the pipe due to the curvature of the invert. The curvature keeps the impacts concentrated along the centerline. In larger-diameter pipes the angle of the curvature is less, allowing the impacts to wander over a larger area, which spreads out the damage and does not create deep grooves.

4.4. MICROBIOLOGICALLY INDUCED CORROSION PROCESSES

4.4.1. Description

MIC results from a complex series of chemical, physical, and biological reactions that naturally occur in all municipal wastewater collection systems to some extent. The root cause of MIC is sulfate in the wastewater being biologically converted to sulfide, which in turn is chemically converted to hydrogen sulfide gas, which is then biologically converted to sulfuric acid. The following discussion describes all of these conversions in more detail and provides methods to calculate the MIC potential of any collection system.

4.4.2. Generation Processes

4.4.2.1. Dissolved Sulfide Generation

Sulfide generation is a biochemical process occurring in the submerged portion of sanitary sewers. Fresh domestic sewage entering a wastewater collection system is usually free of sulfide. Due to conditions that naturally develop within virtually all collection systems, dissolved sulfide soon begins to appear. These conditions are: development of an active biological slime layer below the water surface in pipes; low dissolved oxygen content in the slime layer; long detention times; and warm wastewater temperatures. The sulfide generation and corrosion process is illustrated in Figure 4-10. The first step in the sulfide generation process is the establishment of a slime layer below the water level in a sewer pipe. This slime layer is composed of bacteria and inert solids held together by a biologically secreted polysaccharide "glue" called zooglea. When this biofilm becomes thick enough, dissolved oxygen cannot fully penetrate and an anoxic zone develops within it. Approximately two weeks are required to establish a fully productive slime layer in new concrete pipes.

Within the anoxic zone, sulfate-reducing bacteria use the sulfate ion ($SO_4^{=}$)—a common component of wastewater—as an oxygen source for

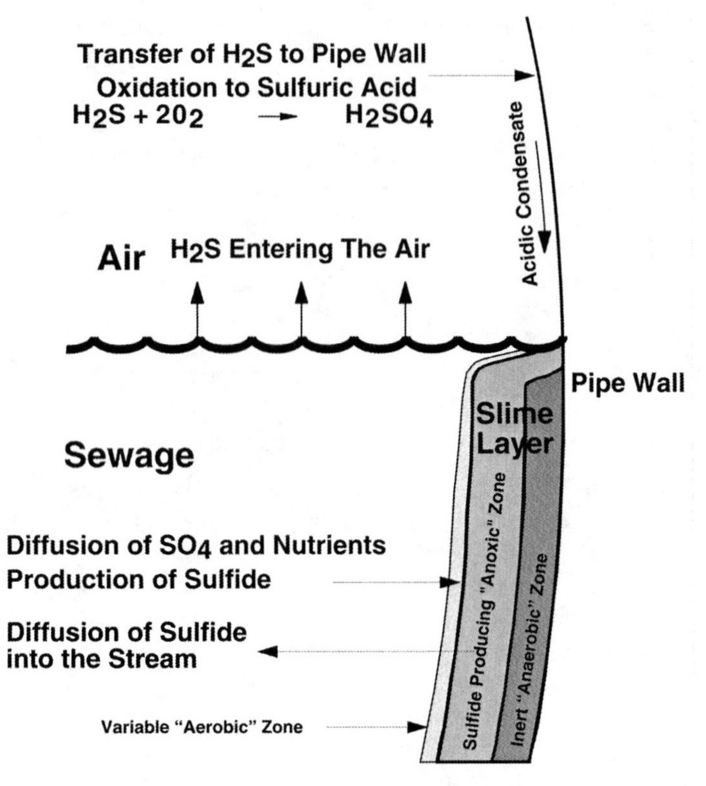

FIGURE 4-10. Sulfide generation in sewers.

the assimilation of organic matter, in the same way dissolved oxygen is used by aerobic bacteria. Sulfate is readily abundant in normal domestic wastewater because it is commonly present in drinking water in more than sufficient quantities to produce sulfide. Even if there were no sulfate in drinking water, the sulfate released by the hydrolysis of sulfur-bearing proteins in sewage is enough to generate enough sulfide to produce severe corrosion. When these bacteria utilize sulfate, the sulfide ion ($S^{=}$) is the by-product. The rate at which sulfide is produced by the slime layer depends on a variety of environmental conditions, including the concentration of organic food source (BOD), wastewater dissolved oxygen concentration, temperature, wastewater velocity, and the normally wetted perimeter of the pipe.

As sulfate is consumed, the sulfide ion by-product is released back into the wastewater stream where it immediately establishes a dynamic chemical equilibrium between four forms of sulfide: the sulfide ion ($S^{=}$), the bisulfide ion (HS^{-}), aqueous hydrogen sulfide [$H_2S_{(aq)}$], and hydrogen sulfide gas [$H_2S_{(g)}$]. The particular dominant sulfide specie is greatly dependent upon wastewater pH, as illustrated in Figure 4-11.

- *Sulfide Ion* ($S^{=}$). The sulfide ion carries a double negative charge, indicating that it can react by giving up two electrons in the outer shell. It is a colorless ion in solution and cannot leave wastewater in this form. It is not present in significant concentrations in the normal pH range of domestic wastewater and does not contribute to odors when in this ionic form.
- *Bisulfide Ion* (HS^-). The bisulfide (or hydrosulfide) ion carries a single negative charge. This is because one of the negative charges of the sulfide ion is taken up by a positively charged hydrogen ion (H^-). The hydrogen ion is obtained from a water molecule (H_2O), leaving one hydroxyl ion (OH^-) behind. It is a colorless, odorless ion that can only exist in solution. It also does not contribute to odors in the ionic form.
- *Aqueous Hydrogen Sulfide* [$H_2S_{(aq)}$]. Hydrogen sulfide can exist as a gas dissolved in water. After reacting with one more hydrogen ion and leaving behind one more hydroxyl ion, the hydrogen sulfide molecule is complete. The polar nature of the hydrogen sulfide molecule makes it soluble in water. In the dissolved aqueous form, hydrogen sulfide does not cause odor. However, this is the only form of sulfide that can leave the aqueous phase to exist as a free

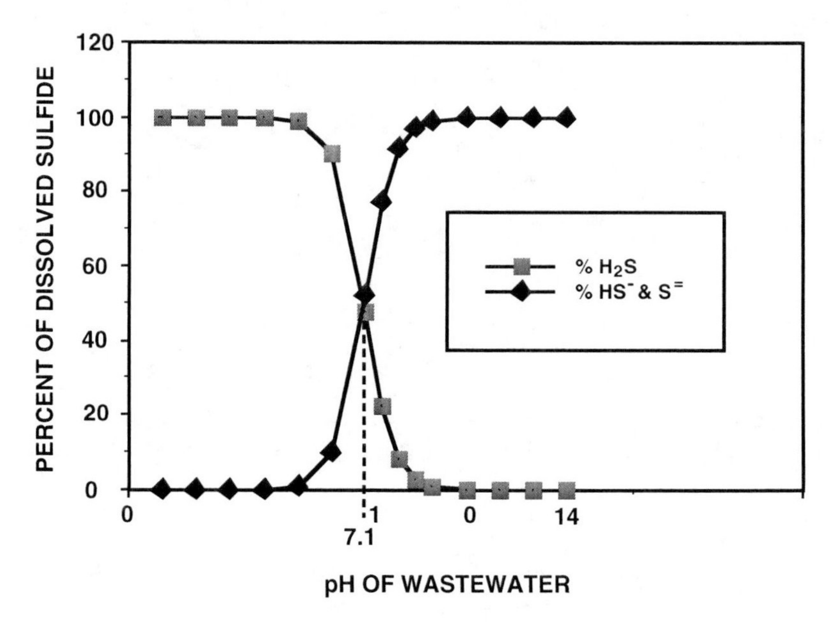

FIGURE 4-11. pH effect of wastewater on sulfide specie present.

gas. The rate at which hydrogen sulfide leaves the aqueous phase is governed by the partial pressure of the gas (Henry's Law), the degree of turbulence of the wastewater, and the pH of the solution. It is important to note that for every mole of sulfate that is biologically converted to hydrogen sulfide gas, two moles of hydroxide alkalinity are produced.

- *Gaseous Hydrogen Sulfide* [$H_2S_{(g)}$]. Once hydrogen sulfide leaves the dissolved phase and enters the gas phase, it can cause odor and corrosion. Hydrogen sulfide gas is a colorless but extremely odorous gas that can be detected by the human sense of smell in concentrations as low as 0.00047 ppm. It is also very hazardous to humans in

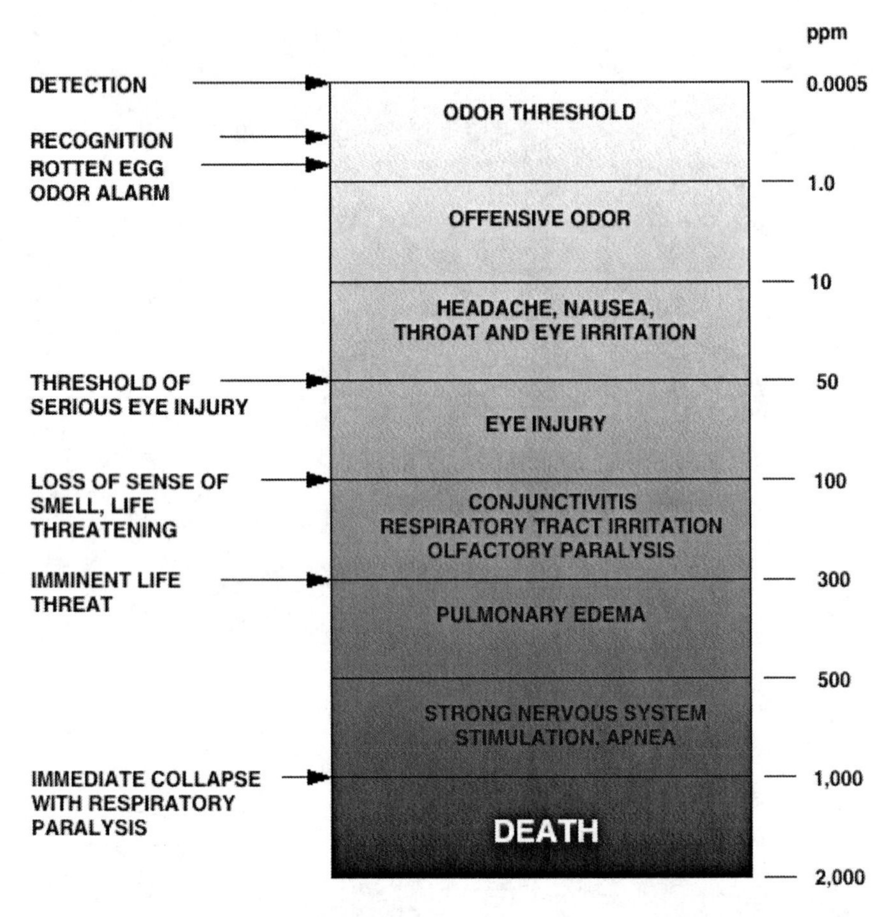

FIGURE 4-12. Hydrogen sulfide gas toxicity spectrum. Source: "Hydrogen sulfide." (1977). Report by the Subcommittee on Hydrogen Sulfide of the Committee on Medical and Biologic Effects of Environmental Pollutants, National Research Council, Washington, D.C.

high concentrations and can cause a number of health-related problems, including death. A hydrogen sulfide odor and toxicity spectrum is presented in Figure 4-12.

In concentrations as low as 10 ppm, hydrogen sulfide gas can cause nausea, headache, and conjunctivitis. Above 100 ppm, it can cause serious breathing problems and loss of the sense of smell, along with burning of the eyes and respiratory tract. Above 300 ppm, death can occur within a few minutes. For these reasons, the Occupational Safety and Health Administration (OSHA) has established an 8-hour, time-weighted, personal exposure limit of 10 ppm. Because it can remove the sense of smell in concentrations above 100 ppm, it is a particularly dangerous gas that tricks its victims into thinking that it is no longer present. It is the cause of death in numerous wastewater accidents every year.

Due to the continuous production of sulfide in wastewater, hydrogen sulfide gas rarely, if ever, re-enters the liquid phase. Sulfide continuously produced by the slime layer replaces that which is lost to the atmosphere as hydrogen sulfide gas in the collection system. In addition, once the hydrogen sulfide gas is released, it usually disperses throughout the sewer environment and never reaches a high enough concentration to be forced back into solution.

The four sulfide chemical species are related according to the following equilibrium:

$$H_2S_{(g)} \Leftrightarrow H_2S_{(aq)} \overset{pKa = 7}{\Leftrightarrow} HS^- \overset{pKa = 14}{\Leftrightarrow} S^{=}$$

Hydrogen Sulfide Gas	Hydrogen Sulfide (dissolved)	Bisulfide Ion	Sulfide Ion

4.4.2.2. Hydrogen Sulfide Release

As indicated by the equilibrium equations above, once hydrogen sulfide is released into the gas phase, the bisulfide ion is immediately transformed into more aqueous hydrogen sulfide to replace that which is lost. Concurrently, sulfide ion is transformed into bisulfide to replace that lost to aqueous hydrogen sulfide. Through this type of continuously shifting equilibrium, it is possible, through stripping, to completely remove all sulfide from wastewater as hydrogen sulfide gas. This is generally not recommended or advantageous due to odor releases and the accelerated corrosion that can take place.

The pKa values above the equilibrium symbols indicate the pH value at which the respective sulfide species on either side of the symbol are in equilibrium (e.g., 50% of the total sulfide on each side of the symbol). The most critical pKa is between the HS^- and $H_2S_{(aq)}$ species where the conversion to the releasable form of hydrogen sulfide is pronounced. Note that the pH value is 7 for this equilibrium, which is in the range of most domestic wastewaters. This means that at the pH of most wastewaters, hydrogen sulfide is very capable of releasing as hydrogen sulfide gas.

4.4.2.3. Biological Sulfuric Acid Generation

Aerobic bacteria, which commonly colonize pipe crowns, walls, and other surfaces above the waterline in wastewater pipes and structures, have the ability to consume hydrogen sulfide gas and oxidize it to sulfuric acid. This process can only take place where there is an adequate supply of hydrogen sulfide gas (>1 ppm), high relative humidity, and atmospheric oxygen. These conditions exist in virtually all wastewater collection systems. The pH of surfaces exposed to severe hydrogen sulfide environments (>50 ppm in air) has been measured as low as 0.3, which is approximately equivalent to a 7% sulfuric acid solution (by weight). The simplified and balanced equation for the biological metabolic process which converts hydrogen sulfide to sulfuric acid is:

$$\underset{\text{Hydrogen Sulfide Gas}}{H_2S_{(g)}} + \underset{\text{Atmospheric Oxygen}}{2\,O_2} \xrightarrow{\text{Bacteria}} \underset{\text{Sulfuric Acid}}{H_2SO_4}$$

4.4.3. Corrosion Processes

4.4.3.1. Metals Corrosion

Most metals will suffer corrosive reactions when exposed to a low-pH environment. A low-pH environment is rich in hydrogen ions (+ charge) that are seeking an electron (− charge) to stabilize their internal charge balance. We know that when metals lose electrons they become oxidized, which reduces their strength. The low-pH environment therefore enhances the corrosion of metals. It is important to note that sacrificial anodes used to protect metals from the galvanic loss of electrons do not protect metals in sewers from oxidation by acids.

Almost all metals used in sewers are susceptible to low-pH corrosion. With the exception of some rare and costly alloys (Monel, Hastelloy C, etc.), only 316 SS can withstand the low-pH environments in sewers. It is important to point out that not all stainless steels are similarly corrosion-resistant. Often 304 SS (also known as "carpenter stainless") is used in sewers where no particular grade of stainless steel is specified. The 304 SS cannot resist low-pH environments as well as 316 SS and will exhibit signs of rusting oxidation very quickly. Similarly, 416 SS does not perform as well as 316 SS in these environments. Only 316 SS can cost-effectively perform in a low-pH environment in sewers. This makes 316 SS the preferred metal to use in the presence of hydrogen sulfide gas in sewers.

Aluminum is also susceptible to low-pH corrosion for the same reasons. The aluminum metal is oxidized back to aluminum sulfate (alum) in the presence of low pH caused by sulfuric acid. There are various alloys of aluminum, some of which offer better low-pH corrosion protection than others. The 6000 series of aluminum alloys (e.g., 6061) is slightly more resistant to acid conditions. Anodizing aluminum hardens the surface but provides only slight extra resistance to acid attack. Aluminum is also susceptible to corrosion in high pH (>10).

4.4.3.2. Concrete Corrosion

The effect of sulfuric acid on surfaces exposed to the sewer environment can be significant. Entire sewer segments have been known to collapse due to loss of structural stability from corrosion. However, the process of concrete corrosion is a step-wise process that can sometimes give misleading impressions. The following discussion briefly describes the general process of concrete corrosion in the presence of a sewer atmosphere containing hydrogen sulfide gas.

Freshly placed concrete has a pH of approximately 12 or 13, depending upon the mix design. This high pH is the result of the formation of calcium hydroxide [$Ca(OH)_2$] as a by-product of the hydration of cement. Calcium hydroxide is a soluble, very caustic crystalline compound that can occupy as much as 25% of the volume of concrete. A surface pH of 12 or 13 will not allow the growth of any bacteria. The pH of the concrete is lowered over time through the effects of carbon dioxide (CO_2) and hydrogen sulfide gas (H_2S). These gases are both known as acid gases because they form relatively weak acid solutions when dissolved in water. CO_2 produces what is called carbonic acid and H_2S produces various sulfur products depending upon conditions. These gases dissolve into the water on the moist surfaces above the sewage flow and react with the calcium hydroxide to slowly reduce the pH of the surface. Eventually, the surface pH is reduced to a level that can support the growth of bacteria.

The time it takes to reduce the pH is a function of the concentration of carbon dioxide and hydrogen sulfide in the sewer atmosphere. It can sometimes take years to lower the pH of concrete from 13 to 9; however, in some severe situations this can be accomplished in a few months.

Once the pH of the concrete is reduced to approximately 9, biological colonization can occur. More than 60 different species of bacteria are known to regularly colonize wastewater pipelines and structures above the water line. Most species of bacteria in the genus *Thiobacillus* have the unique ability to convert hydrogen sulfide gas to sulfuric acid in the presence of oxygen. Because each specie of bacteria can only survive under a specific set of environmental conditions, the particular species inhabiting the colonies changes with time. Since the production of sulfuric acid from hydrogen sulfide is an aerobic biological process, it can only

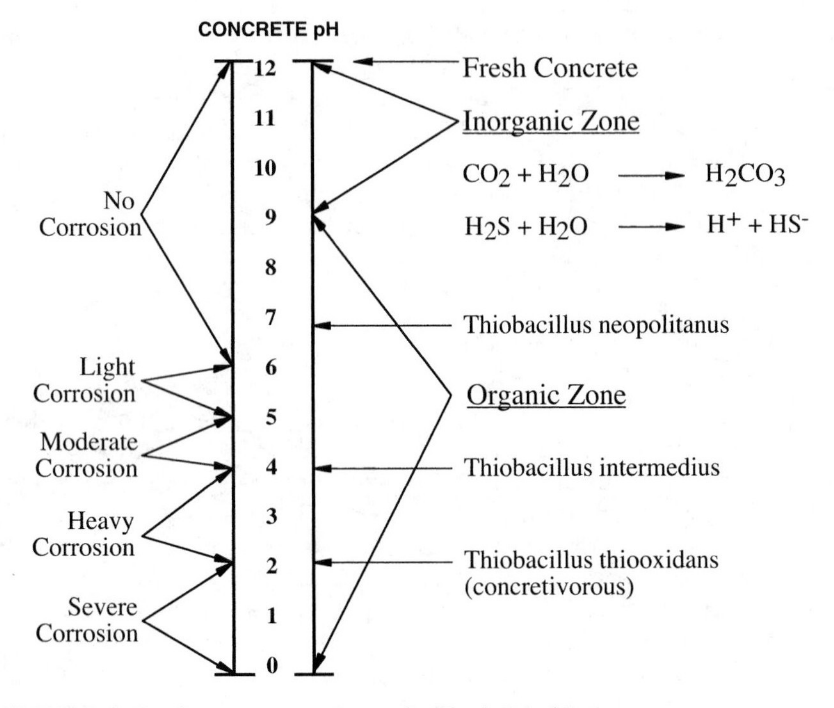

FIGURE 4-13. Concrete corrosion and pH relationship in sewers.

occur on surfaces exposed to atmospheric oxygen. The relationship between inorganic pH reduction and biological pH reduction is illustrated in Figure 4-13.

As a simplified example, one specie of *Thiobacillus* only grows well on surfaces with a pH between 7 and 5.5. However, when the sulfuric acid waste product decreases the pH of the surface below 5.5, the bacteria die off and another specie (which can withstand lower pH ranges) thrives. The succeeding specie grows well on surfaces with a pH between 5.5 and 4. When the acid produced by this specie drops the pH below 4, a new specie takes over. The process of successive colonization continues until species that can survive in extremely low pH conditions take over. One such specie is *Thiobacillus thiooxidans*. This organism has been known to grow well in the laboratory while exposed to a 7% solution of sulfuric acid. This is equivalent to a pH of approximately 0.5. It is not uncommon to measure pH values below 1 on surfaces exposed to aggressive corrosion in wastewater systems. An airborne concentration of hydrogen sulfide gas above 30 ppm is sufficient to generate a surface pH of 1 under some conditions.

Sulfuric acid attacks the matrix of the concrete, which is primarily composed of calcium silicate hydrate gel (CSHG), calcium carbonate from aggregates, and unreacted calcium hydroxide. Although the reaction products are complex and result in the formation of many different compounds, the process can be generally illustrated by the following reactions:

$$H_2SO_4 + Ca,Si \Rightarrow CaSO_4 + Si + 2\,H^+$$
$$H_2SO_4 + CaCO_3 \Rightarrow CaSO_4 + H_2CO_3$$
$$H_2SO_4 + Ca(OH)_2 \Rightarrow CaSO_4 + 2\,H_2O$$

The primary product of concrete decomposition by sulfuric acid is calcium sulfate ($CaSO_4$), commonly known by its mineral name, gypsum. From experience with this material in the form of drywall board, we know that it does not provide any structural support, especially when wet. It is usually present as a pasty white mass on concrete surfaces above the scour line. In areas where diurnal or other high flows intermittently scour the walls above the water line, concrete loss can be particularly rapid. It has been demonstrated that in washing off the "protective" coating of gypsum, fresh surfaces are exposed to acid attack, which accelerates the process.

The color of corroded concrete surfaces can also be various shades of yellow, caused by the direct oxidation of hydrogen sulfide to elemental sulfur. This only occurs where a continuous supply of atmospheric oxygen or other oxidants is available. The upper portions of manholes and junction boxes exposed to high hydrogen sulfide concentrations are often yellow because of the higher oxygen content there. This same phenomenon can be observed around the outlets of odor scrubbers using hypochlorite solutions to treat high concentrations of hydrogen sulfide gas.

The rate of concrete loss is dependent upon a number of factors, of which hydrogen sulfide gas concentration is only one. It is fairly common to see a rate of concrete loss of 1 inch (2.54 cm) per year in high-sulfide environments.

The production of sulfuric acid by biological oxidation of hydrogen sulfide gas can cause significant damage to structures associated with a wastewater collection and treatment system. Both concrete and steel can be attacked and destroyed by long-term exposure to even low concentrations of hydrogen sulfide gas. Manholes, concrete pipe, concrete wet wells, and even force mains and submerged sewers (see Section 4.4.3.2.1., Crown Corrosion) have been damaged and have failed from biogenic sulfuric acid attack. The prevention of this type of corrosion is the major reason for upstream sulfide controls in many wastewater collection systems.

Corrosion of gravity sewer pipe made of cement-bonded materials is not uniform. Lack of uniformity is due in part to the air currents that control the rate of dissolved sulfide release and transfer of H_2S to the pipe wall. The greatest corrosion is generally observed where turbulent force, main discharge conditions exist and for a distance downstream. The distance downstream that corrosion caused by force main discharges can extend is a function of the velocity of the air in the sewer headspace and the rate of biological uptake by acid-forming bacteria. Where steep slopes flatten and air velocities slow, the rate of transfer will increase. Corrosion is also greatest at the downstream soffit of the gravity sewer manhole, as this is where a greater mass of hydrogen sulfide physically impacts the surface. Test specimens hung in a sewer manhole may provide information on the relative corrosion rates of different materials, but they will not indicate how fast a pipe wall will corrode.

There is normally a flow of air down the sewer but, in addition, transverse currents are set up by temperature differences. The pipe wall is normally cooler than the water, especially in the summer when sulfide concentrations are at a maximum. The air that is cooled by the walls moves downward and slightly warmer air rises from the center of the stream surface. As a result, the maximum rate of transfer of H_2S to the pipe wall is at the crown.

Uneven distribution of corrosion also results from the migration of acid-containing condensate down the pipe wall, particularly when there is a high rate of acid production. In the zone that is intermittently washed by the wastewater during diurnal flow variations, the pasty decomposition products are cleaned away. As a result, the pipe wall is laid bare to the attack of the acid when the water level is low. Deeper penetration may therefore be observed in this zone.

Corrosion of force mains occurs when an air pocket exists in the pipe, usually at high points in the line where air release valves are inadequate or nonexistent, or in long downhill stretches. If the air pocket exists for any length of time, sulfide-oxidizing bacteria will produce sulfuric acid and corrosion will occur.

A more dramatic effect of force main-related corrosion may be noticed in manholes and lift station wet wells where force mains discharge. If the discharge of the force main is turbulent, great quantities of hydrogen sulfide are stripped from the wastewater. The resulting high concentrations of H_2S feed the acid-producing bacteria and can cause severe corrosion to surfaces above the normal high water level. The washing effect of the fluctuating wastewater level provides some measure of protection for surfaces between the high- and low-water levels, and corrosion does not usually occur below the low-water level since those surfaces are always submerged.

Corrosion of gravity sewers is also influenced by the moving mass of air in a sewer. As air moves down a gravity sewer, its inertia forces it against the outside of a turn or bend in the pipe. The increased presentation of hydrogen sulfide gas to the outside of a bend causes the production of more acid and increased corrosion at that location. The downstream soffit of manholes is also prone to increased corrosion for the same reason.

4.4.3.2.1. Crown Corrosion

Since the production of sulfuric acid requires oxygen, carbon dioxide, hydrogen sulfide gas, and water, it would typically be assumed that MIC would not be present in a full-flowing gravity sewer. Full pipes typically do not allow gases to be present. However, a recently identified phenomenon called crown corrosion has been found in concrete pipes which intentionally flow full. The process starts when entrained air bubbles in the flow coalesce on the top of the crown and move to a point in the pipe where the bubble can remain in place. This is typically at a joint in RCP or any high spot in the line where hydraulic shear cannot dislodge the air bubble. All of the conditions required for MIC will exist inside this bubble of air. As acid is generated in the bubble, concrete corrodes and makes the air pocket larger. Over time, the pocket can become quite large and even breach the crown of the pipe. Figure 4-14 is a photograph of some crown corrosion in a normally full-flowing gravity pipe. Note that the corrosion starts at a joint and then proceeds upward as the bubble gets larger and more acid is generated. The corrosion does not extend downstream of the

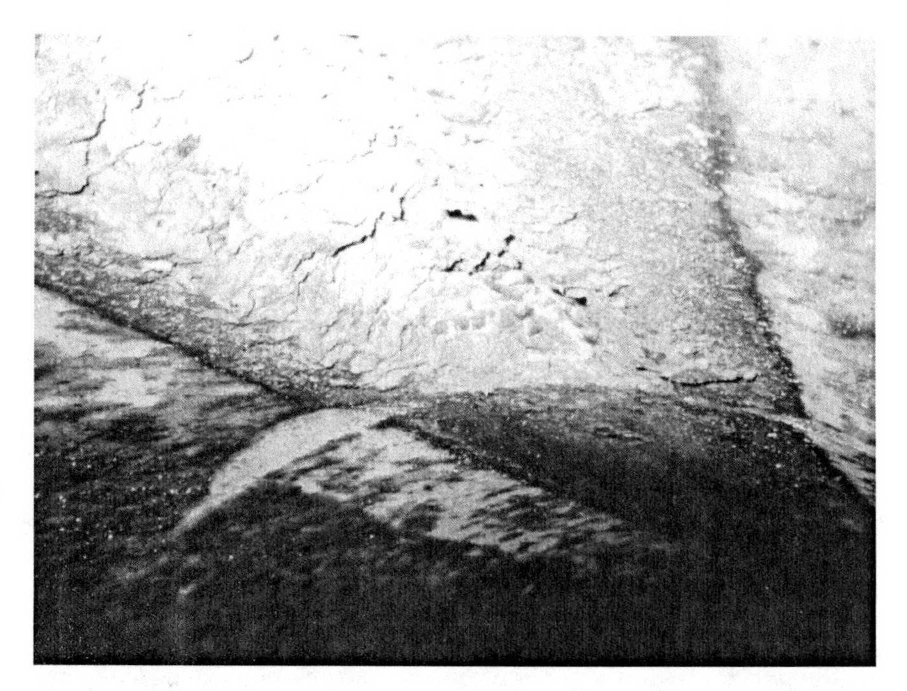

FIGURE 4-14. Typical crown corrosion in a full-flowing pipe.

joint where the bubble originally formed since it is continuously wet and not exposed to gases.

4.4.3.2.2. Corrosion of Cementitious Materials

The corrosion of concrete and other cementitious materials is of primary concern in wastewater systems. Concrete is an extremely versatile and inexpensive construction material, particularly for large hydraulic structures and pipes. Therefore, when this universal building material cannot perform adequately, it presents a significant challenge for the designer.

In general, with conventional concrete mix designs using typical Portland cements, concrete has the ability to withstand moderately low pH surfaces for long periods of time. The generally accepted ranges for corrosion categories and surface pH values are:

- *Severe Corrosion.* This category of concrete corrosion is characterized by significant measurable concrete loss or active corrosion. There is exposed aggregate and occasional exposed reinforcing steel. The original concrete surface is not distinguishable. The surface is covered with soft, pasty corrosion products where active scouring is not present. There is generally a depressed wall pH (<3), indicating active corrosion.
- *Moderate Corrosion.* This category of concrete corrosion is characterized by some concrete loss with aggregate slightly exposed, but the original concrete surface is still distinguishable. The surface may have a thin covering of pasty material which is easily penetrated. There is generally a depressed wall pH (>3, <5), indicating moderately corrosive conditions.
- *Light Corrosion.* This category of concrete corrosion is characterized by a slightly depressed pH (>5, <6) and a concrete surface that can be scratched with a sharp instrument under moderate hand pressure with the removal of some concrete material. The original concrete surface is fully recognizable and aggregate may or may not be exposed.
- *No Corrosion.* This category of concrete corrosion is characterized by normal pH ranges (>6) and a normal concrete surface which cannot be penetrated or removed by a sharp instrument under moderate hand pressure. The surface of the concrete may have biological growth and moisture but the concrete is normal and the aggregate is not exposed.

4.4.3.2.3. Hydrogen Sulfide Gas Concentrations

The principal driving force for MIC in wastewater systems is biogenic sulfuric acid production. The precursor for sulfuric acid is hydrogen sulfide gas. Without hydrogen sulfide gas, there would be no sulfuric acid.

The question remains: How low do the hydrogen sulfide gas concentrations need to be before corrosion stops? Another way of asking this same question is: What is the highest concentration of hydrogen sulfide gas that can be tolerated in a collection system before giving rise to concern about corrosion?

There are two definitive works on the subject. One was a paper produced by the Los Angeles County Sanitation Districts that demonstrated that a H_2S gas concentration of 2 ppm or less did not allow the growth of *Thiobacillus* and therefore prevented serious pipe damage. Once *Thiobacillus* bacteria colonize, significant pipe damage can occur.

Another study from South Africa related that *Thiobacillus* bacteria could exist in concentrations between 2 to 5 ppmv. This tends to make one believe that *Thiobacillus* bacteria become substrate-limited at H_2S gas concentrations below 2 ppmv. Thus, 2 ppmv would be a good number to use for continuous concrete pipe exposure in a sewer environment without producing sulfuric acid.

The problem with using this approach is that the hydrogen sulfide gas concentration in a sewer is never constant. It is always changing—rising or falling. In fact, recent studies on the effect of the Clean Water Act on sulfide production in domestic wastewater show that sulfide concentrations are rising dramatically as a result of regulations.

4.5. CORROSION PREDICTION MODELS

The power of computers has allowed a much faster evaluation of the rather complicated equations required to predict corrosion. The following discusses the development of corrosion models and their limitations.

4.5.1. Corrosion Model Development

The equations most commonly used to estimate the rate of corrosion due to hydrogen sulfide-induced corrosion come from research conducted by the U.S. Environmental Protection Agency (EPA 1992). These empirical equations were originally presented in a joint publication by the EPA, ASCE, and the American Concrete Pipe Association (EPA 1992).

The corrosion prediction model can be used to identify, locate, and define areas susceptible to corrosion and estimate (not predict) rates of corrosion. Reasonable assumptions must be made in order for the model to provide reasonable results. The more data that is based on actual inspection and testing, the more reliable the model results will be. Any reliable dissolved sulfide prediction model can be used to provide average dissolved sulfide concentrations for different wastewaters, at different times of the year, and under different conditions. Model predictions should be confirmed with as much field data as possible.

The average rate of sulfide-related corrosion is calculated through the use of two primary equations. The first equation calculates the sulfide flux (the rate transfer of hydrogen sulfide from the wastewater to the sewer pipe). A common form of the flux equation used in the model is:

$$\text{Flux} = 0.45\,[(s\,v)^{3/8}]\,(j)\,DS\,(b/p')$$

where

- Flux = rate of hydrogen sulfide transfer to walls; g/m^2-hr
- 0.45 = conversion from meters to feet
- s = energy grade line
- v = velocity of stream; ft/sec
- j = factor relating the fraction of dissolved sulfide present as $H_2S_{(g)}$ to pH
- DS = dissolved sulfide concentration; mg/L
- b = stream width; ft
- p' = perimeter of pipe exposed to atmosphere; ft.

Following input of the system physical parameters, the model calculates b and p', as well as velocity and the j factor. The sulfide concentration and wastewater pH are required. The average rate of corrosion of the exposed pipe perimeter is calculated based the sulfide flux. The form of the equation is:

$$C_{avg} = \frac{0.45\,(k)\,\text{flux}}{A}$$

where

- C_{avg} = average rate of corrosion, inches/year
- 0.45 = conversion factor from meters to feet
- k = incomplete acid reaction factor, ranging from 0.3 to 1
- A = concrete alkalinity, expressed as calcium carbonate, equivalent decimal fraction.

The k factor estimates the efficiency of the acid reaction considering the estimated fraction remaining on the wall. The k factor approaches unity for nearly complete reaction. Pipes with frequent flushing or with high humidity and high rates of sulfide production would have lower k values because acid would tend to rinse or drip off the wall before reacting. The A value for granitic aggregate concretes ranges from 0.17 to 0.24; for calcareous aggregate concretes, A ranges from 0.9 to 1.1. The equivalent A value for mortar-lined pipes is 0.4, for asbestos cement, 0.5, and for ferrous pipe a value of 0.5 should be used to account for the attack of the sulfuric acid as well as the direct attack of the hydrogen sulfide gas.

The average rate of corrosion represents the total loss of pipe over the total surface area considered. As discussed above, the crown of the pipe tends to have a greater rate of corrosion than the intermediate perimeter. Corrosion is also higher in areas of greater turbulence because of increased sulfide release. In order to calculate the maximum rate of corrosion, these two factors must be taken into account. The following equation is used to calculate the maximum rate of corrosion:

$$C_{max} = C_{avg} * CCF * TCF$$

where

CCF = crown corrosion factor, typically 1.5 to 2

TCF = turbulence corrosion factor, typically 1 to 2.5 for well-designed dropstructures and 5 to 10 for drops or other turbulent junctions.

Example Analysis

Based upon some field sampling data and some estimated values, a prediction of corrosion rates was performed for the sewers downstream of various force main discharges for a large East coast municipality. Table 4-2 contains corrosion model predictions for a typical collection system. The universal estimated data for all model runs was:

Acid Reaction Factor (k) = 0.85 (indicates moderate reaction)
Concrete Alkalinity (A) = 0.80 (normal for concrete pipe with limestone aggregate)
Crown Corrosion Factor (CCF) = 1.5 (moderate)
Turbulence Correction Factor (TCF) = 5.0 (high for gravity sewers)
Slope = 0.10 (ft/ft)

TABLE 4-2. Example Results of Corrosion Modeling on Gravity Sewers Downstream of Force Mains

Location	Wastewater pH	Sulfide (mg/L)	Average Rate (in./yr)	Maximum Rate (in./yr)	Sewer Diameter (in.)
Horse Pen Pump Station	7.4	2.3	0.0219	0.1642	36
Muirkirk Pump Station	7.4	1.1	0.0086	0.0597	18
Atwood Grinder System	7.9	3.8	0.0038	0.0286	8
Arcola Pump Station	7.5	1.4	0.0048	0.0359	12
Olney Pump Station	7.3	5.6	0.0530	0.3973	36
Squaw Hill Grinder System	7.9	6.8	0.0114	0.0853	8
Seneca Pump Station	7.4	2.0	0.0294	0.2204	48
Willow Ridge Grinder System	7.4	8.5	0.0495	0.3716	12

The information most likely contains some degree of inaccuracy due to estimations; however, it represents a comparison of the relative corrosion rates with most factors held constant. It also indicates the possible magnitude and relative severity of the corrosion problems in common concrete pipe.

4.6. SULFIDE CORROSION CONTROL

4.6.1. Chemical Controls

There are many different chemical control options that could be used effectively to reduce or eliminate odors in collection systems. The majority of these chemicals are used specifically for the treatment of sulfide-related odors and corrosion. Although chemicals have been used very effectively to control collection system odors in many applications, a thorough evaluation of all odor control options must be performed before selecting a chemical treatment for corrosion control. Each case is unique, with site-specific parameters which may or may not make chemical treatment feasible or economical. When an evaluation indicates that chemical control options are viable, the results can be economical and provide effective odor and corrosion control.

In situations where sulfide is causing both odor and corrosion in the collection system, cost/benefit analyses which include the reduction of corrosion often indicate that it is more cost-effective to add chemicals to the collection system than scrub odorous air at the treatment plant. It should be recognized that corrosion and hydrogen sulfide gas concentrations in wastewater collection systems vary both diurnally and seasonally. To achieve efficient and economical odor control using chemicals, regular monitoring of sulfide concentrations and adjustment of chemical dosage rates are required. This requires an investment in training for operations personnel and the necessary sampling and analytical equipment.

Chemical additives come in many forms and react to control odors in several different ways. The following sections briefly discuss the various chemical treatments for odor and corrosion control in greater detail, but do not include their chemistry, process description, or operational and maintenance considerations. The interested reader is directed to Chapter 10 of the Water Environment Federation's Manual of Practice 25—*Control of Odors and Emissions from Wastewater Treatment Plants,* which contains much more information on the specifics of chemicals used for sulfide control in sewers.

Chemical additives work in different ways to stop or reduce sulfide production by bacteria in the slime layer; they do not react with hydrogen sulfide gas directly. This is a valid mechanism, since without sulfide production there will be no hydrogen sulfide gas. Some chemical additives increase the oxygen concentration of the wastewater. We know that oxygen is used preferentially by the slime layer bacteria instead of sulfate when present, hence there is less sulfide production in aerobic wastewater. The oxygen can come in the form of atmospheric air sparged into the wastewater, pure oxygen injection, or chemically bound oxygen found in products such as nitrate solutions.

Some chemical additives combine with the sulfide before it can be released as hydrogen sulfide gas. Iron compounds are most commonly used for this purpose, although zinc, nickel, copper, and many other metal ions will also perform as well as or better than iron. The main difference is the toxicity of zinc and other metals and the fact that the Clean Water Act prohibits them. Iron compounds combine with the dissolved sulfide and form a relatively insoluble precipitate of iron sulfide, which can no longer be released as hydrogen sulfide gas. This is a stoichiometric reaction, so if the sulfide concentration is high it will require more iron solution.

Still other chemical additives are strong oxidants which react with sulfide and other reduced compounds in wastewater. In these reactions, the sulfide is typically oxidized to other sulfur-containing compounds which are not volatile (sulfate, sulfite). Examples of these types of chemicals are hydrogen peroxide, potassium permanganate, and various forms of chlorine. Most of the oxidant chemicals are also hazardous to handle and store.

Due to the large number of chemicals that could be used and their many specific reactions, chemicals used for hydrogen sulfide control in sewers must be selected with assistance from a professional. The subject cannot be treated in a detailed manner in this chapter.

4.6.2. Coatings

The class of products that are applied to concrete surfaces as liquids or semi-solids and then allowed to chemically cure to a final product are called coatings; they generally fall into the categories of epoxies and urethanes. Many different formulations and combinations are available. Evaluation of the many different formulations is not possible within the context of this discussion—they must be evaluated on a case-by-case basis. The construction of these materials often leaves microscopic pinholes in the coating, even when applied under strict supervision and inspection. It is extremely difficult to spray or apply coatings under the unfavorable conditions of sewage, temperature, and humidity found in sewers. The pinholes and other defects formed in the coating provide a route for the acid to penetrate to the concrete and cause failure of the coating's anchor or attachment mechanism. In some instances, the organic components of various coatings have been found to be biodegradable and digested by the aggressive biological environment in a sewer.

Some general guidelines for coatings use for corrosion protection in wastewater collection systems are:

- Surface preparation is critical to achieving a successful coating. A certified corrosion engineer should assist with all phases of coating application, particularly surface preparation.
- An appropriate primer (typically 100% solids epoxy) should always be used. This is particularly critical when coating concrete.
- The topcoat should be applied in two separate coats of equal thickness with appropriate set/cure time between coats. The topcoat should be compatible with the primer system and suitable for the service required.
- The finished coating system should be spark-tested to detect pinholes. All pinholes or defects should be corrected.

In general, coatings with fillers or extenders which serve no corrosion protection function should not be used. An example of an organic extender/filler is coal tar. Coatings also contain varying amounts of silica dust and other inorganic materials as fillers. Coatings should be used sparingly in sewers where visual inspection is difficult and periodic, since failures can occur quickly and cause significant damage. Coatings generally work better in manholes and wet wells where visual inspection is more frequent. The primary concern with coatings is to catch the failure before damage can occur. It is not a question of *will* a coating fail as much as *when* a coating will fail.

4.6.3. Liners

The other general class of concrete protection methods consists of a thick, pinhole-free, manufactured sheet of polyethylene (PE/HDPE), polyvinyl chloride (PVC), or other impervious plastic material that is securely fastened to the concrete surface. The primary difference between coatings and liners is the virtual absence of pinholes in a sheet of plastic manufactured under ideal factory conditions. Another advantage that liners have over coatings is resistance to hydrostatic head and vapor pressures generated by groundwater at great depths. Coatings must anchor to the concrete substrate, which means that hydrostatic and vapor pressures exerted through the porous concrete wall must be 100% physically resisted by the coating material. Coatings have not demonstrated the ability to withstand these types of forces. Practically 100% of inspected coatings under similar conditions have blisters and failures within two years of application. Most liner systems do not work by resisting hydrostatic and vapor pressure effects, but instead allow water and air a relief pathway from the concrete to the sewer environment—this does not defeat the corrosion protection mechanism. The methods of pressure relief provided by different liner systems are often unique and inventive; however, this is an extremely important aspect of liner design.

In 1953 the City of Los Angeles installed a large-diameter concrete pipe with plastic sheet liners attached to it to prevent hydrogen sulfide-related corrosion (Ameron T-Lock™). That 50 year-old sewer is showing every indication that it will perform for another 50 years with zero corrosion. Since no commercially available coating has remained in continuous service without maintenance for this period of time, it is clear that liners provide a long-term advantage that cannot be offered by coatings. For this reason, coatings are not discussed further in this chapter as a viable method for large-diameter sewer corrosion protection.

4.6.3.1. Liner Systems

There are many different types of liners that are intended for many different types of sewer applications. Some liner systems have been developed for small-diameter sewer rehabilitation, such as the fold-and-form type of expandable liners. These liners are mostly PE or polypropylene (PP) and are folded and pulled into the sewer to be rehabilitated. Once in place, the pipe is expanded with air or water pressure to fit tightly against the host pipe. Some grouting of the annular space may be required to prevent dislocation. Other types of liner systems utilize a cured-in-place technology where a circular felt bag full of chemical resins is inserted in the pipe, expanded against the host pipe, and subsequently cured with hot water or air. These liners are only suitable for smaller-diameter sewer rehabilitation since they require significant installation and curing time which necessitates full-flow bypass pumping, and therefore are not appropriate for this subject matter.

Some types of liners, called slip liners, are intended for large-diameter sewer rehabilitation only and do not require full-flow bypass pumping. These liners are thin shells of corrosion-resistant piping [PVC, PE, glass-reinforced plastic (GRP), fiberglass-reinforced plastic (FRP)] which have an outside diameter roughly equal to the inside diameter of the pipe being rehabilitated. By pushing these pipe shells one after the other into a corroded pipe, the sewer is effectively lined against corrosion. Grouting of the annular space secures the sections in place. Slip liners can be designed to provide both corrosion protection and structural enhancement of the deteriorated pipe if necessary.

The following discussion will focus primarily on those liner systems that are appropriate for new concrete sewer construction and concrete sewer rehabilitation. More specifically, the liner systems discussed are those liners that are not intended to provide structural enhancement of the host pipe. Therefore, these liner systems are appropriate only for situations where the host concrete pipe is new or still fully capable of supporting the intended static and dynamic design loads and does not require structural enhancement. Although many of the liner systems discussed in this chapter are capable of providing some additional measure of structural enhancement, designing these systems for this purpose is not the focus of the technology. Identifying liner systems that provide corrosion resistance, security of attachment, and long-term protection of a concrete sewer infrastructure are the primary discussion topics.

4.6.3.2. *Rehabilitation Versus New Construction*

When considering sewer corrosion protection, it is always more economical to provide protection during the original design and installation of the sewer as opposed to rehabilitation. The average cost increase for liner protection systems when installed as part of the original construction ranges from 4% to 10% of the cost of the pipe on an installed basis. Typical rehabilitation costs for large-diameter, deep sewers can range from $2,200 to $3,800 per linear meter (depending upon diameter and method used)—nearly equal to the original cost of installation. However, there is ample need for sewer rehabilitative services and liners are still the preferred choice, and there are several liner systems more adapted to rehabilitation than new construction. Rehabilitating a large-diameter, deep sewer presents construction challenges that are often difficult and costly to overcome. Access shafts must often be dug to allow insertion of construction equipment and materials. Wastewater flow must be accommodated by working wet or bypass pumping. The working environment is a confined space and must be accommodated by the appropriate safety and breathing equipment. Because of the increased difficulty of construction and inspection, the quality of work performed in an active sewer environment is often not equal to that obtainable on the surface.

4.6.3.3. *Selecting and Designing Liner Systems*

Liners are well-suited for both initial protection of new concrete pipe and structures as well as for rehabilitation and protection of existing corroded sewers. Attachment details for liners vary from application to application based upon the specific needs of the particular situation. Different design criteria are used to select liners for concrete rehabilitation purposes, as opposed to new concrete installations. Although many of the criteria, such as acid resistance and security of attachment, are practically identical, some of the most important criteria for these separate purposes are:

Rehabilitation	*Protection*
Flow capacity reduction of method	Years in similar service
Constructability	Joint treatment details (overlaps)
Method of attachment	Hydrostatic head handling
Grout design	Ease of construction (joint welds)
Installation with no flow interruption	Resistance to scour/abrasion
Confined space work	Degree of coverage (270 degrees versus 360 degrees)

Unique conditions exist in large-diameter, deep tunnels that do not exist in smaller sewer lines. When selecting a liner system for deep tunnels, special considerations must be made because of the unique conditions. Among these considerations are:

- High hydrostatic pressure from groundwater due to the depth of the tunnel.
- Surcharged conditions (intentional or unintentional) often exist in deep sewer systems. If an annular space exists between the concrete pipe and the liner, and this space is significant enough to contain water, then the rapid dewatering of the tunnel may impose forces on the liner from the stored water in the annulus.
- The inability to easily bypass flow in a deep tunnel requires work to be done under live flow conditions in a confined space. Pumping wastewater up and around a bypass for construction in a deep tunnel sewer can be prohibitively expensive, or expensive enough to give a competitive price advantage to a rehabilitation method that does not require bypass pumping.
- The diameter of access structures (manholes) in deep tunnels is generally much smaller than the diameter of the tunnel. This restricts the size of the equipment that can be inserted into the tunnel for construction purposes.
- The location of the deep tunnel property limits (easement) may affect the rehabilitation liner system selection, depending on how certain methods may require additional land area for material laydown and construction equipment storage.
- The distance between existing access locations (manholes) affects the liner system selection. Depending upon the rehabilitation liner system, some have more difficult construction staging requirements and maximum distances that can be accommodated.

- Large-diameter, deep sewers under the influence of extensive surface collection systems can experience rapid and dramatic flow variations due to wet weather impacts (see the bullet item Surcharged Conditions, above). These flow variations can impart very high, transient hydraulic forces upon a liner and can also cause pressure waves that induce alternating pressure/vacuum cycles against a liner and its attachment mechanism.

4.6.3.4. *Types of Liner Products Available*

There are several different types of liner products that can be basically classified by material of construction and method of attachment. The basic types of liners, as well as more detailed information that can be used to evaluate each liner system and descriptions of the application processes, are discussed in the following sections.

4.6.3.4.1. Cast-in-Place Liners

These types of liners have knobs, tees, ribs, or other types of protrusion from one side of the liner sheet, which are embedded into concrete or grout during construction. When the forms are removed, the smooth liner surface is revealed to provide an acid barrier. Proprietary examples are T-Lock™ (PVC), Bekaplast™ (PE), Studliner™ (PE), and Agru Sure-Grip™ (PE). These types of liners can be used for either new construction or rehabilitation, although they are most often used in new construction. Some of these liners are gaining acceptance as viable rehabilitation methods due to advances in grouting technology and in-pipe construction techniques [e.g., T-Hab™ (PVC), Danby (PVC), Bekaplast (PE)].

Whether cast-in-place liners are used for new or rehabilitation service, they relieve hydrostatic and vapor pressure by providing a pathway for escape into the sewer without sacrificing corrosion protection. Since the anchors are the only part of the liner embedded in the pipe wall, the flat surface of the liner is held close against the concrete but not tightly adhered to it. This allows weep water and vapor pressure to equalize into the sewer flow without creating an open airspace where gases can accumulate. Figure 4-15 shows a typical T-Lock™ cast-in-place liner detail on a vertical wall. The tee-shaped extensions are usually oriented vertically in structures and horizontally in sewers. Figure 4-16 shows how the conical anchors of Bekaplast(tm) liner secure the liner to the concrete.

Recommendations for Cast-in-Place Liners. Cast-in-place liners work best in sewers when installed on precast concrete pipe sections at the factory. Superior performance with cast-in-place liners has been experienced when the pipe sections are wet-cast and steam-cured, as opposed to dry-cast pipe construction. It is not recommended to install cast-in-place liners

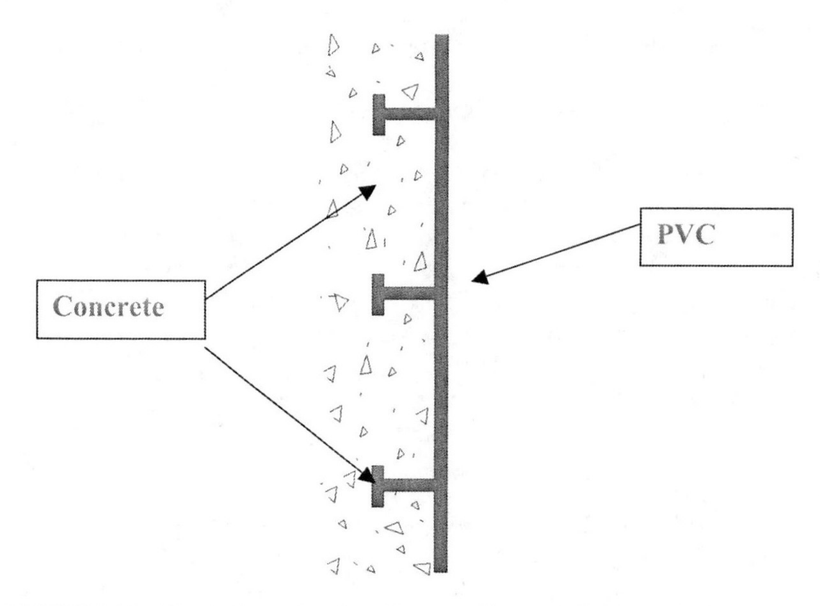

FIGURE 4-15. Typical cast-in-place liner profile on wall (plan view).

FIGURE 4-16. Bekaplast™ attachment using conical anchors. Courtesy of Georg Steuler Industriewerke GmbH, Hors Grenzhausen, Germany.

using the dry-cast method of pipe manufacture due to potential problems with proper embedment of the attachment anchors. Careful attention to the degree of coverage for the liner is recommended. It has often been found that 270 degrees of coverage is generally not sufficient to protect the concrete below the liner during some flow events. From 300 to 320 degrees of coverage is recommended as a minimum to protect pipes. Three hundred sixty-degree coverage is not recommended because it makes walking or standing in the pipe dangerous. Regardless of the circumference of coverage, openings must be provided for relief of the weep water and vapor pressure so the liner is not bulged or pressurized.

When used in a rehabilitation mode, cast-in-place liners must be grouted in place over the prepared surface of the host pipe. This requires an internal form that can pass wastewater during the placement and grouting procedure. Design of the grout mix and the placement technique are critical to the success of any rehabilitation application using cast-in-place liners. The grout must be specially designed for nonshrink characteristics, flow characteristics, and strength. No expense should be spared in providing the best grout mix. In order to prevent the buildup of hydrostatic pressure at the concrete–grout interface and to transmit weep water directly to the grout–liner interface, the grout must have a permeability (to water) equal to or less than the permeability rate of the host concrete pipe. This requires a special mix design for the grout.

In addition to care in installation, care is also required in welding the liner sheets to each other. It is very important that the welding be performed by expert, trained welders and that the final project be inspected by experienced inspectors familiar with the product. Unless the welding is done properly, there will be "cold joints," gas will get behind the liner, and concentrated corrosion will occur. Improper welding and poor inspection techniques have led to failure of sheet liners in some installations.

4.6.3.4.2. Chemically Attached Liners

These liners come in two basic forms. One type has protrusions on one side of a PVC liner which are pressed into freshly applied epoxy mastic on the concrete. The mastic flows around the protrusions to provide the attachment mechanism when cured (Arrow-Lock®). Another form of chemically attached liner is a PVC sheet, flat on both sides, which is pressed into a freshly applied polyurethane mastic (Linabond®). A chemical additive in the PVC sheet allows chemical fusion and cross-polymerization to occur between the urethane and the PVC, resulting in a very secure bond between the two. These liners can be applied over new concrete; however, they are more commonly used in rehabilitation. There are other more secure and economical methods for new construction than chemically attached liners. One of the potential drawbacks to chemically attached liners is the requirement to physically resist *all* hydrostatic and vapor pressure exerted against the liner–host pipe interface. Since the liner is glued to the pipe wall with an epoxy or urethane mastic, the attachment of the epoxy or urethane to the pipe wall is critical to liner performance. The ability to resist anticipated hydrostatic and vapor pressures should be adequately demonstrated. Figure 4-17 shows a typical chemically attached liner profile on a vertical concrete wall.

Recommendations for Chemically Attached Liners. The weakest link in the chemically attached liner process is the reliance on a physical–chemical bond between the epoxy or urethane mastic and the host pipe. This bond is not unlike the bond required for coatings, which have often demonstrated an inability to withstand hydrostatic and vapor pressure forces exerted in a deep sewer environment. The bond is enhanced by using a thick layer of mastic with a high modulus of elasticity so that bridging and hoop stresses are generated to help withstand the hydrostatic forces. Some chemically attached liners use a polymer concrete that has an extremely high modulus of elasticity and should be able to withstand anticipated hydrostatic and vapor forces if properly designed and applied.

Due to the reliance on a mechanical–chemical bond, the preparation of the host pipe surface is very important to a successful application. In addition, the selection of the primer to be used under the mastic or polymer

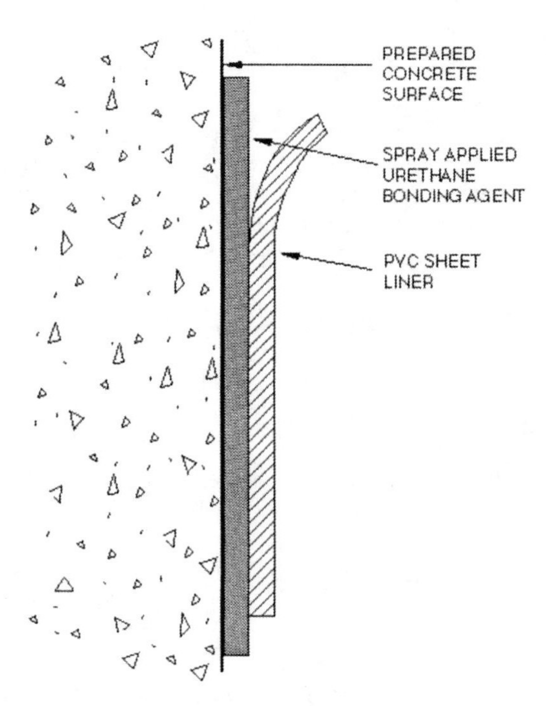

FIGURE 4-17. Typical chemically attached liner profile.

concrete can be critical to the success of the liner/mastic system. It is very difficult for coating systems to resist hydrostatic and vapor forces when applied directly at the concrete–mastic bond. In order to prevent the buildup of forces against the mastic material at this point, a low-viscosity, 100% solids epoxy primer with good wetting characteristics (hydrophilic) is applied just prior to mastic application. The primer drives the interface between hydrostatic forces and the mastic deeper into the matrix of the concrete where the primer has cured. This lessens the dependence upon the mechanical bond with the host concrete surface and makes the interior of the concrete host pipe carry some of the hydrostatic load.

4.6.3.4.3. Mechanically Attached Liners

These liners can be almost any material, although rigid (unplasticized) PVC and HDPE are most common. The most significant variation is the method of attachment, which can vary from small expansion-type anchors to stainless steel batten strips with epoxy grout anchors. The high degree of flexibility offered by mechanically attached liners makes them adaptable to a wide variety of conditions. Attachment systems can be designed that will safely accommodate the anticipated hydraulic and hydrostatic forces. Liner thicknesses can also be easily varied for additional hoop strength to adjust to specific design needs. These liners have been used primarily for new construction, although numerous rehabilitations have been reported with equal success.

Mechanically attached liners, like the cast-in-place liners, do not resist hydrostatic and vapor forces but pass them harmlessly to the sewer environment. Like the cast-in-place liners, the flat inner surface of the liner is held closely against the interior of the concrete pipe but is not fastened to it. This allows water to flow harmlessly into the sewage and not create a localized high-pressure area that could bulge and fail. Figures 4-18 and 4-19 illustrate how mechanically attached liners are secured in a pipe and on a wall.

Scale buildup has been reported behind mechanically attached liners. When groundwater leaches through concrete it picks up minerals, including calcium hydroxide [$Ca(OH)_2$], a by-product of the hydration of cement. The groundwater arriving at the interior of concrete pipe can have a pH as high as 13 or 14, primarily as a result of the dissolved calcium hydroxide. If even a small air space exists between the liner and the concrete wall, this air space will contain carbon dioxide gas (CO_2). Carbon dioxide dissolves into water, producing carbonic acid (H_2CO_3), which then reacts with the calcium hydroxide to form calcium carbonate ($CaCO_3$) and water according to this simple acid–base reaction:

$$CO_2 + H_2O \Leftrightarrow H_2CO_3 \text{ (carbonic acid)}$$
$$H_2CO_3 + Ca(OH)_2 \Rightarrow CaCO_3 + 2\,H_2O$$

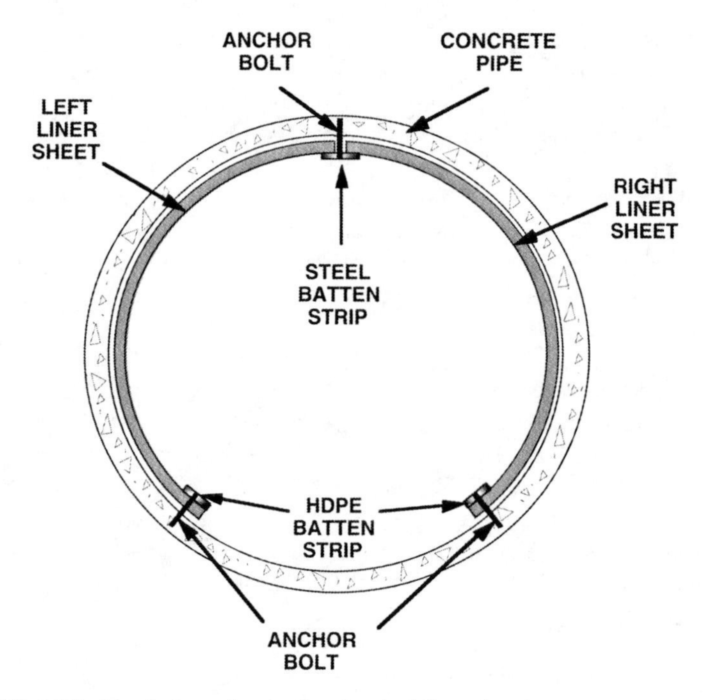

FIGURE 4-18. Typical mechanically attached liner in pipe.

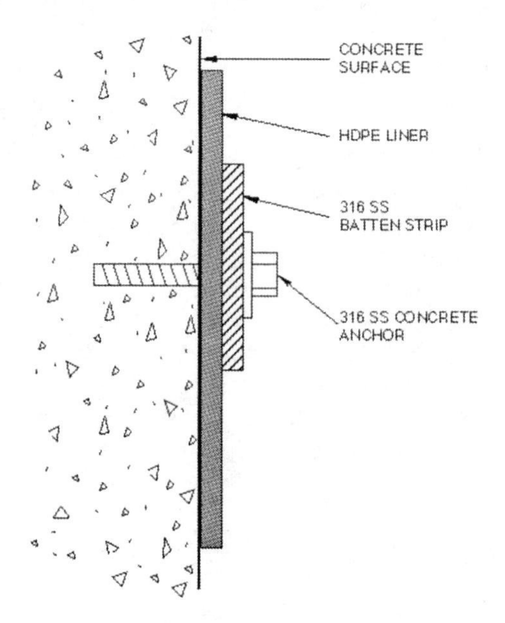

FIGURE 4-19. Typical mechanically attached liner to surface.

The production of calcium carbonate by this reaction is exactly the same process that forms stalactites and stalagmites in caves. Calcium carbonate is deposited on the surfaces between the concrete wall and the liner, forming thin, fragile flakes that often bridge between the liner and the pipe wall. When fluctuating water levels in the pipe cause flexure of the liner, flakes of calcium carbonate fall off and, being heavier than water, accumulate in the lower portion of the annular space between the liner and pipe wall. Significant accumulation of this material can restrict or prevent the movement of water into and out of the annular space, leading to hydrostatic pressure relief problems.

Recommendations for Mechanically Attached Liners. Particular care must be exercised during liner design to provide adequate structural support for the liner (thickness, hoop strength, and attachment security) to withstand the anticipated loads with a safety factor of 2. Serious failures have resulted when these design issues and other anticipated forces and loads upon the liner were not addressed. Designs must consider all hydraulic scenarios, particularly surcharging, in deep sewers and tunnels, and must account for the successful handling of both hydrostatic pressures and hydraulic shear forces, pressure waves, and shear forces. Spacing, type, size, and material of construction of the attachment anchors are critical factors for successful mechanically attached liner installation. Mechanically attached liners should not have spacers intentionally placed between the liner and the pipe wall. Spacers increase the volume of water behind the liner during a surcharge event and can add to forces trying to push the liner off the wall when water levels fall.

4.6.4. Inert Materials

The selection of the proper pipe material for a new large-diameter sewer or tunnel must consider corrosion impacts. When corrosion resistance alone is considered, pipe material selection favors plastic materials such as PE/HDPE, PVC, or various resin-based, composite materials such as GRP or FRP. However, when the cost of producing such large-diameter plastic pipes [>72-inch (1,830-mm) diameter] is considered, it is quickly found that the cheapest material by far is RCP. The relatively low cost and high strength of concrete makes it economically attractive for large-diameter, deep sewers, but the corrosion resistance of concrete is minimal. Concrete is readily attacked and destroyed by strong mineral acids like H_2SO_4, hence the need for liners.

Inert pipe materials are still economical for pipe diameters less than about 72 inches (1,830 mm), which comprises the vast majority of the total service area of the typical collection system. The vast number of plastic pipe types, formulations, and configurations makes a complete discussion in this chapter impossible. However, care should be exercised to determine the specific formulation of plastic pipes, since all plastics are not equal. For instance, PVC pipe is manufactured using a number of different filler materials. This is due to the cost of the resins used and the strength of the particular resin. Fillers are added to provide bulk and mass to the pipe and reduce excessive resin consumption. Some filler materials can be attacked by acidic conditions that leach the fillers from the pipe and reduce the structural capacity.

REFERENCES

U.S. Environmental Protection Agency (EPA). (1992). "Detection, control, and correction of hydrogen sulfide corrosion in existing wastewater systems." EPA 832 R-92-0001, EPA, Washington, D.C.

CHAPTER 5

HYDRAULICS OF SEWERS

5.1. INTRODUCTION

The functional purpose of a sanitary sewer is to safely convey wastewater to its destination. Sanitary sewer analysis and design is sometimes viewed as a simple task. In fact, many and varied local land use, topographical, subterranean, structural, and hydraulics problems combine to make projects very challenging. Computer-based sewer analysis and design systems are powerful tools now commonly used to assist engineers in sewer analysis and design. However, users should be well-informed of and alert to the appropriateness and adequacy of hydraulic principles incorporated into design systems. Although design codes and guidelines are valuable and often provide a safety net against problems resulting from poor engineering and bad judgment, engineers need to be alert to situations where codes and guidelines are inadequate—by being either too restrictive or not restrictive enough.

Prime functional goals for sewers are to (1) carry maximum flows without significant surcharge, and (2) achieve adequate self-cleansing during low-flow periods. At a point in a sanitary sewer, flow rates may vary greatly within a given day and much more over the service life of the sewer (see Chapter 3). Gravity-flow sanitary sewers are usually designed to flow full or nearly full at design peak flow rates. In some cases at extreme peak flows, it is permissible for sanitary sewers to operate as a low-pressure conduit where the hydraulic grade line (water surface) intermittently rises up into manholes to some acceptable height above the crown of the sewer. This type of operation is usually limited to deep sanitary sewers with no service connections at low elevations where backflow into buildings might occur.

Pressure sanitary sewers and force mains are pressure pipelines that carry wastewater in situations where open-channel gravity flow is not possible. Deep tunnels, inverted siphons, and submerged outfalls are other examples of pressure sewers. In such pressure conduits, large flow rate variation often makes solids deposition a significant consideration in design and operation.

Wastewaters contain a wide variety of soluble and solid particles to be conveyed to treatment facilities. Solids buildup can retard and even block the flow, and may foster generation of hydrogen sulfide and methane. In designing sewer systems, possible solids deposition during low-flow periods should always be considered. Wherever possible, designers should use accurate low-flow projections and sound hydraulic principles to design sewers for adequate self-cleansing to minimize solids deposition and avoid longer-term solids buildup. Since self-cleansing is not possible in all cases, sewer cleaning can be considered an acceptable backup position to be used as seldom as possible. The tractive force approach to self-cleansing has been adopted widely internationally and is now advocated by ASCE and the Water Environment Federation (WEF) in the United States. This approach is presented later in this chapter. It is a more refined and accurate approach to self-cleansing design than full-pipe-velocity and related approaches, but its application requires a significant paradigm shift and accompanying incorporation into codes and guidelines.

Sewer hydraulics is a rather complicated subject because of many possible conduit entrance and exit conditions, the range of sewage flows that change over time and by location, water surface profiles, the issue of self-cleansing, and more. Fortunately, steady, uniform flow principles are sufficient for analysis and/or design of most sanitary sewers. All hydraulic principles necessary to deal with all design situations and problems cannot be covered in depth in this chapter. The approach here is to present important hydraulic principles of particular value in sanitary sewer design, providing the engineer with concepts and tools that foster an appropriate analysis and design for the case at hand. For more discussion, see Yen (2001).

5.2. TERMINOLOGY AND SYMBOLS

A = cross-section area, L^2
A_c = cross-section area at critical depth
A_f = total area of closed conduit, L^2
A_n = cross-section area at normal depth, L^2
C_{HW} = Hazen-Williams equation coefficient
D = conduit diameter, L
D_0 = normal depth, L

d = particle diameter, mm
d_m = hydraulic depth = A/T, L
E = specific energy, L
E_m = minimum specific energy, L
F_n = Froude number, dimensionless
F = force
f = friction factor in Darcy-Weisbach equation, dimensionless
G = acceleration of gravity, LT^{-2}
H = total head, L
H_L = head loss, L
h_L = head loss at manhole, L
k = unit conversion factor. In Manning's equation, k = 1 in metric units, and k = 1.486 in English units ($1.486 = \sqrt[3]{3.28}$, where 3.28 is the number of feet in 1 meter)
K = head loss coefficient, dimensionless
n = Manning's roughness coefficient
n_f = Manning's roughness coefficient specifically for pipe full
p = pressure intensity, FL^{-2}
P = wet perimeter of cross section, L
Q = discharge, L^3T^{-1}
Q_{min} = design minimum discharge, L^3T^{-1}
Q_{max} = design maximum discharge, L^3T^{-1}
R_e = Reynolds number, dimensionless
R_f = hydraulic radius for pipe full, L
R_h = hydraulic radius = A/P, L
S = slope of energy grade line, dimensionless
S_0 = slope of invert, dimensionless
S_w = slope of water surface, dimensionless
$\overline{S}$ = average energy grade line slope
T = width of the water surface, L
V = mean velocity (Q/A), LT^{-1}
V_1, V_2 = velocity at upper and lower reaches or different conduit cross sections
V_c = critical velocity, LT^{-1}
V_n = velocity at normal depth
X = distance along conduit or channel, L
Y = height above invert, L
y_1,y_2 = conjugate depths before and after a hydraulic jump, L
y_c = critical depth, L
y_n = normal depth, L
z = height of invert above datum, L
$\bar{z}_1$, $\bar{z}_2$ = distance from the centroid to the free surface at section 1 or 2
β = momentum coefficient, dimensionless
γ = water specific weight, FL^{-3}

Δ = difference operator
θ = angle of sewer invert from horizontal, radians
= angle in radians; see Fig. 5-3
ε = pipe roughness, L
μ = absolute viscosity, FTL^{-2}
ν = kinematic viscosity, L^2T^{-1}
ρ = water density, ML^{-3}
τ_0 = average shear stress between fluid and conduit wall, FL^{-2}
τ_c = critical shear stress, FL^{-2}

5.3. HYDRAULIC PRINCIPLES

5.3.1. Types of Flow

Hydraulically, there is little to distinguish wastewater from stormwater or potable water. Herein, the words *water* or *liquid* imply wastewater, and *conduit* implies any type of sanitary sewer. *Open-channel* flow means that atmospheric pressure (zero gauge pressure) exists at the surface of flow, and *pressure flow* means that water completely fills the conduit and the gauge pressure at the conduit crown is greater than zero. Flows in sewers vary both spatially and temporally; they are described at any point as *unsteady* flow, and from location to location as *spatially varied* flow. However, these variations are of negligible importance in the design of open-channel sewer conduits as long as capacity and self-cleansing are addressed for each reach in the system.

The traditional and usually accepted approach to achieving adequate capacity and self-cleansing design is to assume one-dimensional, incompressible, steady, uniform flow. To accomplish these goals, for each reach in a system good estimates of flow rates are needed for *design maximum flow* (design capacity flow) for capacity evaluation and *design minimum flow* for self-cleansing evaluation.

5.3.2. Equations for Fluid Flow

5.3.2.1. *One-Dimensional Steady Flow*

Equations of motion for one-dimensional steady flow along a conduit are rather simple. Special forms of general equations for one-dimensional, spatially varied, unsteady flow must be used in special cases. The general equations must be used for more complex flows, such as waves in open-channel flow (Sturm 2001); passage from free surface flow to pressure flow; unsteady flow in pressure conduits (Yen 2001); and water hammer (Martin 1999). In sewers that have large, concentrated, extraneous inflow, rapid filling may occur during rain events, with the flow passing rapidly from free surface flow to pressurized flow, possibly resulting in surcharging manholes and structural damage. The use of the full dynamic equations or the Preissman slot model is needed to analyze such situations (Yen 2001; Cunge and Wegner 1964). Calculations for many of the complex cases are now practical, largely due to modern computer speed (see Table 5-8; Zhao 2001; Stein and Young 2001; and Yen 1999).

5.3.2.2. *Continuity Principle*

The continuity principle as commonly used in water flow is based on the general conservation of mass model. For steady flow (constant over time) with no storage changes along a section, inflow equals outflow. For incompressible flow, continuity is often expressed as in Eq. (5-1).

$$Q = V_1A_1 = V_2A_2 = \ldots = V_nA_n \tag{5-1}$$

where Q is the volumetric flow rate, V is the average velocity, and A is the cross-sectional area normal to the velocity vector, V.

5.3.2.3. *Energy Principle*

At any section in a flowing fluid, total energy is the sum of the potential energy—generally consisting of elevation and pressure terms—and the kinetic energy. The most common way of expressing total energy at a point in open-channel flow is given in Eq. (5-2):

$$H = z + y + \frac{V^2}{2g} \tag{5-2}$$

In this formulation, H is energy per unit weight and has units of FL/F, or net units of L (height). H is often referred to as *total head* or *energy line* (*EL*) and the three terms on the right-hand side of Eq. (5-2) are referred to as *elevation head, z, depth head, y,* and *velocity head, $V^2/2g$*. The plot of the sum of z and y along conduit is the free water surface and is also called the *hydraulic grade line* (*HGL*).

More accurately, the depth head term, y, in the equation should be written $y \cos^2 \theta$, where θ is the slope angle of the channel, but θ must be greater than about 2 degrees (slope > 0.035) to change the third significant number in values for that term. Likewise, the velocity head term would be more accurately given as $\alpha V^2/2g$, where α accounts for the effect of the velocity distribution in the cross section on the kinetic energy. For sewer flow, the value of the velocity coefficient, α, ranges from 1.10 to 1.20 (Chow 1959). Given these small deviations from 1.0 and the fact that velocities are normally low in sewers—making the velocity head term rather small compared to the depth term—α is normally assumed to be 1.0.

In pressure conduits, the energy equation is modified to give Eq. (5-3):

$$H = z + \frac{p}{\gamma} + \frac{V^2}{2g} \tag{5-3}$$

where z is the elevation of the cross-section centroid and p/γ is the pressure head at the area centroid of the pressure conduit.

The difference in total head at two different sections is equal to the energy losses between those sections, as given in Eq. (5-4):

$$H_L = \left(z_1 + y_1 + \frac{V_1^2}{2g}\right) - \left(z_2 + y_2 + \frac{V_2^2}{2g}\right) \tag{5-4}$$

Figure 5-1 depicts slope and energy relationships for open-channel flow. For uniform flow, then $S = S_0 = S_W$; for varied flow, they are not equal.

5.3.2.4. *Specific energy, alternate depths, and critical depth*

Specific energy is defined as energy relative to the invert (bottom) of the channel. The common expression for specific energy is given in Eq. (5-5):

$$E = y + \frac{V^2}{2g} \tag{5-5}$$

Specific energy is of no direct value in the hydraulic analysis or design of sewers, although it is of theoretical interest and can be used in deriva-

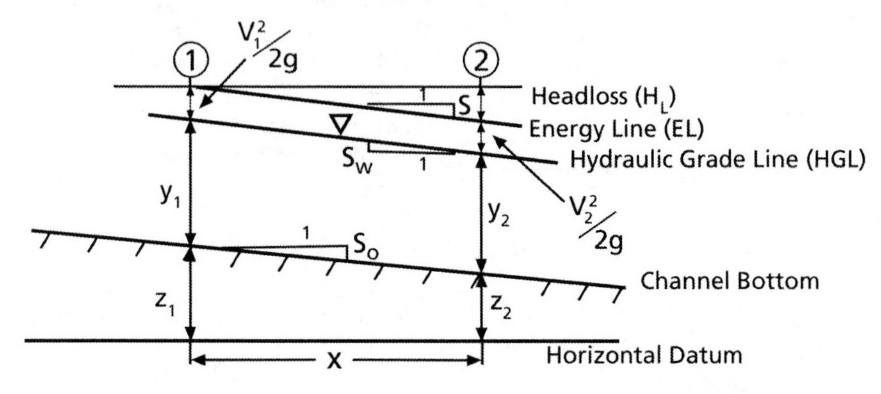

FIGURE 5-1. Slope and energy relationships for open-channel flow.

tion of the critical depth equation or in calculation of water surface profiles, as discussed later in this chapter. Given a channel cross section and a flow rate, a plot of the specific energy versus depth of flow is referred to as a *specific energy curve*. An example for an 8-inch (200-mm) pipe is given in Fig. 5-2. The curve illustrates that, in general, for any depth of flow there is another depth of flow that has the same energy. These two depths are called *alternate* depths. However, in a closed conduit the second depth may not be possible. The fact that two depths exist at the same energy is of limited practical significance in open channels, including sewers, because a different slope and/or roughness would be needed to cause the other depth to occur as a uniform flow depth.

The minimum value point on a specific energy curve identifies the *critical depth*, a singular point at which the alternate depths merge into a single depth. It is of passing interest that, near critical depth, relatively large variations in depth of flow, y, are associated with small variations in energy and flow rate.

In a channel, flow at depths deeper than critical depth (and thus velocities smaller than critical velocity) is referred to as *subcritical* flow, and flow at shallower depths (and thus velocities larger than critical velocity)

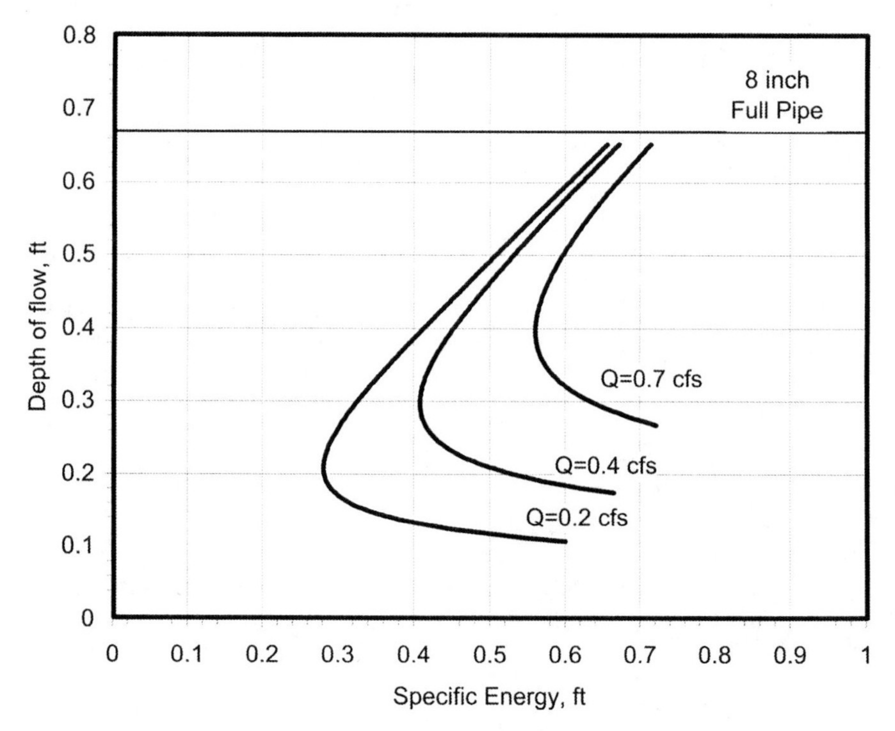

FIGURE 5-2. Examples of specific energy curves for a circular conduit—Q in cfs.

as *supercritical* flow. Two of the more important considerations associated with these flow regimes are (1) when supercritical flow occurs, a hydraulic jump will occur when the flow downstream is subcritical; and (2) when considering water surface curves, the control point (i.e., the starting point depth of a water surface profile calculation) is upstream and the surface curve develops going downstream when the actual depth of flow is in the supercritical range. Conversely, the control point is downstream and the surface curve develops back upstream when in the subcritical range.

For a general cross section, the specific energy [Eq. (5-5)] can be differentiated with respect to y and equated to zero to determine the minimum point. The resulting relationship many be written as Eq. (5-6):

$$\frac{Q^2}{g} = \frac{A_c^3}{T} \tag{5-6}$$

When $V_c A_c$ replaces Q, this equation may also be rewritten as Eq. (5-7):

$$V_c = \sqrt{\frac{gA_c}{T}} \tag{5-7}$$

where the cross-sectional area at critical depth is A_c and width of the water surface is T. The ratio A/T is called the *hydraulic depth*, d_m, so Eq. (5-7) may be modified to give Eq. (5-8):

$$V_c = \sqrt{gd_m} \tag{5-8}$$

The *Froude number*, F_n, may be stated as the ratio of the actual velocity to the critical velocity as given in Eq. (5-9):

$$F_n = \frac{V}{\sqrt{gd_m}} \tag{5-9}$$

For subcritical flow, $F_n < 1$ and for supercritical flow, $F_n > 1$, with $F_n = 1$ at the critical point.

In Eq. (5-9) for either a unit width or for a rectangular channel, d_m becomes depth y_c and at the $F_n = 1$ critical point the equation can be written as Eq. (5-10).

$$\frac{V_c^2}{2g} = \frac{y_c}{2} \tag{5-10}$$

Because the specific energy is the sum of the depth y and the velocity head $V_c^2/2g$, for a rectangular channel the relationship at critical depth becomes Eq. (5-11):

$$E_{min} = \frac{3}{2} y_c \tag{5-11}$$

For complex cross sections (such as circular or elliptical), equations, tables, or plots are often used to determine the hydraulic depth, d_m. For a round conduit, the relationship is given in Eq. (5-12) for the variables as shown in Fig. 5-3.

$$d_m = \frac{D}{4}\left(\frac{\theta - 1/2(\sin 2\theta)}{\sin\theta}\right) \tag{5-12}$$

where θ is in radians and D is the conduit diameter.

5.3.2.5. *Momentum Equation*

For steady flow, Newton's second law, $F = ma$, can be written as $F = (m/\Delta t)\Delta V$ and applied to a fluid control volume (volume between two sections) and expressed as Eq. (5-13):

$$\Sigma F = \rho Q(V_2 - V_1) \tag{5-13}$$

where ΣF is the sum of all forces acting on the control volume, including pressure forces $\int p dA$ at each end section, the weight of the fluid, and the peripheral forces the conduit applies to the fluid to constrain and guide the flow. The density is $\rho = \gamma/g$.

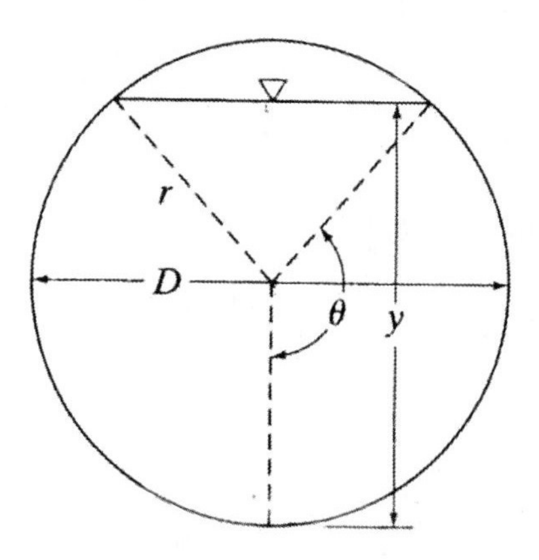

FIGURE 5-3. Circular conduit geometry.

This equation is called the *momentum equation*. It is a vector equation and the forces and velocities have both magnitude and direction. In cases where velocity variations across the cross section are large and/or high accuracy is desired, a β momentum coefficient may be used, replacing V with βV. In sewers, β values range between about 1.03 and 1.07 (Chow 1959). The fact that β values are so near to 1 and are generally about the same at both sections leads to usually assuming they are 1.

Momentum can be used to evaluate a hydraulic jump. Figure 5-4 shows a general hydraulic jump. The depths y_1 and y_2 are called *conjugate depths;* y_2 is also known as the *sequent depth* to y_1.

If conduit friction is small compared to the pressure forces, P_1 and P_2 (this will be the case unless the conduit is extremely rough), the momentum equation [Eq. (5-13)] can be written in one dimension (along the conduit) for a hydraulic jump to give Eq. (5-14):

$$\frac{Q^2}{gA_1} + \bar{z}_1 A_1 = \frac{Q^2}{gA_2} + \bar{z}_2 A_2 \qquad (5\text{-}14)$$

where $\bar{z}_i$ is the distance normal to the free surface to the centroid of the area A. Equation (5-14) is a general form of the one-dimensional hydraulic jump equation for a general conduit cross section.

Hydraulic jumps are interesting but are of little practical value in design of sewer conduits. This is the case since each reach is designed to flow at its normal depth of full pipe or less. If upstream flow is supercritical and downstream is subcritical, the hydraulic jump that develops will simply be the natural transition from the first depth to the second depth. In essentially all cases in sewer flow, jumps are very small and energy loss

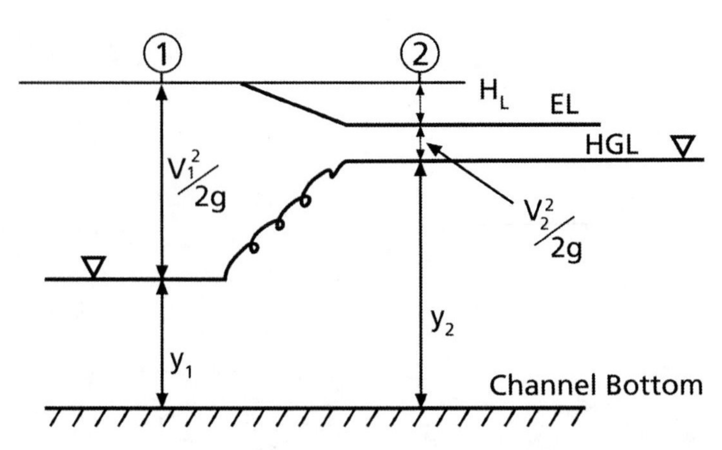

FIGURE 5-4. A hydraulic jump in a channel of small slope.

is likewise very small. In larger sewers [≥24 inches (610 mm) in diameter] they may need to be evaluated. One concern about hydraulic jumps is that the turbulence may localize the release of malodorous and corrosive gases, when present.

5.4. FLOW RESISTANCE

5.4.1. Energy Losses

Conceptually, friction losses are simply the energy converted to heat (lost) due to viscous shear friction along the conduit walls. Since steady, uniform flow is normally assumed for hydraulic design, the loss rate is just the pipe slope. In a pressure conduit of constant diameter, where velocity head is constant, the energy gradient (rate of head loss) is the change in $z + p/\gamma$ over the length and is $\Delta(z + p/\gamma)/L$. For uniform open-channel flow, the energy gradient, S, is equal to the sewer slope, S_0, and the slope of the water surface, S_W. If one takes a water element dx long between two sections and considers (1) static equilibrium for the element, and (2) its weight and shear forces the walls are exerting upon the element, then the average shear stress, τ_0, can be expressed as given in Eq. (5-15):

$$\tau_0 = \gamma R_h \frac{d}{dx}\left(\frac{p}{\gamma} + z\right) \tag{5-15}$$

For open-channel, uniform flow, the last part of the equation can be replaced by conduit slope to give Eq. (5-16):

$$\tau_0 = \gamma R_h S_0 \tag{5-16}$$

where R_h is the *hydraulic radius* and is defined as the ratio of area A to wetted perimeter P for the cross section, as given in Eq. (5-17):

$$R_h = \frac{A}{P} \tag{5-17}$$

For a full circular conduit, $R_h = D/4$.

Equation (5-16) is also known as the *tractive force equation;* it gives the average shear stress exerted on the channel by the flowing water. This τ_0 can be used to evaluate self-cleansing in conduits as presented in more detail later in this chapter. The shear stress necessary to move a design particle down the sewer is entered into the equation to determine the minimum sewer slope needed for self-cleansing for the given diameter and design minimum flow rate, Q_{min}.

5.4.2. Energy Loss Equations

When open-channel flow is steady and uniform, slopes of the water surface and energy grade line are equal to the slope of the conduit. Various empirical equations have been developed over the years to relate slope (energy loss) to channel size and material and average velocity. Continuity is then used ($Q = VA$) to determine the flow rate. For uniform flow and a given discharge, there is just one depth at which the flow equation is satisfied—the *normal depth*, y_n.

The Chézy equation was probably the first flow friction formula. It was developed by Antoine Chézy around 1768. Nearly 100 years later, Kutter's equation established a mathematical relationship for the Chézy constant. It was published about 1869 and received wide acceptance for estimating open-channel flows. Both the Chézy and the Kutter equations are now rarely used in practical applications and are mentioned here for historical interest.

5.4.2.1. The Manning Equation

The Manning equation is widely used and is one of the best open-channel hydraulics equations. It was developed by the Irish engineer Robert Manning and first published in 1890 (Yen 1992a). Over time, it has largely replaced the Chézy and Kutter equations in engineering practice because of its relative simplicity. The Manning equation can be written as:

$$Q = \frac{k}{n}(AR^{2/3}S^{1/2}) \tag{5-18}$$

where $k = 1.486$ for BG units (Imperial units) and 1.000 for SI units [1.486 is a unit conversion factor, the cubic root of 3.28 (i.e., the number of feet in a meter), so the Manning n is the same for both BG and SI units]. The variable n is called the *Manning roughness coefficient*. Although the n value is commonly perceived as just a function of conduit roughness, it is actually a function of several other factors (Yen 1992b). This perception of n being just a function of the conduit roughness is acceptable in cases where it is sufficient to have velocity and flow rate accurate to within perhaps 20% or so of true values. If greater accuracy in the Manning n estimate (and hence the discharge) is desired, then other factors have to be considered. These will be presented later in this chapter.

Usual ranges of n for various conduit materials are listed in Table 5-1. A more complete table can be found in Chow (1959).

Figure 5-5 is a nomograph for the solution of the Manning equation for circular pipes flowing full. An example of its use is illustrated by the broken lines. The nomograph can be used for other shapes of closed conduits and open channels if the discharge scale is ignored and the diameter scale is taken as four times the hydraulic radius of the actual cross section.

TABLE 5-1. Typical Values for Manning Equation n

Conduit Material	Manning n[a]
Closed Conduits	
Asbestos-cement pipe	0.11–0.015
Brick	0.13–0.017
Cast iron pipe	
Cement and seal-coated	0.011–0.015
Concrete (monolithic)	
Smooth forms	0.012–0.014
Rough forms	0.015–0.017
Concrete pipe	0.011–0.015
Corrugated Metal Pipe ½ in. (13 mm) × 2⅔ in. (68 mm) Corrugations	
Plain	
Paved invert	0.022–0.026
Spun asphalt	0.018–0.022
Plastic Pipe (Smooth)	0.011–0.015
Polyethylene	0.011–0.015
Polyvinyl Chloride	0.009[b]
Vitrified Clay Pipe	0.010[b]
Vitrified Clay Liner Plates	0.011–0.015
	0.011–0.020
Open Channels—lined	
Asphalt	0.013–0.017
Brick	0.012–0.018
Concrete	0.011–0.020

[a]Modified from American Society of Civil Engineers (ASCE). (1982). "Gravity sewer design and construction." ASCE Manual and Reports of Engineering Practice No. 60, ASCE, Reston, Va., unless otherwise noted.
[b]French, R. H. (2001). "Hydraulics of open channel flow." Stormwater collection systems design handbook, L. W. Mays, ed., McGraw-Hill, New York, with permission.

For example, if Q is given with depth to be determined, then the choice of S and n will indicate a point on the transfer line. By pivoting from this point, sets of values of $4R_h$ and V may be obtained. By successive trials, a compatible set of values for the given Q and the cross section may be found, which furnishes the correct depth of flow, y. Such approaches are now largely replaced by computer programs, available in numerous forms for both calculators and digital computers, which can make such calculations with relative ease.

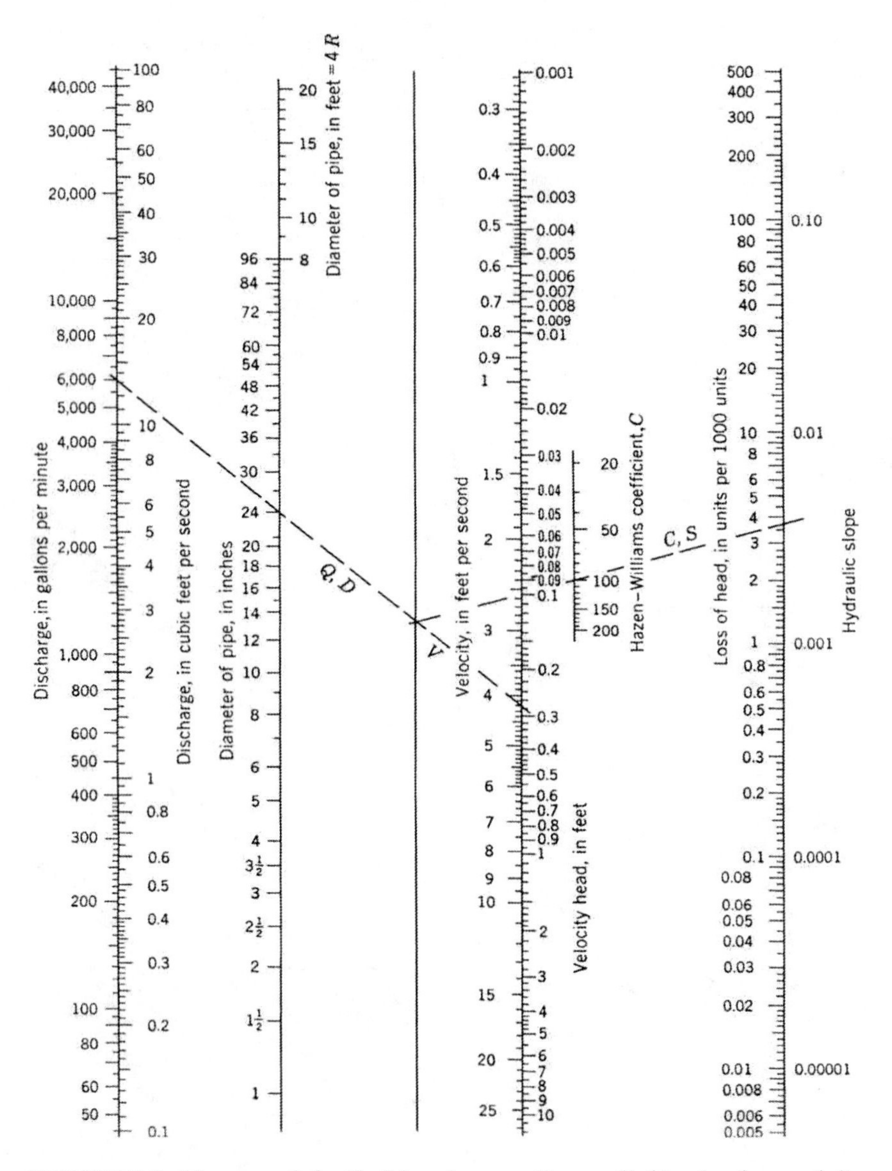

FIGURE 5-5. Nomograph for the Manning equation applied to circular conduits.

5.4.2.2. The Darcy-Weisbach Equation

The equation most widely used for all Newtonian fluid flow is the Darcy-Weisbach equation, which can be stated as given in Eq. (5-19):

$$S = \frac{H_L}{L} = \frac{f}{D}\frac{V^2}{2g} \tag{5-19}$$

where H_L is energy loss in a uniform conduit of length L and diameter D, and f is the friction coefficient. The f factor is a function of the Reynolds number and the relative roughness. Traditionally, f values are determined from the Moody diagram shown in Fig. 5-6 or from Eq. (5-20), the Colebrook-White formula (Colebrook 1939):

$$\frac{1}{\sqrt{f}} = -2.0\log\left[\frac{\varepsilon}{14.83R_h} + \frac{2.52}{4R_e\sqrt{f}}\right] \tag{5-20}$$

This formula is applicable for Reynolds numbers of $R_e = VR_h/\nu > 4{,}000$.

When viscous effects are negligible (high Reynolds numbers, R_e approximately $>10^4$), Eq. (5-20) becomes Eq. (5-21):

$$\frac{1}{\sqrt{f}} = -2.0\log\left[\frac{\varepsilon}{14.83R_h}\right] \tag{5-21}$$

Since Eq. (5-20) is implicit in terms of f, many explicit relationships have been proposed. Among these is the Swamee-Jain (Swamee and Jain 1976) equation, written as Eq. (5-22):

$$f = \frac{1.325}{\left[\log\left(\frac{\varepsilon}{3.7D} + \frac{5.74}{R_e^{0.90}}\right)\right]^2} \tag{5-22}$$

The apparent conduit roughness, ε, is sometimes referred to as the *hydraulic roughness* or else the *effective absolute roughness*. For open-channel flow, D is replaced with $4R_h$. The dimensionless Reynolds number is formulated as given in Eqs. (5-23) and (5-24):

$$R_e = \frac{DV}{\nu} \tag{5-23}$$

$$R_e = \frac{4R_hV}{\nu} \tag{5-24}$$

where ν is the kinematic viscosity. Equation (5-24) is used for a general cross section.

Introducing S in Eq. (5-19) into Eq. (5-16), and noting that $R_h = D/4$, yields Eq. (5-25):

$$\tau_0 = \frac{\rho f V^2}{8} \tag{5-25}$$

where $\rho = \gamma/g$ is the fluid density.

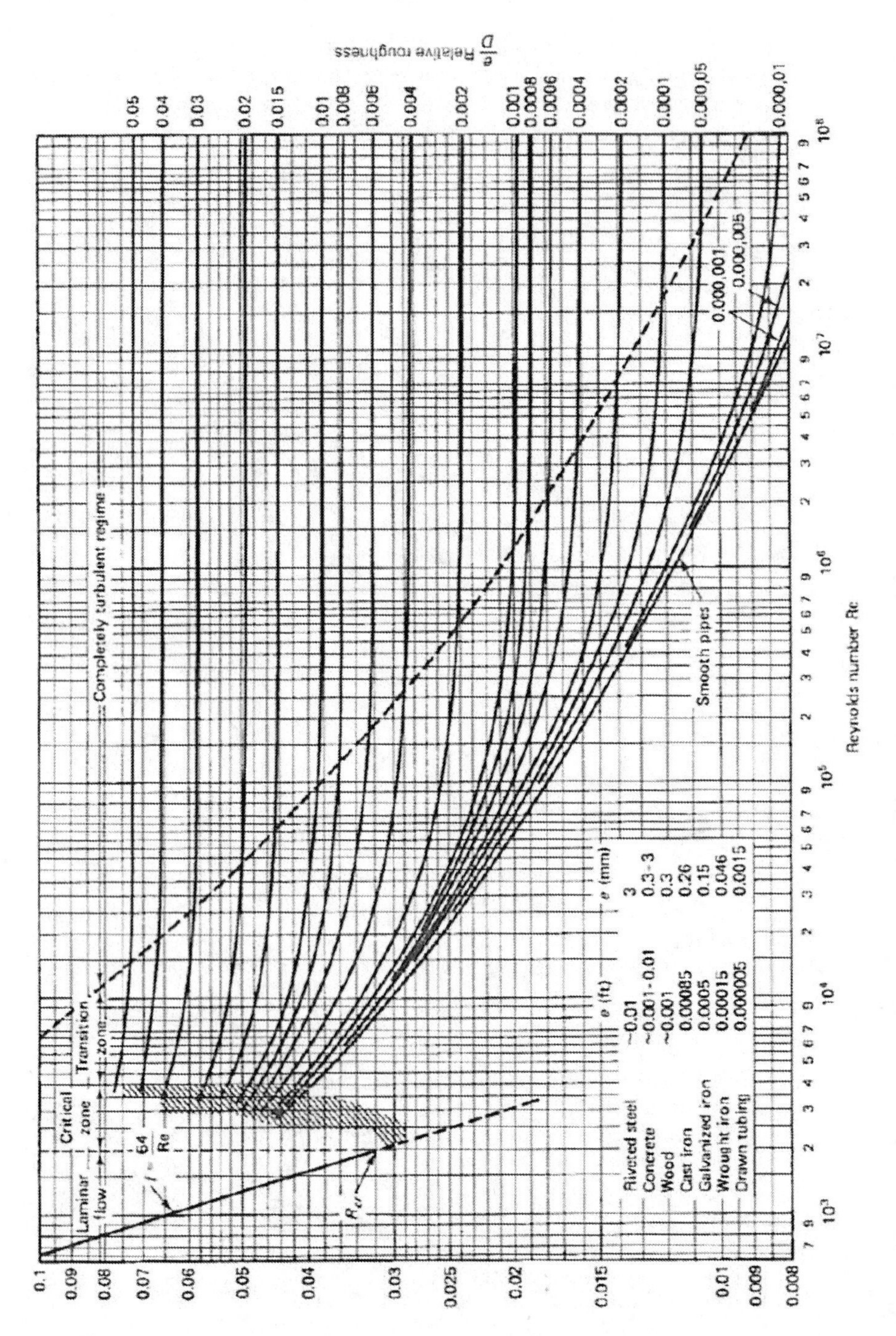

FIGURE 5-6. Moody diagram. Source: L. F. Moody. (1944). Trans ASME, Vol. 66.

It is important to note that most applied water hydraulics design situations, including pressure flow and open-channel flow, fall into the Reynolds number range of 10^5 to 10^6 and relative roughness range of 10^{-4} to 10^{--6} (0.0001 to 0.000001). As can be found in Fig. 5-6, this zone is near to or on the hydraulically smooth boundary. In this zone, rather large differences in conduit roughness cause only slight differences in f values for the normal range of manufactured and constructed channel or pipe roughness used in sanitary sewers. This outcome results from the presence of a laminar sublayer near the conduit wall that is thick enough in this zone that the physical roughness is largely or completely submerged within this sublayer, and the resistance is essentially fluid flowing on a fluid boundary, which is the smoothest that the boundary can be.

The Darcy-Weisbach equation is applicable to both pressure and open-channel flow, but in civil engineering practice it has been used sparingly and usually for pressure flow. Going forward into the future, the Darcy-Weisbach equation should be used much more, perhaps to the point of being rather exclusive, for both pressure and open-channel applications. The reason for this transition to more general use is that modern calculators and computers make the Darcy-Weisbach equation just as easy to apply as other less accurate and more limited equations. Ironically, such equations were initially developed and used because, in precomputer times, they were less unwieldy than the Darcy-Weisbach equation. Some argue that the Darcy-Weisbach equation cannot be used in sewer design since buildup of a slime layer in sewers has an unknown effect on the f value—that is, the ε/D or $\varepsilon/(4R_h)$ is hard to estimate. As addressed in the preceding paragraph, this is a negligible concern since ε value variation over a reasonable range results in little change in the f value.

5.4.2.3. *The Hazen-Williams Equation*

The Hazen-Williams equation (5-26) has been used as a practical equation for water flow in pressure conduits, similar to the use of the Manning equation for open-channel flow. Both equations may be used for both pressure and open-channel flows as long as velocities are moderate and the conduit diameter is not too small.

$$Q = kC_{HW}AR_h^{0.63}S^{0.54} \tag{5-26}$$

where $k = 1.32$ in BG units and 0.850 in SI units. The nomenclature is the same as used above; C_{HW} is the Hazen-Williams coefficient, which is also considered to depend mainly on just conduit roughness.

Values for the coefficient vary from about 40 for a badly tuberculated, smaller-diameter pipe to about 160 for a very smooth pipe. Normal application values range from about 100 to 140. However, the Hazen-Williams equation is being used less because, again, the more accurate and versatile

Darcy-Weisbach equation is now easily used on calculators and computers. Although the Hazen-Williams equation can be used for open-channel calculations, such use is rare. If it is used for partially full conduits, the hydraulic elements in Fig. 5-7 can be used as it is in the other equations. Values of C_{HW} can be found in Lansey and El-Shorbagy (2001) and many other sources.

5.4.2.4. *Partial Depth Calculations*

Fluid flow equations are used for conduits of all shapes flowing either full or partly full. For conduits with complex cross sections, equations, tables, or graphs must be available that give relationships between size and depth of flow and the area and hydraulic radius. For a circular shape (as shown in Fig. 5-3), the equations relating the pertinent variables are given as Eqs. (5-27), (5-28), (5-29), and (5-30). Since the depth, y, is more convenient than θ to measure or use directly, Eq. (5-28) can be used to determine θ for a given conduit diameter and flow depth. This θ, measured in radians, can then be used in Eqs. (5-29) and (5-30) to calculate the partially full area and hydraulic radius.

$$y = \frac{D}{2}(1 - \cos\theta) \tag{5-27}$$

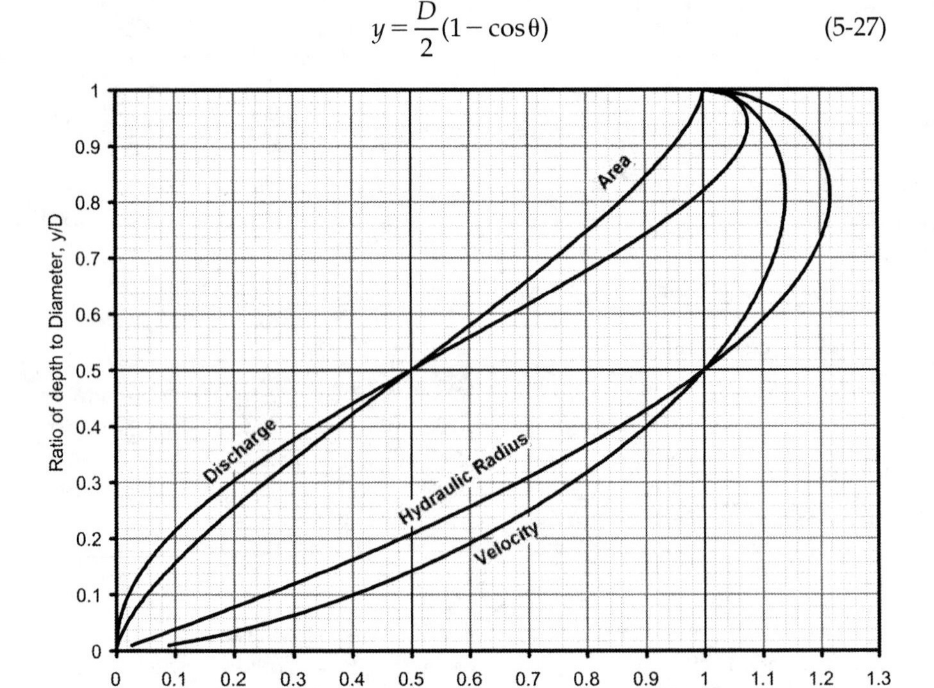

FIGURE 5-7. Hydraulic elements plot for circular conduits. Modified from: T. R. Camp. (1946). "Design of sewers to facilitate flow." Sew. Works Jour. 18(3).

$$\theta = \cos^{-1}\left(1 - 2\frac{y}{D}\right) \quad (5\text{-}28)$$

$$A = \frac{D^2}{4}\left(\theta - \frac{1}{2}\sin 2\theta\right) \quad (5\text{-}29)$$

$$R_h = \frac{D}{4}\left(1 - \frac{\sin 2\theta}{2\theta}\right) \quad (5\text{-}30)$$

Simultaneous solution of these geometric elements and the selected flow equation is rather unwieldy, but is much easier now via calculators and computers. To bypass using these equations, hydraulic elements graphs are still used at times to facilitate applying flow equations to partially full flow. Figure 5-7 graphically depicts the geometric as well as velocity and flow rate relationships between full pipe and partial depth flows. Figure 5-7 is used by calculating the full pipe area, the hydraulic radius, and full-pipe velocity and flow rate using the selected flow equation—usually the Manning equation. Depending on which variables are known, values are identified on the appropriate curves and unknowns are calculated from the associated ratio values taken from the plot.

For example, for $y/D = 0.8$, which corresponds to the depth of maximum velocity, one reads from Fig. 5-7 that $A = 0.86A_f$ and $R_h = 1.22R_f$. Noting that $A_f = \pi D^2/4$ and $R_f = D/4$, then $AR_h^{2/3} = 0.306D^{8/3}$ and Manning's equation becomes:

$$Q = 0.306\frac{k}{n}D^{8/3}S^{1/2} \quad (5\text{-}30A)$$

from which

$$D = 1.559\left[\frac{nQ}{kS^{1/2}}\right]^{3/8} \quad (5\text{-}30B)$$

This relation can be used for the determination of the diameter for a given Q, n, and S so that the flow occurs at 80% depth.

For full pipe,

$$D = 1.548\left[\frac{nQ}{kS^{1/2}}\right]^{3/8} \quad (5\text{-}30C)$$

If the engineer wants to design for some other partial depth of flow, then A and R_h values for the partial depth desired could be put into Eq. (5-30A) to yield the desired equation for the calculated diameter for the given Q, n, and S values.

5.4.3. Manning *n* Values from the Darcy-Weisbach Equation

The Darcy-Weisbach equation is generally accepted as the most accurate equation for fluid flow. If the Manning and Darcy-Weisbach equations are combined (Yen 1992b) by setting $S = H_L/L$ and $D = 4R_h$ in the Darcy-Weisbach equation, the result is Eq. (5-31):

$$n = kR_h^{1/6} f^{1/2} \tag{5-31}$$

where $k = 0.0926$ for BG units and 0.1129 for SI units.

This relationship indicates dependence of n on hydraulic radius and f (diameter, velocity, viscosity, and pipe roughness.) Equations (5-22), (5-24), and (5-31) can be used to evaluate these relationships. For full-pipe flow and the range of variables D: 6 to 48 inches (153 to 1,220 mm); V: 1.5 to 12 fps (0.5 to 3.5 mps) for constant values for $\varepsilon = 0.0001$ ft (0.03 mm); and ν at 60 °F (15 °C), calculated n values range from the lowest of $n = 0.0076$ for a 6-inch- (153-mm)-diameter conduit at 12 fps (3.7 mps), to the highest of $n = 0.0106$ for a 48-inch- (1,220-mm)-diameter conduit at 1.5 fps (0.5 mps) (Merritt 1998). Figure 5-8 illustrates the relationships between n value, velocity, roughness, and diameter. Water temperature also affects n value via viscosity in the Reynolds number, but the effect is relatively small over the 50 °F to 75 °F (10 °C to 23 °C) temperature range normally

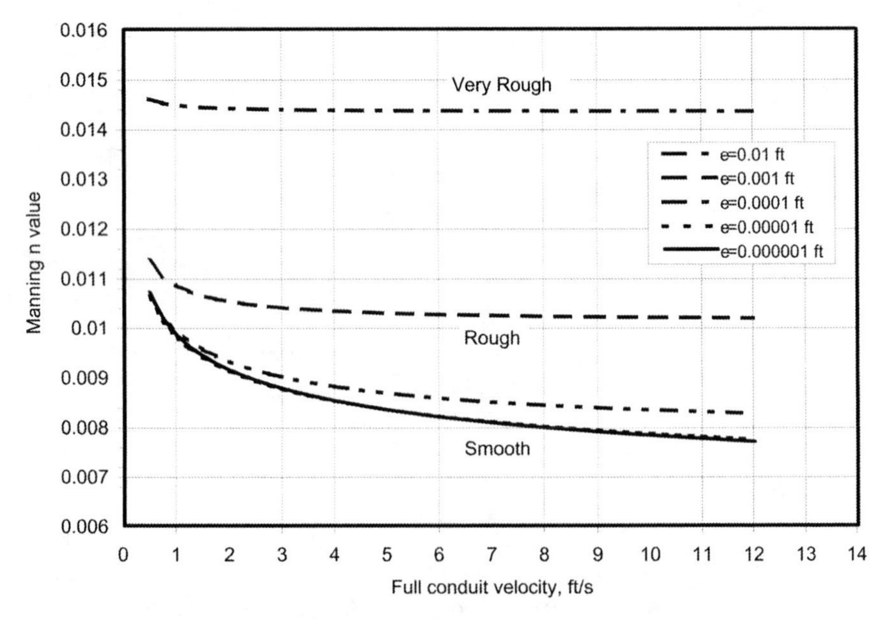

FIGURE 5-8. Manning n *versus velocity for selected roughness values for an 8-inch (200-mm) pipe.*

encountered in sewage. Likewise, for the normal range of sewer conduit roughness—from about 0.0001 ft (0.03 mm), a value for smooth-finish concrete, to about 0.000005 ft (0.0015 mm) for plastics—roughness differences cause relatively very small differences in n. However, for increasing roughness values [greater than about 0.0001 ft (0.03 mm)], n values begin to increase very significantly, as these increasing ε values move the flow regime away from the hydraulically smooth boundary and ε/D values become important.

5.4.4. Variation in Manning *n* with Partial Flow Depths

If Eqs. (5-27) through (5-31) are used for partial depths to evaluate n variation with depth of flow, results show essentially no variation in n until the partial flow ratio (y/D) is less than about 15%, at which the n/n_f ratio begins to increase. The n increase is not significant at two significant digits until the conduit is less than about 10% full (Merritt 1998). This observation is supported by experimental findings of others (e.g., Schmidt 1959; Tullis 1986) and indicates that n value variation with depth of flow in circular sewer conduits [as tentatively postulated by Camp (1946)] and included in recent editions of this Manual is in error. As a practical matter, constant or slightly variable n with depth of flow is of little consequence in hydraulic design of sewers as long as an appropriate n value is used for capacity calculations.

5.4.5. Manning *n* Recommended Values

Most sewer design codes and reference sources for n are based on observations and judgments made nearly 100 years ago. They typically recommend a Manning $n = 0.013$ be used for all sanitary sewer design. Subsequent empirical results from full-scale setups and operating sewers using modern pipe and joints report that n actually ranges from about 0.008 to 0.011 (May 1986; Straub et al. 1960; Tullis 1986). These values are about the same as those calculated from Darcy-Weisbach relationships. It should be noted that values calculated from the Darcy-Weisbach relationships are empirical (not theoretical), since they are based on actual water flow experiments in open channels. Thus, it appears that for operating sewers Manning n values are very close to values predicted by the Darcy-Weisbach relationships, although they range from about 15% to 40% less than the traditional value of 0.013.

Proponents of higher Manning n values argue that in operating sewers a higher n value covers possible significant flow retardation, which can be caused by pipe misalignment and joint irregularities, interior corrosion or coating buildup, cracks and breaks, protruding or interfering laterals, sediment buildup, etc. Although these factors do cause higher operating *n* values, when they are moderate [in combination, in the smaller diameters of perhaps less than 24 inches (610 mm)], they cause far less increase than exhibited by the traditional 0.013 value. For larger diameters, values are closer to the traditional 0.013 value. Designers should recognize that with good installation and maintenance, actual pipe capacities will be larger than calculated values resulting from use of a traditional high *n* value. Regulatory agencies are encouraged to allow informed discretion in selecting *n* values. Values given in Table 5-2 are suggested (Haestad et al. 2004). Agencies are encouraged to adopt similar values into their codes and guidelines.

The *Extra Care* values are based on the Darcy-Weisbach equation; *Typical* values are 15% higher; and *Substandard* values are 30% higher. The use of *n* values given in Table 5-2 acknowledges the significant variation of *n* with pipe diameter, while ignoring the smaller variations caused by velocity, relative roughness, and temperature. A cautionary note: If the actual conduit roughness is greater than about 0.0001 ft (0.03 mm) (rougher than smooth-finish concrete), *n* values begin to increase significantly, as shown in Fig. 5-8; however, even then the presence of a normal, thin slime film moderates the increased roughness.

5.5. SELF-CLEANSING IN SANITARY SEWERS

5.5.1. Sanitary Sewer Solids

Solids encountered in sanitary wastewater can be classified in the following categories (Ashley and Hvitved-Jacobsen 2002):

1. Large fecal and other organic matter.
2. Paper, rags, and miscellaneous sewage litter (tampons, sanitary towels, etc). Garbage grinders also contribute higher organic solid loadings. Other substances may be added from industrial sources.
3. Fine fecal and other organic particles.

TABLE 5-2 Suggested Values of Manning *n* for Sewer Design Calculations

	Pipe Diameters in inches										
Condition	6.	8.	10.	12.	15.	18.	24.	30.	36.	48.	60.
Extra Care:	0.0092	0.0093	0.0095	0.0096	0.0097	0.0098	0.0100	0.0102	0.0103	0.0105	0.0107
Typical:	0.0106	0.0107	0.0109	0.0110	0.0112	0.0113	0.0115	0.0117	0.0118	0.0121	0.0123
Substandard:	0.0120	0.0121	0.0123	0.0125	0.0126	0.0127	0.0130	0.0133	0.0134	0.0137	0.0139

Note: Extra care values are calculated from the Darcy-Weisbach equation for 60 °F, 2 fps velocity, $\varepsilon = 0.0001$ ft. Typical values are 15% higher than Extra Care values; Substandard values are 30% higher than Extra Care values.

After Haestad, M., et al. (2004). Wastewater collection system modeling and design, Haestad Press, Waterbury, Conn., with permission.

In the United States, the concentration of suspended solids in domestic sewage varies from 100 to 350 mg/L (Metcalf & Eddy, Inc. 2003). Typical sanitary sewage contains about 210 mg/L of suspended matter, of which about 160 mg/L is mineral and 50 mg/L is organic (Tchobanoglous et al. 2003). Table 5-3 lists sewer sediment characteristics as classified in the UK.

Self-cleansing sewers are capable of transporting most solids to the desired point, usually a treatment facility. When slopes are relatively flat, and thus the pipe's self-cleansing power is marginal, large variations in sewage flow in a reach may result in solids deposition at times of low flows. These deposits are usually eroded and transported later at times of higher flows. As noted in Chapter 4, organic solids deposited during low flows may trigger the generation of hydrogen sulfide and methane gas from anaerobic decomposition This condition, if persistent, can create hazardous conditions, corrode concrete and iron pipes, and make any overflow or exfiltration from incontinent sewers more objectionable. Excess sedimentation may result in clogging, possible surcharge, and more persistent anaerobic decomposition with the attendant problems mentioned above. These factors accentuate the necessity of constructing self-cleansing sewers wherever possible.

5.5.2. Traditional Approach to Self-Cleansing

Traditionally, an experience-based criterion of 2 fps (0.61 mps) for full-pipe flow was adopted and incorporated into most codes and guidelines as the minimum velocity requirement for self-cleansing. This criterion established a rather simple way to calculate minimum slopes based on full-pipe calculations, but 2 fps was not originally intended to be the required velocity for shallow depths of flow; it was intended just for full pipe. The intent was that the lower velocities at shallow depths of flow would still avoid serious sedimentation problems.

TABLE 5-3. U.K. Sewer Sediment Characteristics

Type	Normal Transport Mode	Concentration (mg/L)			Median Particle Size d_{50} (μm)			Specific Gravity		
		Low	Med	High	Low	Med	High	Low	Med	High
Sanitary Solids	Suspension	100	350	500	10	40	60	1.01	1.4	1.6
Stormwater Solids	Suspension	50	350	1,000	20	60	100	1.1	2.0	2.5
Grit	Bedload	10	50	200	300	750	1,000	2.3	2.6	2.7

Butler, D. et al. (1996a). "Sediment transport in sewers—Part 1. Background." *Proc., Inst. Civil Engrs., Water, Maritime and Energy, 118*, 103–112, with permission.

Hager (1999) recommends a minimum velocity between 2 and 2.3 fps (0.6 to 0.7 mps). The 1982 edition of this Manual (ASCE 1982) recommends a minimum full-pipe velocity of 2 fps and indicates a preferred velocity as high as 3 fps (0.91 mps) at maximum discharge, whenever possible, to scour the sands and grit that may have accumulated in the sewer. That Manual also stated that an actual velocity of 1 fps (0.30 mps) at low flows only prevents deposition of the lighter sewage solids but serious deposition of sand and gravel is possible. The tractive force approach given in Section 5.5.3 in this Manual is now recommended since it addresses self-cleansing more accurately. However, during the conversion period some projects may require an analysis of the velocity at actual depth of flow. The procedure to check the velocity for a given Manning n and given maximum discharge Q_{max} and slope is as follows:

1. Estimate the diameter to carry the maximum discharge. For full-pipe condition, use Eq. (5-30C). For maximum velocity occurring at 80% of the depth, the sewer diameter will be given by Eq. (5-30B).
2. Round up the calculated diameter to the next commercially available size.
3. Calculate the left-hand side of the following equation:

$$\frac{nQ}{kS^{1/2}} = AR_h^{2/3} \tag{5-31A}$$

   and solve for the normal depth y_n at the maximum discharge. This can be done by computer program, graphically (Chow 1959; Sturm 2001), or by trial and error. For the latter method, select trial values of y/D, calculate θ successively from (5-28), A from (5-29), R_h from (5-30), and $AR_h^{2/3}$ until the latter value matches the calculated left-hand side of the above equation.
4. For the normal depth obtained in the previous step, calculate $V = Q/A$ (which will be close to the full-pipe velocity) and verify that this velocity satisfies the prescribed 2 fps minimum velocity criterion so that the sewer is self-cleansing at approximately full-pipe discharge (or 80% depth). If not, the slope may have to be increased until a satisfactory velocity is obtained.

This 2 fps (0.61 mps) criterion was also commonly combined with n = 0.013 in the Manning equation to generate minimum allowable slopes for each conduit size. For example, Table 5-4 uses these values to calculate the minimum slopes contained in the widely followed Great Lakes Upper Mississippi River Basin (GLUMRB) standards. Such guidelines have generally resulted in adequate self-cleansing, but result in higher self-cleansing

TABLE 5-4. Recommended Minimum Sewer Slopes, Per Ten-State Standards

Pipe Diameter		Slope, ft/100 ft (m/100 m)	
Inch	mm Equivalent	Calculated[a]	GLUMRB[b]
6	152	0.49	—
8	203	0.34	0.40
10	254	0.25	0.28
12	305	0.2	0.22
15	381	0.15	0.15
18	457	0.12	0.12
21	533	0.093	0.10
24	610	0.077	0.08
27	686	0.066	0.067
30	762	0.057	0.058
33	838	0.05	0.052
36	915	0.050 (2.1)[c]	0.046
39	991	0.050 (2.2)[c]	0.041
42	1,067	0.050 (2.3)[c]	0.037

[a]Calculated with the Manning equation, $n = 0.013$, velocity = 2 fps (0.61 mps).
[b]Great Lakes Upper Mississippi River Board (1997).
[c]Recommended that 0.050 be the minimum slope for large-diameter pipe. Number in parentheses indicates velocity in fps.

power than is needed in many cases (particularly in smaller-diameter pipes in which low-flow depths are deeper than about 20% full), and in inadequate self-cleansing power in others (particularly in larger-diameter pipes flowing less than about 30% full and smaller-diameter pipes flowing less than about 20% full).

Yao (1974) and others have shown that the use of the same full-pipe, minimum-velocity criterion for all conduit sizes often results in overdesign (steeper slopes than needed) of small-diameter pipes and underdesign (flatter slopes than needed) of larger ones. A problem exists with traditional criteria since velocity magnitude alone is a poor indicator of self-cleansing power. Nevertheless, these guidelines are conservative enough that they result in adequate self-cleansing in most sewers. The breakpoint between small and large sewers used here is about 18 inches (457 mm), but the transition is continuous with the smallest diameter having the most conservative minimum slope and the largest diameter having the most troublesome minimum slope, if based on a full-pipe velocity of 2 fps (0.61 mps).

The reliance on full-pipe minimum velocity as a self-cleansing requirement was also widely used internationally, but now tractive force criteria are being increasingly used. Since sediment deposition problems still occur in some sewers designed for about 2 fps (0.61 mps), particularly with increasing sewer size, often the response was to require steeper and steeper minimum slopes to achieve higher and higher full-pipe velocities. By the 1990s this trend generally culminated with some countries promoting minimum slopes that give full-pipe velocities of some 2.5 to 3.5 fps (0.75 to 1.1 mps) for sanitary sewers and 3.5 to 5 fps (1.1 to 1.5 mps) for storm sewers (Nalluri and Ghani 1996). The British Construction Industry Research and Information Association (CIRIA) (Ackers et al. 1996) approach is based on a many-year effort to identify full-pipe velocities that would nearly always ensure self-cleansing for a range of commonly encountered conditions. This approach is outlined in the appendix at the end of this chapter.

More recently, some countries have also begun making a transition to a tractive force design approach that considers actual low flows, which then allows flatter minimum slopes for sewers with relatively large minimum flow rates and requires steeper minimum slopes for those with relatively small minimum flow rates.

According to Ackers et al. (1996), conditions in a sewer should be sufficient to:

- Transport a given concentration of fine-grained or lower-density particles in suspension.
- Transport coarser particles as bedload (temporary small depth of deposition may be acceptable).
- Erode deposits from the bed which may exhibit cohesivelike properties.

In residential sewersheds, bedload sediment transport is usually the controlling transport mechanism in sanitary sewers as well as storm sewers. Peak flows associated with diurnal and weekly flow rate fluctuations are usually sufficient to cause bedload movement and prevent buildup of cohesive deposits. Therefore, to achieve self-cleansing a conduit needs to periodically achieve a critical shear stress sufficient to transport the bedload before the bedload has had sufficient time to congeal or become cohesive. This concept underpins the approach presented in following paragraphs. Note that for very large collectors and combined sewers, a more sophisticated identification of the sediment load may be desirable (Ota and Nalluri 2003); however, the approach given below yields essentially the same results unless the wastewater contains an unusually large quantity of larger sediment and/or has an unusually high cohesive propensity.

5.5.3. Tractive Force Approach to Self-Cleansing

ASCE and WEF now advocate a transition to the tractive force approach for self-cleansing design. Minimum velocity and minimum slope guidelines that are similar to those in Table 5-4 will likely continue to be used for some time; however, they are inferior to the tractive force approach for setting minimum slopes for self-cleansing. Proper application of the tractive force method strongly depends on selection of an appropriate design sediment particle and good, realistic estimates of *design minimum flow rates*, Q_{min}, for each reach designed. See Chapter 3 for information on determining Q_{min} values.

The average shear stress exerted by flowing fluid on a conduit can be expressed as Eq. (5-16) ($\tau_0 = \gamma R_h S$). Although this shear stress has also been referred to as tractive tension, drag shear, and others, the term *tractive force* will be used herein. Tractive force design, as developed in this chapter, is based on the concept that if a conduit is to adequately transport solids, there exists a design discrete-grit particle that must be transported often enough that it does not accumulate into a permanent sediment layer along the bottom. All other solids are more easily transported than this design particle. In concept, at the design low-flow Q_{min} this design particle is not suspended by turbulence, as are most of the larger, less dense, and smaller, equally dense particles. Instead, it is dragged along the conduit invert as *moving bedload.* This occurs since the shear stress acting on it is larger than the maximum frictional resistance it can develop against the conduit invert.

The design particle does not have to be continuously moved along the conduit as long as it is moved often enough that slight accumulations, which might begin to entrap some organic debris, do not congeal to the point where their resistance to movement is greater than the design shear stress.

5.5.4. Sediment Transport Capacity—Magnitude of Tractive Forces Required

Various efforts have been made to identify the tractive forces required to transport various sizes and types of particles in sewer conduits (Medina and Vega 2002). However, for sanitary sewers, more data on sediment quantities, characteristics, and movement vis-à-vis tractive force will help to fine-tune future refinements to this approach. For example, data can be developed regarding the slight differences in tractive force values necessary to ensure movement of the design sediments as a function of other parameters, such as actual representative sediment analysis of the wastewater to be transported or the presence of unusually large amounts of certain cohesive materials.

Some applicable results come from experiments done by Raths and McCauley (1962). A least-squares fit of their results (Haestad et al. 2004) yields Eq. (5-32):

$$\tau_c = kd^{0.277} \tag{5-32}$$

where $k = 0.0181$ for BG units (τ_c in lb/ft^2) and 0.867 for SI units (τ_c in N/m^2), and d is the nominal diameter (in mm) for a discrete mineral particle having a specific gravity of 2.7. Actual domestic sewage was used in their experiments in a full-scale, 8-inch- (200-mm)-diameter vitrified clay pipe with slip seal joints. The experimental sewer was constructed at the Mason, Michigan Sewage Treatment Plant.[1] It is likely that pipe with a new-condition interior surface rougher than vitrified clay may initially exhibit slightly higher frictional resistance to bedload sand particles; however, the very small, sharp projections (which might cause slightly higher resistance along the pipe invert) are likely to be quickly sanded away by the bedload sediments in sewage. Therefore, differences in commercial sewer pipe roughness as affecting particle friction may be largely discounted and the experimental sewer pipes used should be essentially the same in physical-friction resistance as other commercial pipes.

In Raths's and McCauley's experiments, sand particles sized between 0.2 mm and 9.5 mm with an average specific gravity (SG) of 2.7 were fed into the experimental sewer line through top tees located at 8-ft (2.4-m) intervals. The largest particle size to pass through the pipe without deposition was identified for a number of slopes and flow rates. The sewage used as the carrying water in these experiments is considered typical for residential areas in the United States.

Equation (5-16) can also be applied to existing sewers to calculate the shear stress for a particular flow rate. Then Eq. (5-32) can be used to determine the largest particle that will move down the conduit as bedload for the given flow rate. The experiment from which Eq. (5-32) was developed addresses only the bedload sediment for sewage with typical suspended solids levels and typical solids cohesion. Therefore, if the sewage in a sewer system has significantly adverse combinations of very high organic solids load and larger sediment loads, a slight decrease in the bedload sediment carrying capacity might occur.

5.5.4.1. Appropriate Design Particle Size

Consensus seems to be converging toward a 1-mm design particle for typical sewage; this size is in about the middle of sizes that have been sug-

---

[1]Mason is a growing suburban town of 7,200 people (2005) located minutes from Lansing and East Lansing, Michigan.

gested. From Eq. (5-32), a 1-mm design particle is associated with a critical shear stress of 0.0181 lb/ft^2 (0.867 N/m^2). Reasons for using a 1-mm particle include (1) essentially all of the grit is smaller than this size in typical sewage, and (2) shear stresses generated in smaller-diameter sewers—say between 6 and 12 inches (152 to 305 mm) in diameter—placed at traditional minimum slopes transport a 1-mm particle at about 20% relative flow depth. Conventional wisdom from the past indicates that small sewers at traditional minimum slopes generally have no significant sedimentation problems if they regularly flow greater than 20% full. In sewers where larger quantities of larger grit occur, it would be judicious to increase the design particle size somewhat, maybe to 1.5 to 2 mm, or even larger in extreme cases where an extraordinary amount of larger grit is expected in the sewer.

Note that when using Eqs. (5-16) and (5-32), pipes flowing full at traditional minimum slopes [based on 2 fps (0.61 mps)] would transport a 10-mm particle in an 8-inch- (200-mm)-diameter conduit at its minimum slope of 0.00334, decreasing to a 1-mm particle in a 60-inch- (1,500-mm)-diameter conduit at its minimum slope of 0.00023. It is impractical in most cases, where a sewer needs to be as flat as possible, to make minimum slopes steep enough to transport very large grit particles at extremely shallow depths of flow. For example, to transport 10-mm particles in an 8-inch- (200-mm)-diameter pipe at 10% full would require a slope four times steeper (or 0.0134), since the hydraulic radius at 10% full is about one-quarter of that for a full pipe.

Figures 5-9 through 5-12 are similar to Fig. 2-22 in Haestad et al. (2004). These graphs give the relationships between design minimum flow rates and necessary minimum slopes to achieve self-cleansing for 1-mm, 2.7-SG particles in various pipe sizes. Figures 5-9 and 5-11 are based on n values 1.15 times the values calculated from the Darcy-Weisbach relationships—"good" sewer conditions as referenced in Table 5-2. Figures 5-10 and 5-12 are based on a constant Manning n value of 0.013. Equations that may be used in lieu of Figs. 5-9 through 5-12 are given in Tables 5-5 and 5-6. These equations are limited to the range $0.1 < \textit{partial depth} < 0.5$; Q_{min} is the design minimum flow rate as presented in Chapter 3.

It is important to note that if tractive force is applied as outlined, it addresses self-cleansing for the projected worst week in the life of the sewer. All other times would be better (higher flows/higher tractive force and greater self-cleansing power). By the end of the design life, normal long-term growth conditions would substantially increase flow rates for most pipes in a system, resulting in substantially higher self-cleansing power than early in its life, since the Q_{max} value will be at least several times larger than the design Q_{min} in each sewer reach.

5.6. DESIGN COMPUTATIONS

5.6.1. Capacity Design

Each reach in a sewer must have a size and slope sufficient to (1) not violate minimum or maximum sewer depths, (2) carry its design capacity flow rate, and (3) be self-cleansing. The capacity requirement is designed by using the selected flow equation and the design capacity flow rate, Q_{max}. Usually the Manning equation (Eq. (5-18) is used, but the Darcy-Weisbach equation is an acceptable replacement in determining the appropriate n value for the Manning equation. Table 5-2 earlier in this chapter gives n values calculated from the Darcy-Weisbach equation. The given Q_{max}, selected size (usually diameter), and selected n value are used to calculate the necessary full-pipe slope to carry Q_{max}. Some codes require that the capacity design be at some percentage of full pipe. It seems more rational to determine the best possible estimate for Q_{max}, including any safety margin selected, and use full-pipe design. However, a percentage flow requirement may be accommodated by using Fig. 5-7 to determine the flow ratio for the percent full needed and increasing the original Q_{max} by dividing it by this decimal value. The modified Q_{max} can then be used for the full-pipe design.

The pipe slope necessary to go from the upstream invert depth to the minimum depth at the next manhole is then calculated and compared to the necessary pipe slope for capacity. The steeper of the two slopes is selected at this point. Other sizes may be considered for the reach, as reasonable, with each size being identified with its slope.

Maximum velocity criteria are also important, since extremely high velocities are troublesome for a number of reasons, including being a hazard to workers and to the structural integrity of the systems, particularly at manholes. Full-pipe maximum velocity limits for normal sewer conduits in the 10 to 15 fps (3 to 4.5 mps) range are often found in codes, although if the flow goes straight through at essentially the upstream grade, somewhat higher values would be acceptable. It is judicious to use lower maximum velocity values and take extra precautions in cases where flows are increasingly larger.

5.6.2. Self-Cleansing Design

After a diameter and its slope for capacity are determined for a reach, the slope needs to be checked to determine whether a steeper slope is required for self-cleansing. If it is, the self-cleansing slope becomes the design pipe slope and the reach is controlled by self-cleansing rather than by capacity or sewer minimum invert depth constraints. If the engineer is required to use the traditional minimum slope approach, the slope found in Section 5.6.1 immediately above is compared to the required value and the larger slope is used for the reach.

If the engineer uses the recommended tractive force approach, the same check and outcome procedure is followed, except the self-cleansing slope is determined as outlined in the following sections.

5.6.2.1. *Procedure for Calculating Minimum Sewer Slope Using Tractive Force Design for Self-Cleansing*

Several different approaches might be used for self-cleansing based on the relationships given earlier in this chapter. These approaches are summarized below.

The recommended approach is to use the equations from the variable n column of Table 5-5 or Table 5-6 to calculate the needed S_{min} for the design Q_{min}. These equations result from a least-squares fit for the Q_{min} versus S_{min} data for a depth range of 0.10 to 0.40 full. The equations have correlation coefficients above 99%. They are for a 1-mm, 2.7-SG design particle.

A second approach is to use the appropriate curve from Figs. 5-9, 5-10, 5-11, or 5-12. These curves give essentially the same results as the equations in Tables 5-5 and 5-6, but with a moderate loss in accuracy due to the graphical nature of this approach.

A third approach can be used for any desired design particle (tractive force), not just the 1-mm particle used for the tables and figures provided here. It is applied as follows:

1. Determine the design minimum flow rate, Q_{min}, for the reach (see Chapter 3 for procedures to establish Q_{min}). A good estimate of design minimum flow, Q_{min}, for each reach is crucial to accurately achieving the self-cleansing objective
2. Select a design particle size, normally between about 0.5 and 2 mm (1 mm is suggested as appropriate for most areas) and use Eq. (5-32) to calculate the critical shear stress as applicable to your design situation. For a suggested 1-mm particle, τ_c is 0.0181 lb/ft^2 (0.867 N/m^2).
3. Select a trial pipe size.
4. Select an appropriate Manning n value. Use $n = 0.013$ if required; otherwise select an appropriate value from Table 5-2.
5. Iterate on depth of flow, y—select a y value (less than half full) and solve Eqs. (5-27), (5-28), and (5-30) for R_h. Put R_h into Eq. (5-16), solve for S, and calculate A using Eq. (5-29). Calculate flow rate, Q, using Eq. (5-26) or another flow equation. Continue to iterate on y until Q equals the design minimum flow rate, Q_{min}. The associated slope, S_{min}, is the minimum slope for self-cleansing for the reach.

For any of the three approaches used, the resulting S_{min} value is then compared to the S_{max} value from capacity calculations above. S_{min} replaces

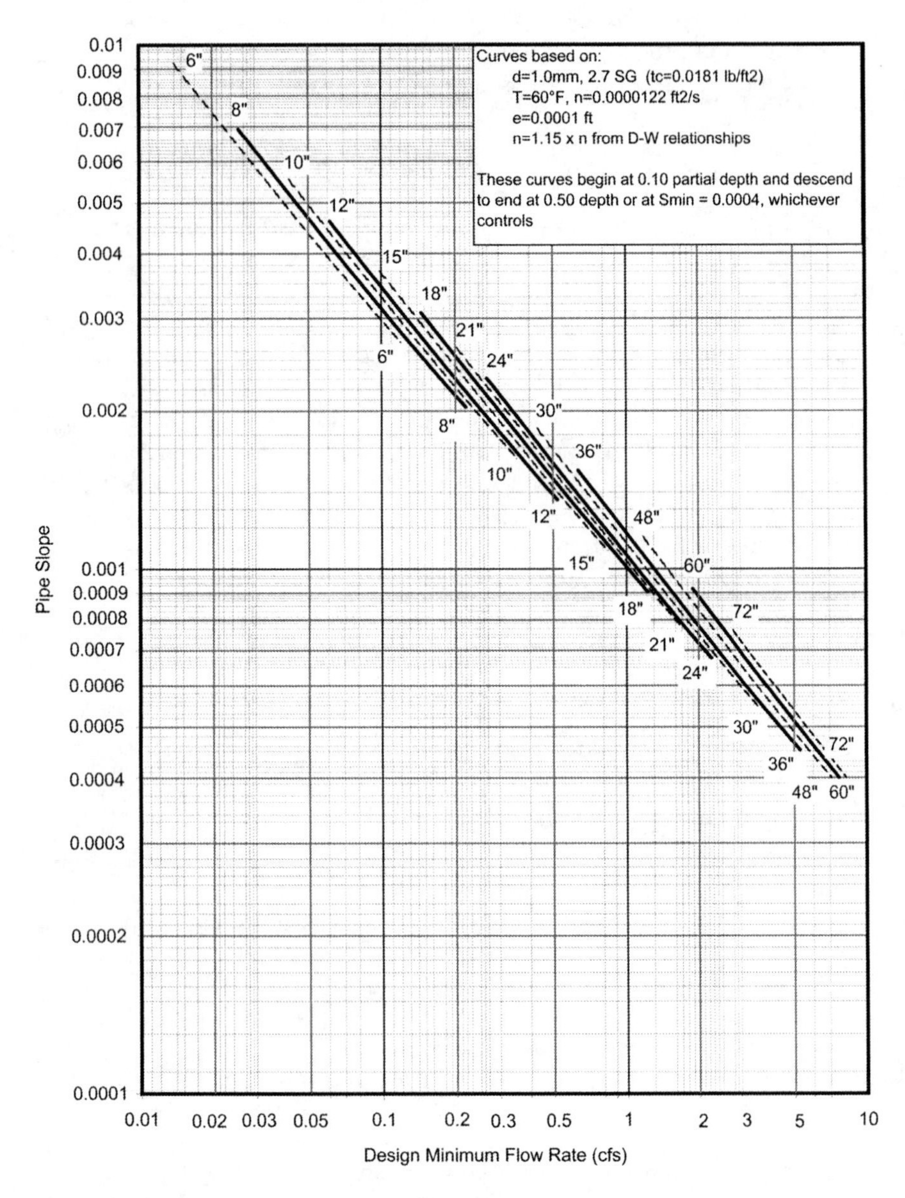

FIGURE 5-9. Self-cleansing slopes as a function of Q_{min} *– calculated Manning* n *(BG units).*

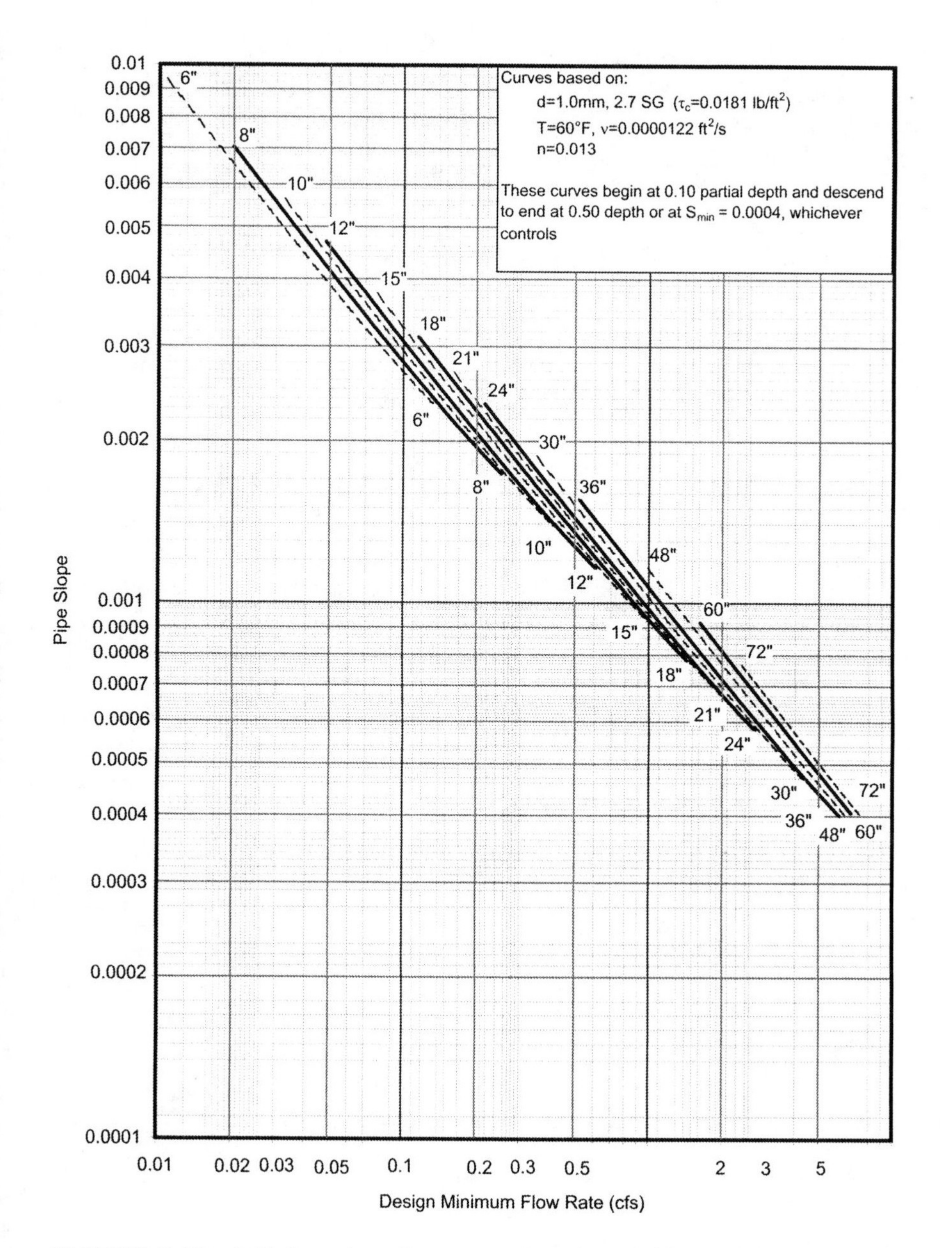

FIGURE 5-10. Self-cleansing slopes as a function of Q_{min} – *Manning* n = *0.013 (BG units).*

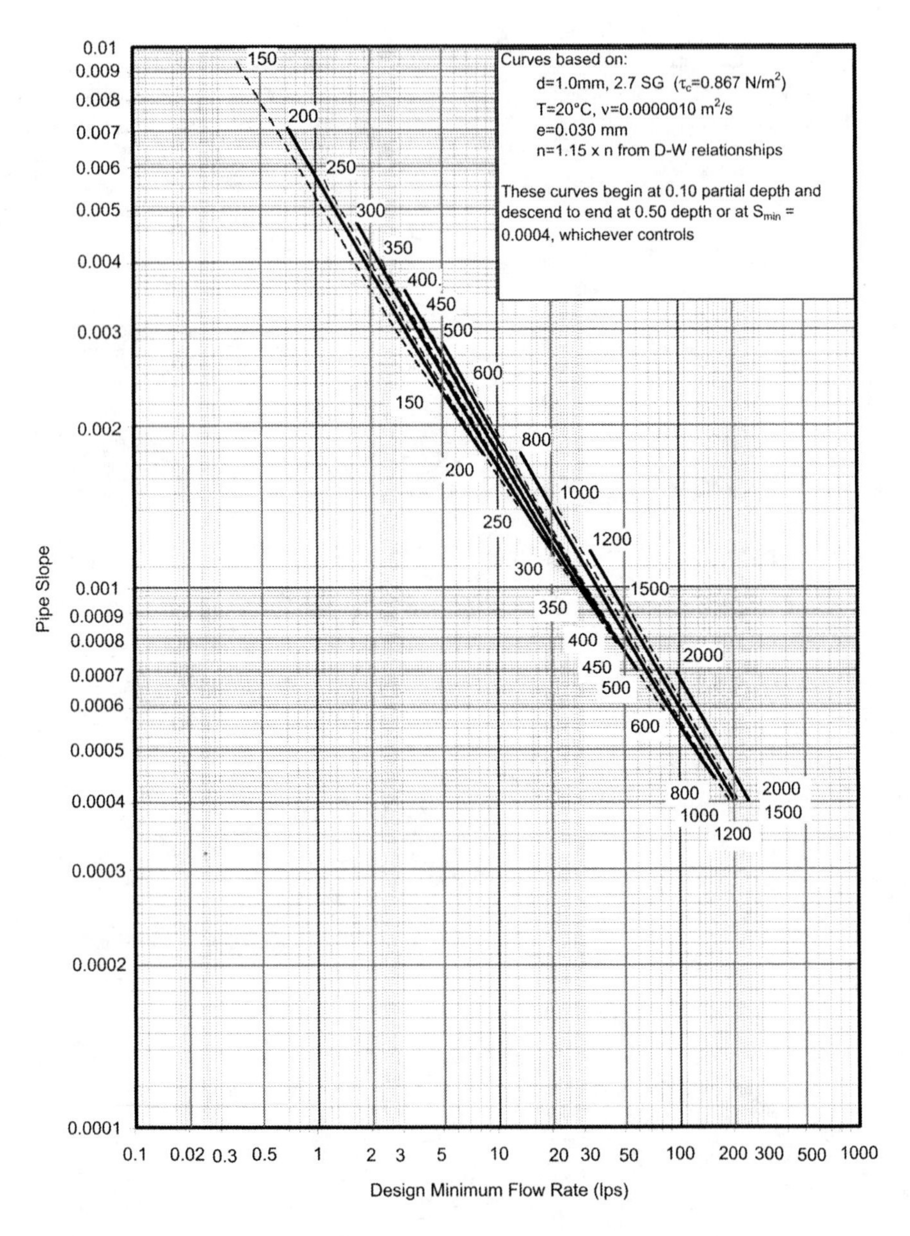

FIGURE 5-11. Self-cleansing slopes as a function of Q_{min} *– Manning* n *calculated (SI units).*

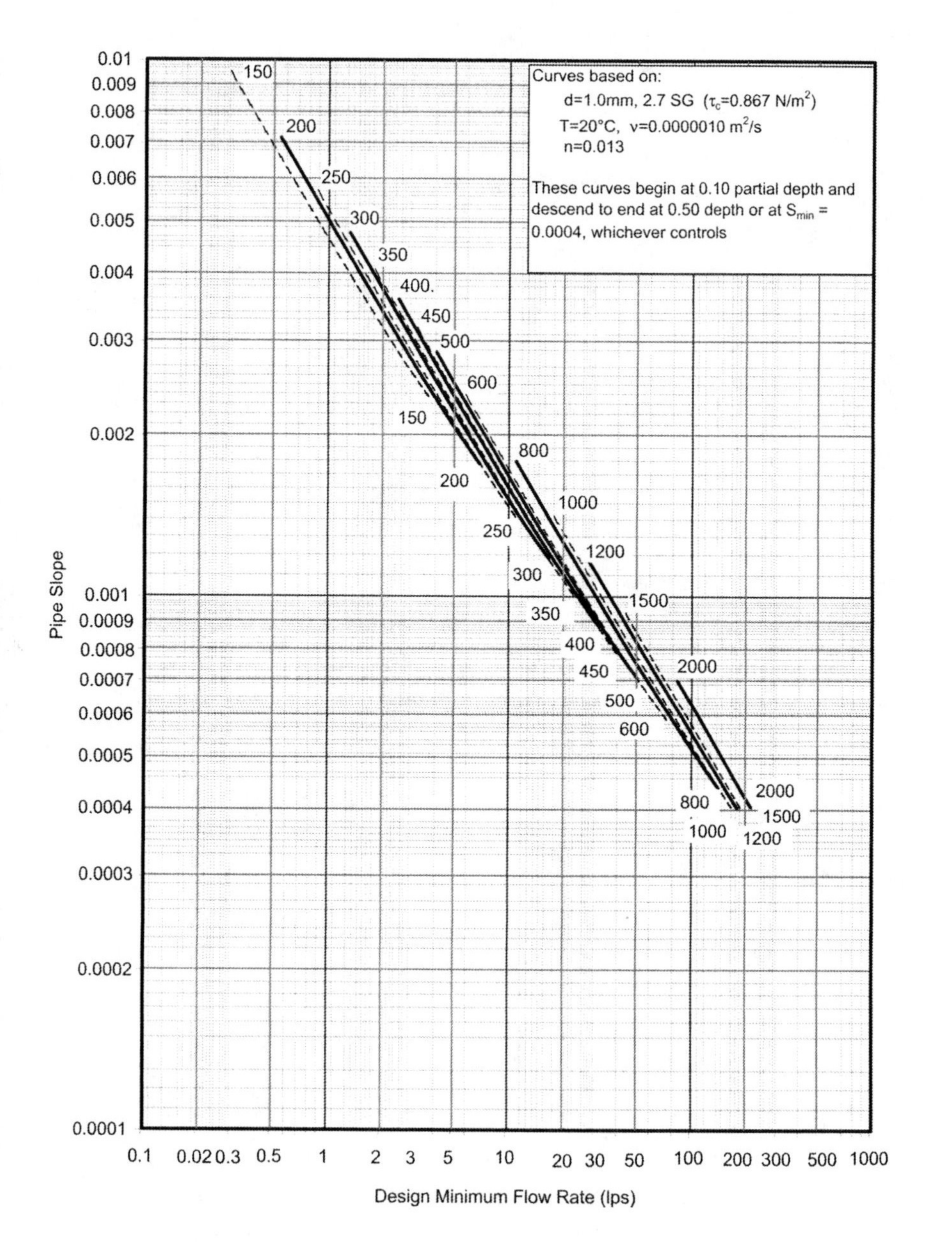

FIGURE 5-12. Self-cleansing slopes as a function of Q_{min} *– Manning* n = *0.013 (SI units).*

TABLE 5-5. Design Equations for S_{min} – BG Units

	Q in cfs	
Diameter	Variable n value*	$n = 0.013$
6 inch	$S_{min} = 0.000801Q_{min}^{-0.5687}$	$S_{min} = 0.000729Q_{min}^{-0.5600}$
8 inch	$S_{min} = 0.000848Q_{min}^{-0.5707}$	$S_{min} = 0.000776Q_{min}^{-0.5600}$
10 inch	$S_{min} = 0.000887Q_{min}^{-0.5721}$	$S_{min} = 0.000813Q_{min}^{-0.5600}$
12 inch	$S_{min} = 0.000921Q_{min}^{-0.5731}$	$S_{min} = 0.000846Q_{min}^{-0.5600}$
15 inch	$S_{min} = 0.000966Q_{min}^{-0.5744}$	$S_{min} = 0.000887Q_{min}^{-0.5600}$
18 inch	$S_{min} = 0.001004Q_{min}^{-0.5754}$	$S_{min} = 0.0009221Q_{min}^{-0.5600}$
21 inch	$S_{min} = 0.001038Q_{min}^{-0.5761}$	$S_{min} = 0.0009529Q_{min}^{-0.5600}$
24 inch	$S_{min} = 0.001069Q_{min}^{-0.5768}$	$S_{min} = 0.0009804Q_{min}^{-0.5600}$
27 inch	$S_{min} = 0.001097Q_{min}^{-0.5774}$	$S_{min} = 0.001005Q_{min}^{-0.5600}$
30 inch	$S_{min} = 0.001123Q_{min}^{-0.5778}$	$S_{min} = 0.001028Q_{min}^{-0.5600}$
36 inch	$S_{min} = 0.001169Q_{min}^{-0.5787}$	$S_{min} = 0.001069Q_{min}^{-0.5600}$
42 inch	$S_{min} = 0.001212Q_{min}^{-0.5812}$	$S_{min} = 0.001106Q_{min}^{-0.5617}$
48 inch	$S_{min} = 0.001255Q_{min}^{-0.5865}$	$S_{min} = 0.001143Q_{min}^{-0.5665}$
54 inch	$S_{min} = 0.001296Q_{min}^{-0.5905}$	$S_{min} = 0.001177Q_{min}^{-0.5700}$
60 inch	$S_{min} = 0.001334Q_{min}^{-0.5936}$	$S_{min} = 0.001210Q_{min}^{-0.5728}$
72 inch	$S_{min} = 0.001405Q_{min}^{-0.5975}$	$S_{min} = 0.001271Q_{min}^{-0.5763}$

*Manning n value calculated from Darcy-Weisbach relationships for 60 °F and $e = 0.0001$ ft, then multiplied by 1.15 to give n values identified as "typical" in Table 5-2.

S_{max} if it is larger than S_{max}. The resulting slope is that to be used for the pipe reach considered. Its value might have been determined by (1) slope to satisfy minimum depth needs, (2) slope needed for peak flow capacity, or (3) slope needed for self-cleansing. The largest slope of these three slopes is the design slope.

5.6.2.2.1. Alternate Tractive Force Procedure

This procedure does not use the empirical relationship [Eq. (5-32)] or a design particle size but, instead, assumes a minimum critical shear stress for self-cleansing. Hager (1999) and Yao (1974) recommend a critical shear stress τ_c of 0.04 lb/ft^2 (2 Pa) for self-cleansing in separate sanitary sewers. This shear-stress level is about what occurs in an 8-inch (200-mm) pipe when flowing from 45% full to full at 2-fps (0.6-mps) velocities.

Table 5-6. Design Equations for S_{min} – SI Units

Diameter	Q in lps: Variable n value*	$n = 0.013$
150 mm	$S_{min} = 0.00538Q_{min}^{-0.5691}$	$S_{min} = 0.00473Q_{min}^{-0.5596}$
200 mm	$S_{min} = 0.00574Q_{min}^{-0.5712}$	$S_{min} = 0.00503Q_{min}^{-0.5599}$
250 mm	$S_{min} = 0.00603Q_{min}^{-0.5725}$	$S_{min} = 0.00527Q_{min}^{-0.5599}$
300 mm	$S_{min} = 0.00629Q_{min}^{-0.5736}$	$S_{min} = 0.00548Q_{min}^{-0.5599}$
350 mm	$S_{min} = 0.00651Q_{min}^{-0.5744}$	$S_{min} = 0.00566Q_{min}^{-0.5599}$
400 mm	$S_{min} = 0.00672Q_{min}^{-0.5751}$	$S_{min} = 0.00583Q_{min}^{-0.5600}$
450 mm	$S_{min} = 0.00690Q_{min}^{-0.5757}$	$S_{min} = 0.00598Q_{min}^{-0.5600}$
500 mm	$S_{min} = 0.00707Q_{min}^{-0.5763}$	$S_{min} = 0.00611Q_{min}^{-0.5600}$
600 mm	$S_{min} = 0.00738Q_{min}^{-0.5772}$	$S_{min} = 0.00635Q_{min}^{-0.5600}$
800 mm	$S_{min} = 0.00790Q_{min}^{-0.5786}$	$S_{min} = 0.00676Q_{min}^{-0.5600}$
1000 mm	$S_{min} = 0.00833Q_{min}^{-0.5795}$	$S_{min} = 0.00676Q_{min}^{-0.5600}$
1200 mm	$S_{min} = 0.00893Q_{min}^{-0.5863}$	$S_{min} = 0.00755Q_{min}^{-0.5661}$
1500 mm	$S_{min} = 0.00972Q_{min}^{-0.5934}$	$S_{min} = 0.00817Q_{min}^{-0.5723}$
2000 mm	$S_{min} = 0.01078Q_{min}^{-0.5998}$	$S_{min} = 0.00898Q_{min}^{-0.5778}$

*Manning n value calculated from Darcy-Weisbach relationships for 20 °C and $e = 0.000031$ m, then multiplied by 1.15 to give n values identified as "typical" in Table 5-2.

After inserting this critical shear stress in Eq. (5-10), the slope becomes:

$$S_o = \frac{\tau_c}{\gamma R_h}$$

With this value of the slope, Manning's equation for the velocity V_c, corresponding to the critical shear stress τ_c, is:

$$V_c = \frac{k}{n}\left(\frac{\tau_c}{\rho g}\right)^{1/2} R_h^{1/6} \tag{5-32A}$$

This equation can be rewritten in a dimensionless form as:

$$V_c^* = \frac{V_c n}{k}\left(\frac{\rho g}{\tau_c}\right)^{1/2} \frac{1}{D^{1/6}} = \left(\frac{R_h}{D}\right)^{1/6} \tag{5-32B}$$

The left-hand side may be thought of as a dimensionless self-cleansing velocity V_c^* corresponding to the critical shear stress, τ_c. Figure 5-12A

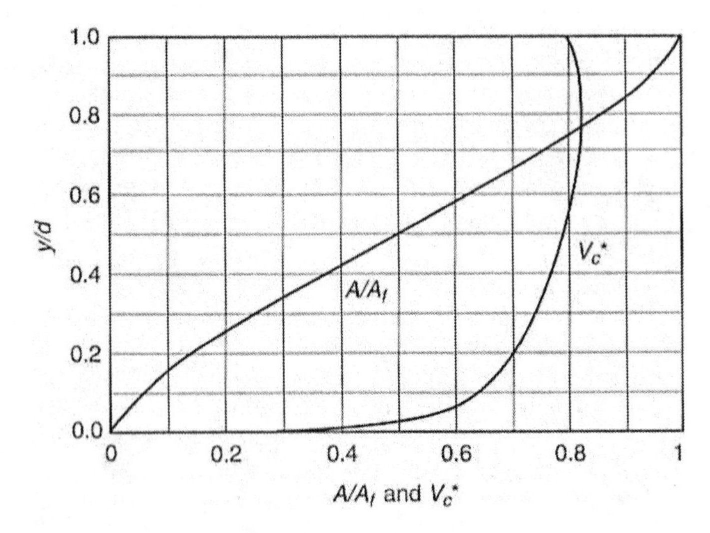

FIGURE 5-12A. Plot of V_c^* *and* A/A_f *versus* y/D.
Reproduced with permission from the McGraw-Hill Companies. Figure 4.11 page 124 in Terry W. Sturm, *Open Channel Hydraulics*, 2001.

shows a plot of V_c^* versus y/D. It can be seen that V_c^* is approximately equal to 0.8 for $y/D \geq 0.4$. Also shown in the figure is a plot of the ratio of the flow area at a given depth to the pipe full area, A/A_f versus y/D.

The calculation procedure for checking whether the design discharge is self-cleansing is as follows:

1. For the given design discharge, trial diameter, and slope, calculate the normal depth. This can be done by computer, graphically (Chow 1959; Sturm 2001), or by trial and error. For the latter method, select trial values of y/D_0, calculate successively θ from Eq. (5-28), A from Eq. (5-29), R_h from Eq. (5-30), and $AR_h^{2/3}$ until the latter value matches $(nQ)/(kS^{1/2})$ [see Eq. (5-31A)].
2. For the calculated normal depth, calculate the flow cross-section area A and the velocity $V = Q/A$.
3. Calculate V_c from Eq. (5-32A) using $\tau_c = 0.04\ \text{lb/ft}^2$ or 2 Pa, or read V_c^* from Fig. 5-12A. If the actual velocity V is equal to or greater than the velocity V_c corresponding to the critical shear stress, the sewer is self-cleansing. If this condition is not satisfied, the slope may be increased to increase the actual flow velocity and the associated shear stress.

It is not recommended that this approach be applied to depths of flow less than about 45% full since it would require a downward adjustment of the shear-stress τ_c value with increasingly shallower depths to avoid requiring unnecessarily high actual velocities and unduly steep slopes at shallower and shallower depths of flow. The Q_{min} and design-particle approach covered above addresses these shallower depths of flow more accurately. The main result of using this alternate approach would be increasingly steeper minimum slopes with increasing pipe sizes as compared to the traditional minimum slopes based on full-pipe flow at 2 fps (0.6 mps).

5.6.3. Observations on the Tractive Force Relationships

As shown in the figures above, when design minimum flow rates are larger than about 0.04 cfs for 6-inch pipe (0.001 m^3/s for 152-mm pipe), 0.06 cfs for 8-inch pipe (0.002 m^3/s for 200-mm pipe), 0.12 cfs for 10-inch pipe (0.003 m^3/s for 254-mm pipe), and so forth, the required S_{min} values will be increasingly smaller than traditional minimum slopes; the curves start at 0.10 y/D values and end with 0.50 y/D (half-full) values. Note that a substantial Q_{min} range for each pipe size requires steeper slopes than the traditional values for full-pipe, 2-fps (0.6-mps) velocity. The curves are not extended below 0.10 relative depths, since flow rates are relatively so small and flow equation accuracy so uncertain that the results are misleading. In such situations, the pipe should be placed as steep as is feasible, preferably as steep as indicated for 0.10 full point; the reach should be flagged; and the pipe cleaned more frequently if planned monitoring indicates it is needed.

Additional important observations from the tractive force approach and Figs. 5-9 through 5-12 include:

- If a larger pipe is selected in order to gain a smaller minimum slope according to a code, the self-cleansing situation is actually worse than if a smaller size were used at the larger pipe's minimum slope. For example, for $Q_{min} = 0.06$ cfs, an 8-inch-diameter pipe would require a 0.004 slope for self-cleansing, whereas a 10-inch-diameter pipe would need an ~0.006 slope rather than the 0.0028 minimum slope required by the code. Based on Eq. (5-32), the 10-inch-diameter pipe would only carry a 0.3-mm design sand particle at the 0.0028 slope when the flow is 0.06 cfs, whereas an 8-inch-diameter pipe at the same slope and flow rate would handle a 1-mm particle.
- S_{min} values are quite sensitive to changes in Q_{min}, especially for the smaller diameters and lower flow rates. This fact highlights the need to establish the best possible estimates for Q_{min} so that the S_{min} values selected are likewise appropriate.
- Relatively speaking, traditional minimum slopes for larger diameters do not self-cleanse as well as those for smaller diameters. As observed above, Yao (1974) and Merritt (1998) have also shown that the use of the same full-pipe, minimum-velocity criterion for all

sizes often results in slopes steeper than needed for small-diameter pipes and slopes flatter than needed for larger ones.

- An incidental observation is that one of the main motivations for the use of egg-shaped sewers, or other type of narrowed cross section at the bottom, is that the hydraulic radius stays relatively larger with decreasing flow rates. Therefore, such shapes self-cleanse better at low flows than does a circular conduit with the same total cross-section area and design minimum flow rate.

5.7. HYDRAULIC CONTINUITY THROUGH MANHOLES

Flow through new manholes should be designed to pass through smoothly to minimize turbulence and energy loss. Changes in slope, direction, diameter, and multiple tributaries often combine to require a drop into the outflowing pipe invert elevation in order to maintain a smooth energy line and water surface through the manhole, and not cause backup in tributary conduits. Figure 5-13 shows a schematic diagram of entering and exiting pipes.

When the energy equation, Eq. (5-2), is applied from an inflowing conduit to the outflowing conduit, the result is Eq. (5-34), which gives the necessary continuity drop, Δz, through the manhole:

$$\Delta z = (y_2 - y_1) + \left(\frac{V_2^2}{2g} - \frac{V_1^2}{2g} \right) + h_L \tag{5-34}$$

where y_1 and y_2 are normal depths of flow in the conduits. Hydraulics programs, which can be used to quickly calculate normal depths and

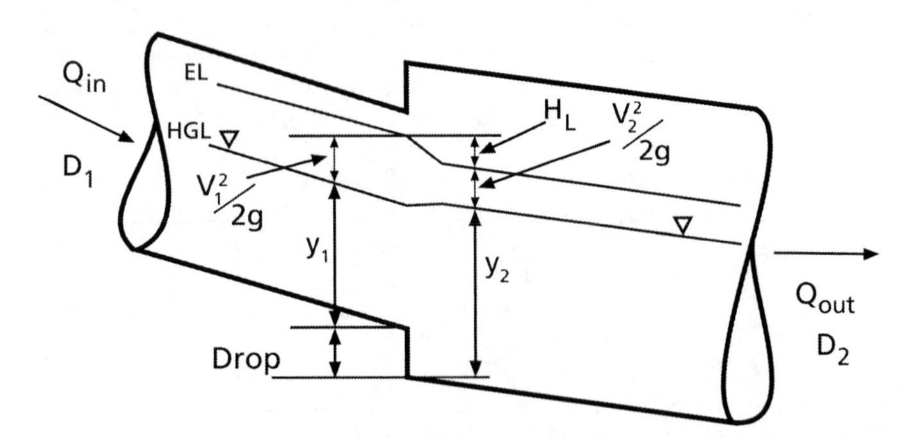

FIGURE 5-13. Hydraulic continuity through a manhole.

velocity heads, facilitate calculations for circular conduits that are quite laborious if done by hand.

If two or more tributary conduits join at the same manhole, then each one should be analyzed separately and the lowest calculated elevation for the exit invert should be used. This continuity analysis should be done for design capacity flow rates since these largest flow rates are most likely to cause backup into entering conduits. The setting most likely to require fairly large drops (sometimes exceeding a foot) is a manhole with two or three joining tributaries and/or a transition from steep to flat slopes. Most drops are fairly small [usually less than 0.10 ft (3 cm)] and rises are not allowed because of possible grit accumulation in the depression.

Some codes and guidelines call for routine manhole drops of some 0.10 to 0.30 ft (3 to 9 cm), usually as a function of the number of inflowing tributaries. Continuity calculations using Eq. (5-34) usually result in significantly smaller values, but sometimes these values are larger. Using arbitrary drops when not needed hydraulically is of negligible concern when terrain is relatively steep and sewers are near minimum depth; however, in relatively flat terrain unnecessary drops sometimes force sewers much deeper than needed because the unneeded drops accumulate downstream. Manhole head losses are reviewed below. For curved and streamlined manhole channels, flow through manholes experiences very small head losses, typically less than 0.02 ft (0.6 cm). When flow depths are considerably greater than about half-full for design capacity flow rates, flows may begin overtopping their manhole channels and larger head losses occur. However, this is usually of little concern and significantly larger drops are not justified, since these high flows are of short duration—just for a few minutes or so during the highest flows of the diurnal cycle. When non-streamlined manholes are used, more detailed techniques should be used to determine manhole head losses. See, for example, Haestad et al. (2004).

If manhole drops are not needed for continuity reasons and head losses are negligible, it may still be prudent to use a drop through manholes of about 0.05 ft (1.5 cm), both to accommodate some small head loss and to act as a construction tolerance so that slight horizontal misalignment of the manhole will not result in an outlet at a higher elevation than the lowest inlet. Even then such drops are not needed when the sewer is, and is very likely to remain, essentially a flow-through pipe at the manhole.

5.8. HEAD LOSS IN MANHOLES

Local energy losses are often estimated using Eq. (5-35) or (5-36):

$$h_L = K\frac{V_2^2}{2g} \tag{5-35}$$

or

$$h_L = K\frac{(\Delta V)^2}{2g} \tag{5-36}$$

For contoured flow-through manholes where each inflow is guided through channels cast into the manhole base, head loss is very small—typically less than 0.02 ft (0.6 cm). When nonstreamlined manholes are used, a more detailed evaluation is needed to estimate the head loss (see Haestad et al. 2004). As a worst-case scenario, the upper bound on a head loss estimate would assume a total loss of the incoming kinetic energy; therefore, the downstream kinetic energy must come from potential energy (depth and/or elevation drop), written as $\Delta(y + z) = V^2/2g$. This represents a change of only 0.06 ft for 2 fps and 0.14 ft for 3 fps; however, since this change is proportional to V^2, it increases rapidly with higher velocities to about 1 ft for 8 fps.

5.9. WATER SURFACE PROFILES

Non-uniform flow (varying depth with distance) in open channels can be categorized as *gradually varied flow* (GVF) or *rapidly varied flow* (RVF). GVF is used to calculate water surface profiles for a conduit. For GVF, slopes of the energy line and water surface are curved and not parallel. Before the advent of digital computers, great emphasis was placed on various mathematical integration techniques for water surface dy/dx relationships. With modern computers, finite step methods are used with ease, which (for relatively small Δy) steps closely approximate integration of the differential equation describing the water surface curvature. One common form of the GVF equation is given in Eq. (5-37).

$$\Delta x = \frac{E_1 - E_2}{\bar{S} - S_0} \tag{5-37}$$

where E is the specific energy for selected depths y_1 and y_2 along a channel, and $\bar{S}$ is the average energy line slope between the two y depths, which will be a calculated distance Δx apart.

A selected flow friction equation, such as the Manning equation, is used to calculate the $\bar{S}$ value, either by averaging the two S values calculated at the ends of the section as if they were each uniform flow or, more commonly, by using average areas, hydraulic radii, and velocities to calculate $\bar{S}$. When increasingly smaller y increments are used in the calculation, the two approaches converge. The answers are only slightly different for fairly large y increments.

Water surface profiles for GVF are classified relative to critical depth, y_c, and the normal depth, y_n, from the uniform flow equation used. Each channel section is given an alpha code based on its uniform flow regime: Mild (M) if subcritical; Critical (C) if at critical (a singular point); Steep (S) if supercritical; and Horizontal (H) or Adverse (A) if the bottom slope is negative. The second identifier is one of three zones based on the actual water depth in the section. Zone 1 is where actual water depth is above both critical and normal depths. Zone 2 is where actual water depth is between critical and normal depths. Zone 3 is below both. Essentially all textbooks and hydraulics reference books have diagrams showing the shapes of all possible GVF surface profiles.

To calculate profiles, one must start at a point of known water depth as the initial y_1. This point is referred to as a *control point* (CP). If the actual water depth at the CP is shallower than the critical depth, the curve will proceed downstream. If the actual water depth at the CP is deeper than the critical depth, the curve will go back upstream.

The most common profiles of consequence in sewer conduits are M1, where the water depth downstream is greater than normal depth and the M1 curve goes back upstream to join the normal depth asymptotically; M2, where the flow in the conduit is subcritical and depth downstream is lower than the normal depth, such as discharging into a tank or pond that is lower; and S1, where conditions are essentially the same as for the M1 curve above except the conduit is steep and the S1 curve goes back upstream to where a hydraulic jump occurs to join the curve. In this latter case, the depth of the S1 curve is the sequent depth to the upstream supercritical depth coming into the jump. Since in sewer conduits velocities along M1 and S1 reaches are quite low, M1 and S1 curves are essentially horizontal lines back upstream from the control point of known depth. In many cases M2 curves will draw down the surface considerably, with the maximum drawdown being critical depth near the outlet of the pipe if the depth at the end is at or below critical depth.

5.10. SERVICE LATERAL SLOPES

Minimum slopes for service laterals range from about 0.0075 (0.75%) to 0.02 (2%). When placed and bedded precisely and carefully, a 0.0075 slope is acceptable when 2- to 4-gal (7.5- to 15-L) flush toilets are used in the building or when larger buildings or several buildings are connected to the lateral. A larger minimum slope value of up to 0.010 is recommended where low-flush toilets [0.5 to 2 gal per flush (2 to 7.5 L)] are used. Larger minimum slopes up to 0.02 are recommended where adequate construction inspection is not planned or conducted. Tractive force principles show that smaller pipes self-cleanse better than larger pipes. However,

3- or 4-inch- (76- to 102-mm)-diameter pipes are likely the smallest feasible sizes for other reasons. Maintenance equipment limitations might result in a 6-inch (152-mm) minimum diameter conduit in some cases. If larger lateral pipes [greater than 6-inch- (152-mm)-diameter] are used, it is suggested that slopes perhaps 50% larger be used for each of the situations stated above if the lateral does not service at least several buildings.

The main issue in the hydraulic design of service laterals is the minimum slope necessary to achieve self-cleansing. Since street sewer depth is often determined by depths necessary for gravity flow in service laterals from buildings and other facilities, shallower street sewers are possible, relatively speaking, when these controlling laterals are shallower. Substantial cost savings might also be gained from relatively shallower placement of laterals themselves as allowed by flatter minimum lateral slopes. The hydraulics of flow in laterals is more difficult to address since it is unsteady, non-uniform flow. Such flow can be stated mathematically, but large variations in daily peak flow from service to service and the use level, and flows in facilities served are subject to relatively unpredictable, major changes over time. Rarely is capacity a concern, since sizes smaller than 4 inches (102 mm) in diameter are seldom used. For perspective: A 4-inch- (12-mm)-diameter line placed at 0.0075 slope would have the capacity to handle peak flow from about 50 houses—based on four people per house, 100 gpcd (gallons per capita per day) (375 Lpcd) (liters per capita per day), and a peaking factor of about 6.

5.11. PARTIAL LISTING OF CURRENT SEWER ANALYSIS/DESIGN SOFTWARE

Table 5-7 contains a partial listing of currently available software. Zao (2001) focuses on public domain U.S. models and a few related proprietary models. Stein and Young (2001) discuss *HYDRA*, a storm and sanitary design and analysis software that is part of the *HYDRAIN* package supported by the U.S. Federal Highway Administration. Yen (1999) gives further details on the *Storm Water Management Model* (*SWMM*) originally supported by the EPA.

APPENDIX

5.A.1. Critical Velocity/Critical Stress

The threshold of sediment movement is usually defined in terms of a critical erosion velocity or a critical shear stress. For full-pipe design for an 8-inch- (200-mm)-diameter pipe, a full-pipe critical velocity of 2 ft/s (0.6 m/s) gives a shear stress of about 0.042 lb/ft^2 (2N/m^2). To achieve this value in larger diameters, the full-pipe velocity is increasingly larger,

TABLE 5-7. Sewer Analysis Software[a]

Model Name	Hydraulic Solution	Routing Level	Time Domain	Sediment Buildup	Multiple Sediment Fractions	Pollutant Transport	Bedload Transport	Source
SWMM RUNOFF Block	Simplified; Nonlinear reservoir	Gutters Pipes	Continuous and Single Events	Linear and nonlinear buildup and washoff Soil loss; Rating curve	No	First-order washoff	No	EPA
SWMM TRANSPORT Block	Simplified; Kinematic wave	Channels, Pipes, Junctions, Pump Stations; Outfall structures, Infiltration; *No* backwater *No* surcharge	Continuous and Single Events	Routes pollutants generated by *runoff* through sewer system	No	Completely mixed reactor, First-order decay (in storage)	No	EPA
SWMM EXTRAN Block	Full	Channels, Pipes, Junctions, Pump Stations; Outfall structures, Backwater surcharge	Continuous and Single Events	No	No	No	No	EPA

(continued)

TABLE 5-7. *(Continued)*

Model Name	Hydraulic Solution	RoutingLevel	Time Domain	Sediment Buildup	Multiple Sediment Fractions	Pollutant Transport	Bedload Transport	Source
STORM	Simplified	Simple storage; Storage/ Treatment	Continuous and Single Events Dry-weather flow	Linear buildup and washoff; Soil loss	No	Six pollutants Mass balance	No	COE
HSPF	Simplified	Simple storage and Channel routing	Continuous and Single Events	Linear buildup and washoff; Soil loss	No	Completely mixed	No	EPA USGS
HYDRAIN/ HYDRA	Simplified/full	Sewers, grade, curb, slotted inlets	Single Event	No	No	No	No	FHWA
HYDRO-WORKS	Full				2		Total load	Wallingford, UK
MOUSE/ MOUSETRAP	Full			Yes	Yes		Yes	DHI, Denmark

[a] *HYDROWORKS* and *MOUSE* are commercial software. They were developed in Europe and have been used in the United States for some time. Their listing does not imply endorsement. A number of commercially available software provide input interfaces that allow direct input into *SWMM* from GIS data and provide output interfaces with excellent graphics for reporting. Among these one may cite (without endorsement): *SEWERCAD* from Haestad (Waterbury, Conn.), *PCSWMM* from CHI (Guelph, Ontario, Canada), *MIKE SWMM* from DHI (Trevose, Penn.), *XPSWMM* from XP Software (Portland, Ore.) and *HYDRA* from Pizer (Seattle, Wash.).

increasing to 3.0 fps (0.9 mps) for a 60-inch- (1,500-mm)-diameter pipe. Because of the organic nature of the particles and the presence of organic slimes and grease, sanitary sediments—particularly old deposits—may exhibit a cohesivelike strength. Incipient movement shear stresses of 2.5 N/m^2 have been observed in synthetic surficial material and 6 to 7 N/m^2 in granular consolidated deposits (Ashley and Verbanck 1996). However, once the cohesive structure is disrupted, the stripped particles are transported as noncohesive sediments (Nalluri and Alvarez 1992).

Once the boundary shear stress exceeds the critical shear stress, the particles are entrained and transported in suspension or as bedload. Sanitary solids are normally transported in suspension, with the larger grit being transported as bedload.

Table 5-8 lists the minimum shear stress values from several countries.

5.A.2. A UK Approach to Self-Cleansing Sewer Design

In the UK, CIRIA developed a methodology for the design of self-cleansing sewers based on sediment transport principles (Ackers et al. 1996). Summaries of the procedure can be found in Butler et al. (2003); Butler et al. (1996a; 1996b); Delleur (2001); and Arthur et al. (1999). The following is based on Butler et al. (2003) focusing on sanitary sewers.

The CIRIA study indicates that for self-cleaning performance over a full range of sediment bed conditions, sewers should be designed so that each pipe satisfies the following three design criteria:

1. Transport a minimum concentration of suspended particles.
2. Transport the coarser, inorganic grit material as bedload at a rate sufficient to limit the depth of deposition to a specified proportion of the diameter.

TABLE 5-8. Minimum Shear Stress Criteria

Reference	Country	Sewer Type	Minimum Shear Stress, N/m^2	Pipe Conditions
Yao (1974)	United States	Sanitary Storm	1.0–2.0 3.0–4.0	Full
Maguire (cited in Nalluri and Ghani, 1996)	UK	Sanitary	6.2	Full
Bischop (1976; cited in Nalluri and Ghani, 1996)	Germany	Sanitary	2.5	Full

Adapted from Nalluri, C., and Ghani, A. A. (1996). "Design options for self-cleansing storm sewers." *Water Sci. & Tech.*, 33(9), with permission.

3. Erode cohesive particles from the deposited bed. This typically requires a bed shear stress $\geq 2.0\ N/m^2$, assuming a particle size of 1 mm and an effective roughness of 1.2 mm.

The design procedure considers the hydraulic roughness of the sediment bed as related to sediment particle size and sediment movement as related to flow velocity and roughness. The methodology is complex. Potential users should refer to the references cited above. However, minimum design velocities for high and medium solids concentration in sanitary sewers (SaH and SaM, respectively) were obtained using a simplified procedure and standard values. The result is shown in Fig. 5-14. This figure is a general guide, not intended as a design tool.

Figure 5-14 also illustrates the three design criteria in terms of the required full-pipe velocities. The curve labeled Sa-H corresponds to criterion 1, the bottom curve Sa-M corresponds to criterion 3, and the remaining curves correspond to criterion 2. In general, smaller pipe design will be based on criterion 3, whereas larger pipes will be designed according to criteria 1 or 2.

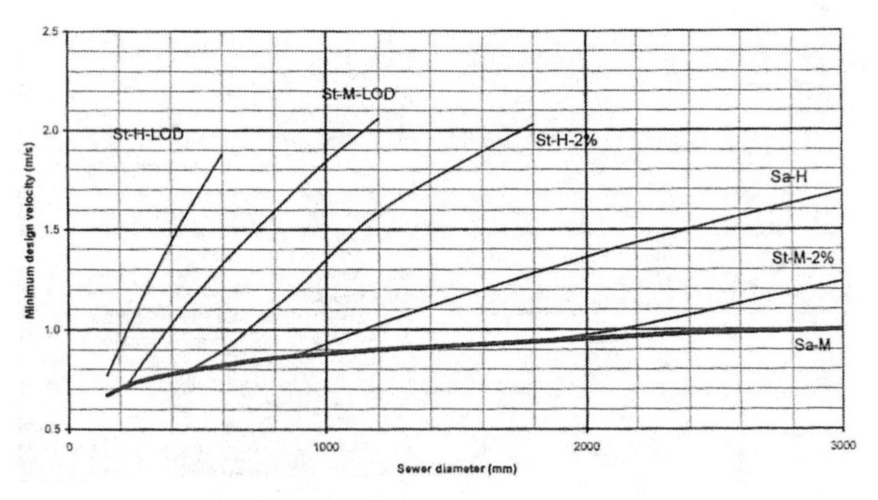

FIGURE 5-14. British CIRIA minimum full-pipe design velocities—by simplified construction industry research and information procedure.
Sewer types: Sa, sanitary; St, storm. Sediment loads: M, medium; H, high. Deposition criteria: LOD, limit-of-deposition; 2%, allowable deposition depth. Butler, D., May, R., and Ackers, J. (2003). "Self-cleansing sewer design based on sediment transport principles." *J. Hyd. Engrg.* (ASCE) 129(4), 276–282.

Comments on the CIRIA Approach

A shortcoming of the CIRIA figure approach in Fig. 5-14 is that it is an envelope approach based on full-pipe velocities and, as such, does not acknowledge actual design minimum flow rates, Q_{min}. The net result is to significantly increase all minimum slopes rather than just in those situations where very small Q_{min} values actually occur. Since the CIRIA approach does not allow flatter slopes when Q_{min} values are relatively large, one of the major benefits of the tractive force and Q_{min} approach is negated. The CIRIA approach results in many sewers being placed much steeper than actually needed, unless one assumes that a cohesive, congealed sediment bed must be eroded—this is the condition that self-cleansing in a sanitary sewer is aimed at preventing, Steeper slopes means deeper sewers and more pumping in many situations.

The SaM velocities in Fig. 5-14 result in a full-pipe tractive force of about 0.045 lb/ft^2 (2 Pa) for all conduit sizes. Based on Eq. (5-32), this tractive force value would move a design particle of about 20 mm in nominal diameter. Assuming that a 1-mm particle is the actual design particle, the slopes associated with the lower curve CIRIA velocities are steeper than needed, unless at Q_{min} a pipe flows less than about 15% full.

The SaH curve velocities have the same inherent problem as mentioned above. However, since the SaH curve applies only to larger-diameter sewers, the significantly higher velocities are gained with fairly small elevation changes. For example, a 60-inch- (1,500-mm)-diameter sewer would need a slope of about 0.00045 to achieve 2 fps, but an increase up to a slope of about 0.00080 to achieve the suggested full-pipe velocity of 3.8 fps. Although this about doubles the slope, it represents an increased elevation change of only about 1.8 ft per mile—for comparison, based on a 1-mm design particle, the tractive force approach herein would require a slope of 0.00093 if the pipe flowed 10% full at its Q_{min}, or 0.00048 for a larger Q_{min} at 20% full.

REFERENCES

Ackers, J. C., Butler, D., and May, R. W. P. (1996). "Design of sewers to control sediment problems." Construction Industry Research and Information Association (CIRIA), Rep. No. R141, CIRIA, London.

American Society of Civil Engineers (ASCE). (1982). "Gravity sewer design and construction." *ASCE Manual and Reports of Engineering Practice No. 60*, ASCE, Reston, Va.

Arthur, S. A. R., Tait, S., and Nalluri, C. (1999). "Sediment transport in sewers—A step towards the design of sewers to control sediment problems." *Proc., Inst. Civil Engrs., Water, Maritime and Energy, 136*, 9–19.

Ashley, R. M., and Hvitved-Jacobsen, T. (2002). "Management of sewer sediments." *Wet-weather flow in the urban watershed technology and management*, R. Field and D. Sullivan, eds, CRC Press/Lewis Publishers, Boca Raton, Fla.

Ashley, R. M., and Verbanck, M. A. (1996). "Mechanics of sewer sediment erosion and transport." *J. Hyd. Res.*, 34(6), 753–770.

Butler, D., May, R. W. P., and Ackers, J. V. (1996a). "Sediment transport in sewers—Part 1. Background." *Proc., Inst. Civil Engrs., Water, Maritime and Energy, 118*, 103–112.

Butler D., May, R. W. P., and Ackers, J. C. (1996b). "Sediment transport in sewers—Part 2." *Proc., Inst. Civil Engrs., Water, Maritime and Energy, 118*, 113–120.

Butler, D., May, R., and Ackers, J. (2003). "Self-cleansing sewer design based on sediment transport principles." *J. Hyd. Engrg. (ASCE), 129*(4), 276–282.

Camp, T. R. (1946). "Design of sewers to facilitate flow." *Sewage Works J., 18*, 3.

Chow, V. T. (1959). Open channel hydraulics, McGraw-Hill, New York.

Colebrook, C. F. (1939). "Turbulent flow in pipes with particular reference to the transition region between the smooth and rough pipe laws." *J. Inst. Civil Engrs., London, 11*, 133–156.

Cunge, J. A., and Wegner, M. (1964). "Intégration Numérique des Équations d'Ecoulement de Barré de Saint Venant par un Schéma Implicite de Différences Finies: Application au Cas d'une Galerie Tantôt en Charge, Tantôt à Surface Libre." ["Numerical integration of the St. Venant flow equations by an implicit finite difference scheme: Application to the case of a conduit sometimes pressurized, sometimes with free surface."] *La Houille Blanche, 1*, 33–39.

Delleur, J. W. (2001). "Sediment movement in drainage systems." *Stormwater collection systems handbook*, L. W. Mays, ed., McGraw-Hill, New York, 14.1–14.25.

Haestad, M., Walski, T. M., Barnard, T. E., Harold, E., Merritt, L. B., Walker, N., and Whitman, B. E. (2004). *Wastewater collection system modeling and design*, Haestad Press, Waterbury, Conn.

Hager, W. H. (1999). *Wastewater hydraulics*, Springer-Verlag, Berlin.

Lansey, K., and El-Shorbagy, W. (2001). "Design of pumps and pump facilities." *Stormwater collection systems design handbook*, L. W. Mays, ed., McGraw-Hill, New York, Table 12.1.

Martin, G. S. (1999). "Hydraulic transient design for pipeline systems." *Hydraulic design handbook*, L. W. Mays, ed., McGraw-Hill, New York.

May, D. K. (1986). *A study of Manning's coefficient for commercial concrete and plastic pipes*, T. Bench Hydraulics Laboratory, University of Alberta, Edmonton, Alberta, Canada.

Medina, L. R., and Vega, F. A. (2002). "Diseño de colectores de alcantarillado aplicando el criterio de tension tractriz." ["Sewer collector design applying the tractive force criterion."] *Ingenieria Hidraulica y Ambiental, 23*(1), 50–55.

Merritt, L. B. (1998). "Manning roughness coefficient in sewer design." *Sewer Design Paper No. 98-2*, Brigham Young University, Provo, Utah.

Metcalf & Eddy, Inc. (2003). *Wastewater engineering: Treatment and reuse*, fourth ed., revised by G. Tchobanoglous, F. L. Burton, and H. D. Stensel, McGraw-Hill, New York.

Moody, L. F. (1944). *Transactions ASME, 66*, American Society of Mechanical Engineers (ASME), New York.

Nalluri, C., and Alvarez, E. M. (1992). "The influence of cohesion on sediment behaviour." *Water Sci. & Tech., 25*(8), 151–164.

Nalluri, C., and Ghani, A. A. (1996). "Design options for self-cleansing storm sewers." *Water Sci. & Tech., 33*(9), 215–220.

Ota, J., and Nalluri, C. (2003). "Urban storm sewer design: Approach in consideration of sediments." *J. Hyd. Engrg. (ASCE), 129*(4), 291–297. Discussion and closure: July, 2004, 718–721.

Raths, C. W., and McCauley, R. R. (1962). "Deposition in a sanitary sewer." *Water and Sewage Works*, 192–197.

Schmidt, O. J. (1959). "Measurements of Manning's coefficient." *Sewage and Industrial Waste, 31*(9), 995.

Stein, S., and Young, K. (2001). "Storm drain design and analysis using HYDRA." *Stormwater collection systems design and analysis handbook*, L. Mays, ed., McGraw-Hill, New York.

Straub, L. G., Bowers, C. E., and Puch, M. (1960). "Resistance to flow in two types of concrete pipes." Tech. Paper No. 22, Series B, St. Anthony Falls Hydraulic Laboratory, University of Minnesota, Minneapolis, Minn.

Sturm, T. W. (2001). *Open channel hydraulics*, McGraw-Hill, New York.

Swamee, P. K., and Jain, A. K. (1976). "Explicit equations for pipe-flow problems." *J. Hyd. Div. (ASCE), 102*(HY5),657–664.

Tchobanoglous, G., Burton, F. L., and Stensel, D. (2003). *Wastewater engineering: Treatment and reuse*, fourth ed., McGraw-Hill New York.

Tullis, J. P. (1986). "Friction factor tests on concrete pipes." Hydraulic Rep. No. 157, Water Research Laboratory, Utah State University, Logan, Utah.

Yao, K. M. (1974). "Sewer line design based on critical shear stress." *J. Env. Engrg., 100*(EE2), 507-520.

Yen, B. C., ed. (1992a). Channel flow resistance: Centennial of Manning's formula, Water Resources Publications, Littleton, Colo.

Yen, B. C. (1992b). "Dimensionally homogeneous Manning's formula." *J. Hyd. Engrg. (ASCE), 118*(9), 1326–1332.

Yen, B. C. (1999). "Hydraulic design of urban drainage systems." *Hydraulic design handbook*, L. Mays, ed., McGraw-Hill, New York.

Yen, B. C. (2001). "Hydraulics of sewer systems." *Stormwater collection systems design handbook*, L. W. Mays ed., McGraw-Hill, New York.

Zhao, B. (2001). "Computer models for stormwater systems design." *Stormwater collection systems design handbook*, L. W. Mays, ed., McGraw-Hill, New York.

CHAPTER 6

DESIGN OF SANITARY SEWER SYSTEMS

6.1. INTRODUCTION

The design of sanitary sewers requires the engineer to meet multiple objectives simultaneously. Sanitary sewers transporting wastewater from one location to another must be installed deep enough below the ground surface to receive flows from the tributary area. However, excessive depths increase construction costs. The pipe material used must exhibit characteristics of resistance to corrosion and structural strength sufficient to carry earth backfill load and any impact and live loads. The size and slope or gradient of the sanitary sewer must be adequate for the flow rate to be carried with a cleansing power sufficient to inhibit deposition of solids, thereby being self-cleansing. The type of sewer pipe selected should be made based on consideration of the characteristics of the wastewater being transported and installation conditions. Manholes, junction chambers, and other structures should minimize turbulence and head loss and prevent deposition of solids. Anticipated service life, economy of maintenance, safety to personnel and the public, and convenience for connection during the useful life of the sewer must be considered.

The objective of design is to provide a sewer system at the lowest annual cost compatible with its function, while providing sufficient durability for the design period. The construction cost is usually proportional to the degree of complexity of the system. Therefore, each of these factors must be weighed and evaluated together with the owner's priorities.

6.2. ENERGY CONCEPTS OF SEWER SYSTEMS

A gravity sanitary sewer system provides a means of transportation for wastewater that utilizes the potential energy resulting from the difference in elevation of its upstream and downstream ends. Energy losses due to free falls, sharp bends, or turbulent junctions must be held to a minimum if the sewer is to operate properly at a minimum slope and depth, unless the system serves an area with a relatively steep general gradient where more than adequate energy is available.

Generally, the total available potential energy is utilized to maintain proper flow velocities in the sewers with minimum hydraulic head loss. Chapter 5 deals with the energy concept in the design of sanitary sewers and provides design guidelines for many of the situations typically encountered.

Where excess elevation differences exist, it may become necessary to dissipate excess potential energy. Where the differences in elevation are insufficient to permit gravity flow, external energy must be added to the system by pumps. Although pumps are often necessary, they add an additional degree of complexity and cost to the design. The consequences of a sanitary sewer overflow due to mechanical or electrical failure in a pumping station must be considered. If a pumping station outage would result in pollution of a waterbody or in any way affect the health and safety of a community, the higher cost of a gravity system may be justified. Other consequences may include enforcement action by state or federal agencies due to the overflow.

6.3. COMBINED VERSUS SEPARATE SEWERS

Many existing sewer systems collect both wastewater and stormwater. In such systems, stormwater, up to a design limit, is channeled with sanitary wastewater to the treatment plant. When combined flows exceed the sewer's capacity, combined sewer overflows (CSOs) occur and wastewater, along with stormwater, is discharged to receiving waters. When stormwater is transported with wastewater to the treatment plant, pumping and treatment costs are increased and treatment problems may occur. Combined sewer systems are no longer designed except as limited extensions or replacements for existing combined systems.

Studies by the U.S. Public Health Service (USPHS 1964) and the U.S. Environmental Protection Agency (EPA 1970) show that combined stormwater and wastewater overflows introduce large quantities of polluting materials into the nation's waters. As a result of these issues and other concerns, the EPA in 1994 adopted a CSO control policy (EPA 1994). This policy includes minimum controls that sharply limit the use of combined sewer systems and essentially preclude their use in new construction.

6.4. LAYOUT OF SYSTEM

The design engineer begins a sanitary sewer system layout by selecting an outlet, determining the tributary area, locating trunk and main sewers, and determining the need for and location of pumping stations and force mains (NTIS 1977). The outlet is located according to the circumstances of the particular project. Thus, a system may discharge to a treatment plant, a pumping station, or a trunk or main sanitary sewer.

Preliminary layouts can be made largely from topographic maps and other pertinent information. In general, sanitary sewers will have a gradient in the same direction as street or ground surfaces and will be connected by trunk or main sewers. Some considerations which may affect the exact location are other existing utilities, traffic conditions, the type and extent of pavement encountered, and the availability of rights-of-way (ROWs).

Sewer system drainage district boundaries usually conform to watershed or drainage basin areas, with ridgelines, or high points, separating the areas. It is desirable to have drainage area boundaries follow property lines so that any single lot or property is tributary to a single system. This is particularly important when assessments for sewer connections are made against property served. The boundaries of subdistricts within an assessment district may be fixed on the basis of topography, economy of sanitary sewer layout, or other practical considerations.

Trunk, main, and intercepting sanitary sewers are located at the lower elevations in a given area. Not all sewer systems will have trunk or intercepting sewers, with their prevalence being dependent upon topographical and construction limitations. The location of these sewers will depend, similarly to the branch sewers, on available ROWs and the presence of low-lying areas in which these sewers can be located.

Consideration should be given to future needs. Each part of the system, as well as the entire system, should be designed to serve not only the present tributary area but also be compatible with an overall plan to serve an entire drainage area unless this is impractical for economic reasons.

Many states prohibit sanitary sewers within a certain radius of public water supplies. When sanitary sewers are located close to public water supplies, it is common practice to use pressure-type sewer pipe, concrete encasement of the sewer pipe, or sewer pipe with joints which meet stringent infiltration/exfiltration requirements. The so-called Ten-States Standards (GLUMRB 2004), as well as most building codes, prohibit sanitary sewer installation in the same trench with water mains. The design engineer should check local health and environmental regulations for site-specific requirements. For example, some states have reduced the separation requirements below the limits specified in the Ten-States Standards. In the absence of other criteria, the Ten-States Standards require 10-ft (3-m) horizontal separation, measured outside to outside between water mains and sewer mains, or 18 inches (450 mm) of vertical separation, with the water main above the sewer main.

Manholes should be located at the junctions of sanitary sewers and at any change in grade, pipe size, or alignment, except in curved alignments (see Section 5). Also, manholes should be placed at locations that provide ready access to the sewer for preventive maintenance and emergency service. Typically, in the absence of other data, the manholes should be spaced at distances not greater than 400 ft (120 m) for sewers 15 inches (375 mm) or less, and 500 ft (150 m) for sewers 18 to 30 inches (450 to 750 mm), except that distances up to 600 ft (185 m) may be acceptable where adequate modern cleaning equipment for such spacing is provided. Greater spacing, up to 1,000 ft (300 m), may be acceptable in larger sewers. Although longer distances may be acceptable, this must be coordinated with the capabilities of the utility's cleaning equipment. In any situation, it is very important to consult with the utility's operational staff to ensure that the design provides proper access for system maintenance.

On the other hand, manholes should not be located in any low area, such as a swale or gutter, where there will be a concentrated flow of water over the top that could cause excessive inflow. This may require a few additional manholes to be constructed but it will enable the operating agency to provide better service during the life of the sewer. Inaccessible manholes are of little or no value to the operating agency. If manholes must be constructed in these areas, careful attention must be paid to design to ensure that the manholes are watertight and protected against flotation.

Street intersections are common locations for manholes. When a manhole is not necessary for a present or future junction, it is better placed outside the pavement of a street intersection, but within the street ROW, to minimize issues during potential future road repairs and to provide room for maintenance personnel.

A terminal manhole at the upper end of a sanitary sewer should be placed in the street ROW so that the manhole and sewer are accessible for maintenance and emergency service. Unless a sufficient access easement is available, it should not be located inside the property line of the last property served. Thus, for proper location of the terminal manhole, the sewer should be extended to the street ROW if necessary.

Sanitary sewer manholes should not be located or constructed in a way that allows surface water to enter. When this is not possible, watertight manhole covers or interior liners should be specified. Interior liners can be either nonmetallic or metallic, with metallic liners preferable in high-traffic areas. Manholes not in the pavement, especially in open country, should have their rims set above grade to avoid the inflow of stormwater and to simplify field location.

Many communities require tees or wyes as part of new sewers when they are constructed. When a sewer is constructed under a proposed street, wye connections should be made with a stub extended beyond the curb or edge of pavement line. If the sewer is located on one side of the street, wyes should be installed and a stub extended across the street to serve property on the other side to avoid tearing up the pavement when the building service line is constructed. Connections to sanitary sewers should be made only with an experienced crew using equipment specifically designed for tapping a sewer. Some cities provide this service with their own staff and charge for each connection made. The authority to make these connections should be carefully controlled so that only authorized, experienced personnel make the connections.

The most common location of a sanitary sewer is at or near the center of a street or alley. A single sanitary sewer then serves both sides of the street with approximately the same length of house connection sewer. In an exceptionally wide street, such as a boulevard, it may be more economical to install a sanitary sewer on each side. In such a case, the sanitary sewer may be located outside the curb, between the curb and the sidewalk or ROW line, or under the sidewalk. Normally, sidewalk locations are used only where other locations are not possible. Locating sanitary sewers within the gutter area is least desirable because of the possibility of stormwater inflow through manhole covers.

Sometimes a sanitary sewer must be located in an easement or ROW (e.g., at back property lines) to serve parallel rows of houses and residential developments without alleys. Easements must be of sufficient width to allow access for maintenance equipment and agreements must provide the right of access for construction, inspection, maintenance, and repair. Local ordinances should provide that all aboveground obstructions belonging to other utilities, such as utility poles, gas meters, and telephone junction boxes, be located as close as possible to one edge of the easement, not in the center. They should also discourage or prohibit, insofar as possible, the planting of trees and shrubs, the construction of fences or retaining walls, or any other aboveground obstruction that would interfere with access to the entire length of the line. The ordinances should indicate where these private obstructions may be located within the easement and that they are placed there at the risk and expense of the property owner. Replacement of the obstructions that are removed to permit access should not be the responsibility of the sewer utility.

Despite such provisions, access to sewer lines located along property lines sometimes becomes difficult or impossible. Sewer maintenance becomes irritating to both the property owners and the utility, and the costs of maintenance may increase significantly. Sewer locations in streets or other public properties are therefore greatly preferred.

6.5. CURVED SANITARY SEWERS

The design and installation of large-diameter sanitary sewers laid on curves is a generally accepted practice in sanitary sewer design (FHA, undated). Several benefits are derived from this type of installation: The installation of curved sanitary sewers will result in economies over straight-run sewers by eliminating manholes needed at each abrupt change of direction. The installation of sanitary sewers parallel to or on the centerline of a curved street makes it easier to avoid other utilities. Such installations will allow manhole locations away from street intersections. The design of curved main or trunk sanitary sewers allows the engineer to follow topographic contours for the desired alignment and simplifies the maintenance of a uniform gradient. Inspection and maintenance requirements generally determine minimum diameters of curved sewers.

6.5.1. Rigid Pipe

The installation of rigid sewer pipe on a curve is accomplished by deflecting the pipe joint from the normal straight position. The maximum permissible deflection must be limited so that satisfactory pipe joint performance is not affected. When rigid sewer pipe is installed on a curve, it is advisable that the manufacturer of the sewer pipe be consulted and the maximum allowable pipe joint deflection determined.

The radius of curvature that may be obtained by deflection of straight sewer pipe is a function of the deflection angle per pipe joint (joint openings), the diameter of the sewer pipe, and the length of the sewer pipe sections.

The radius of curvature (R) is computed by the equation:

$$R = \frac{L}{2\left[\tan\frac{1}{2}\left(\frac{\Delta}{N}\right)\right]} \tag{6-1}$$

where

R = radius of curvature, in ft (m)

L = length of sewer pipe sections measured along centerline, in ft (m)

Δ = total central (or deflection) angle of the curve, in degrees (radians)

N = number of sewer pipe sections

Δ/N = total central or deflection angle for each sewer pipe, in degrees (radians).

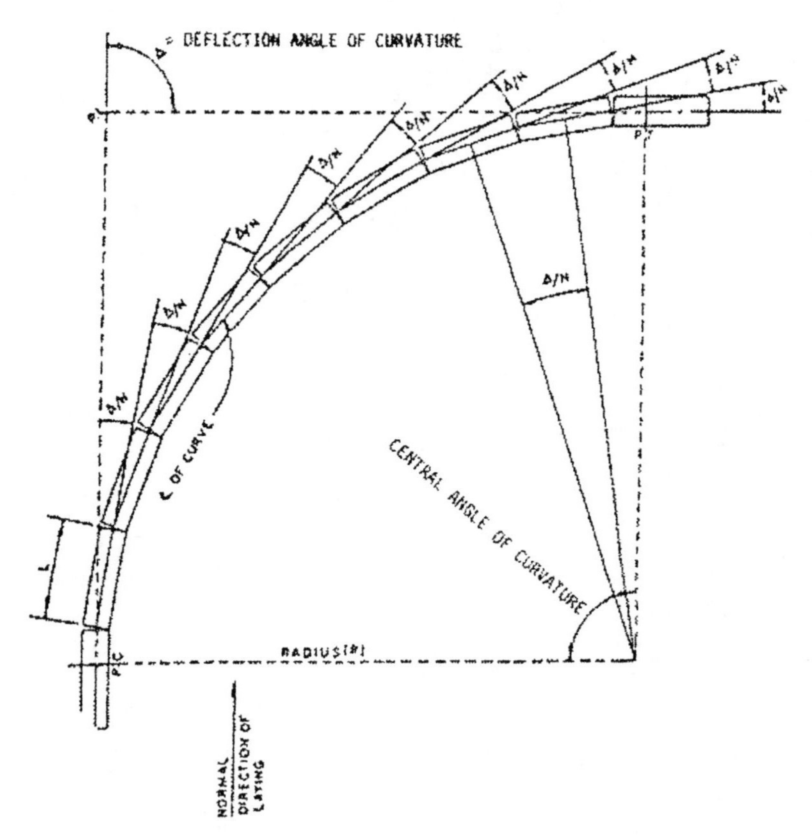

FIGURE 6-1. Deflected straight sewer pipe. Source: "Feasibility of curved alignment for residential sanitary sewers." Federal Housing Administration Report 704, U.S. Government Printing Office, Washington, D.C.

From Fig. 6-1, the angular deflection from the tangent of the circle, $\frac{1}{2}(\delta/N)$, is further defined as:

$$\frac{1}{2}\left(\frac{\Delta}{N}\right)=\sin^{-1}\left(\frac{Deflection}{2B_c}\right) \tag{6-2}$$

where *Deflection* = joint opening, in ft (m) and B_c = outside sewer pipe diameter, in ft (m).

Figure 6-2 shows sewer pipe installed on a curved alignment using straight sewer pipe. As an alternate to deflecting the straight sewer pipe, the desired radius of curvature may be based on the fabrication of radius sewer pipe. Radius sewer pipe, also referred to as beveled or mitered pipe, incorporates the deflection angle into the sewer pipe joint. The sewer pipe is manufactured with one side of the pipe shortened. The amount of shortening or drop is dependent on manufacturing feasibility.

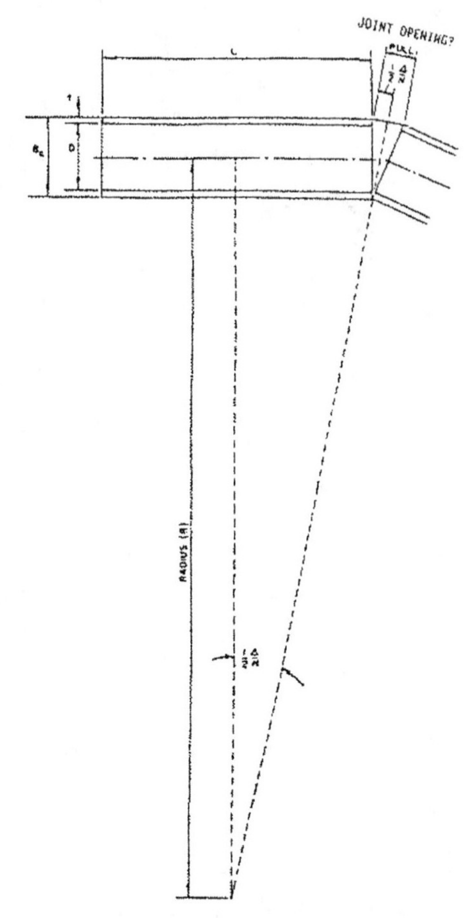

FIGURE 6-2. Curved alignment using deflected straight sewer pipe. Source: "Feasibility of curved alignment for residential sanitary sewers." Federal Housing Administration Report 704, U.S. Government Printing Office, Washington, D.C.

With the possibility of greater deflection angles per joint, shorter radii can be obtained with radius sewer pipe than with deflected straight sewer pipe.

The radius of curvature, which may be obtained with radius sewer pipe, is a function of the deflection angle per joint, the diameter of the sewer pipe, the length of sewer pipe section, and the wall thickness. It is computed from the equation:

$$R = \frac{L}{\tan\left(\frac{\Delta}{N}\right)} - \left(\frac{D}{2} + t\right) \qquad \text{(6-3)}$$

where

R = radius of curvature

δ = total central or deflection angle of curve, in degrees (radians)

N = number of radius sewer pipe sections

L = standard sewer pipe length being used, in ft (m)

D = inside diameter of sewer pipe, in ft (m)

t = wall thickness of the sewer pipe, in ft (m).

From Fig. 6-3, the radius of curvature (R) can be expressed in terms of the bevel and is given by the equation:

$$R = \frac{L(D+2t)}{Drop} - \left(\frac{D}{2}+t\right) \quad \text{or} \quad \left(\frac{L}{Drop} - \frac{1}{2}\right)B_c \tag{6-4}$$

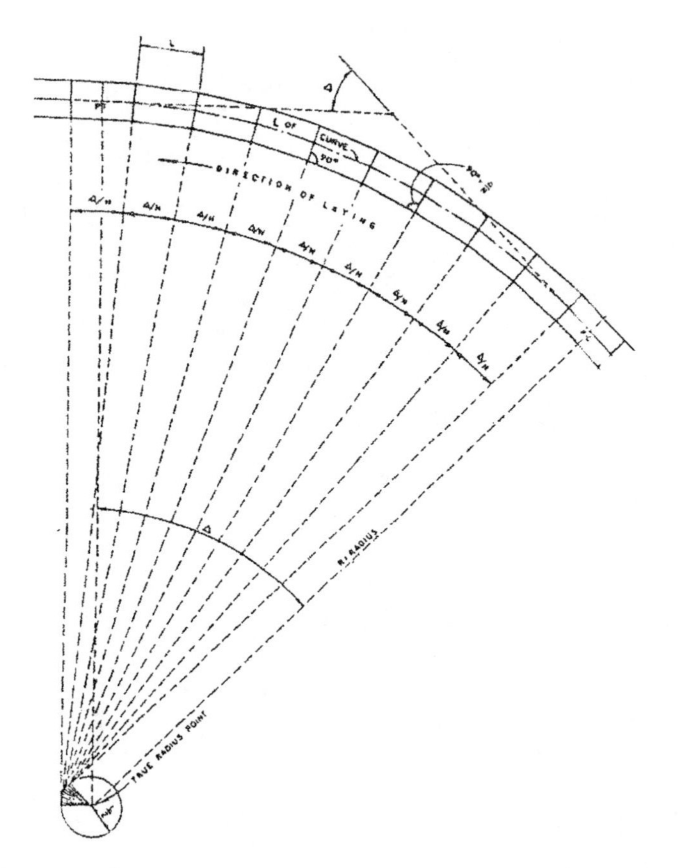

FIGURE 6-3. Radius sewer pipe. Source: "Feasibility of curved alignment for residential sanitary sewers." Federal Housing Administration Report 704, U.S. Government Printing Office, Washington, D.C.

where B_c = outside diameter of the pipe, in ft (m) and *Drop* = length pipe is shortened on one side, in ft (m).

It is essential to coordinate the design of curved sanitary sewers with the radius pipe manufacturer. Figure 6-4 shows sewer pipe installed on a curved alignment using radius sewer pipe. The minimum radius of curvature obtained from Eqs. (6-1) and (6-3) is within a range of accuracy that will enable the sewer pipe to be readily installed on the required alignment.

6.5.2. Flexible Pipe

The installation of flexible sewer pipe on a curve is accomplished by controlled longitudinal bending of the pipe and deflection of the pipe joint.

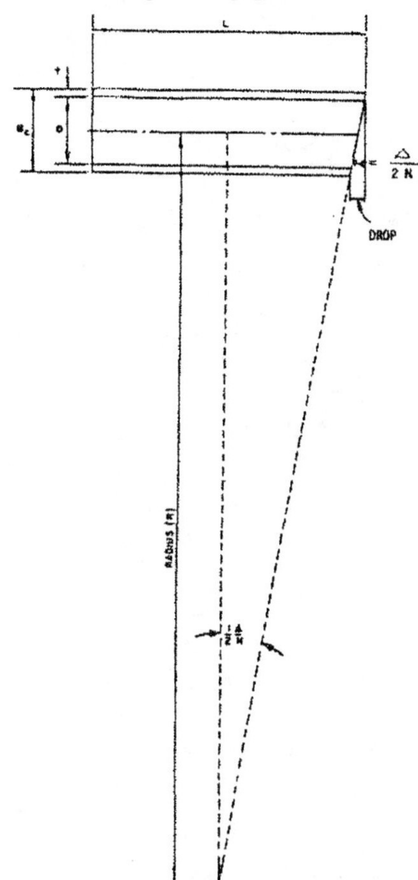

FIGURE 6-4. Curved alignment using radius sewer pipe. Source: "Feasibility of curved alignment for residential sanitary sewers." Federal Housing Administration Report 704, U.S. Government Printing Office, Washington, D.C.

It is recommended that any necessary deflection be obtained through axial joint deflection. Permissible joint deflection may be significant when gasketed joints, which are designed for deflection, are provided on thermoplastic pipe. Solvent cement or fusion joints provide no flexibility.

When flexible sewer pipe must be installed on a curve, it is advisable that the manufacturer of the pipe be consulted on the minimum radius of bending. When considering the deflection of the pipe between joints, the system can be analyzed based on available mathematical relationships for longitudinal bending of pressurized tubes (Reinhart 1961; Reissner 1959). One critical limit to bending of flexible pipe is long-term flexural stress.

The equation for allowable bending stress (S_b) in pascals (lb/ft^2) is:

$$S_b = (HDB)\frac{T}{F} \tag{6-5}$$

where

HDB = hydrostatic design basis of pipe, in lb/ft^2 (pascals)
F = safety factor (2 is suggested for nonpressure pipe)
T = temperature rating factor [1 at 73.4 °F (23 °C)].

Also to be considered is the mathematical relationship between stress and moment induced by longitudinal bending of pipes:

$$M = \frac{S_b I}{c} \tag{6-6}$$

where

M = bending moment, in ft-lbs (n-m)
S_b = allowable bending stress, in lb/ft^2 (pascals)
c = distance from extreme fiber to neutral axis = OD (outside diameter)/2, in ft (m)
I = moment of inertia, in ft (m) to fourth power.

With reference to Fig. 6-5, and assuming that the bent length of pipe conforms to a circular arc after backfilling and installation, the minimum radius of the bending circle (R_b) can be found by Timoshenko's equation (Timoshenko 1948):

$$R_b = \frac{EI}{M} \tag{6-7}$$

where E = Young's modulus of elasticity, in lb/ft^2 (pascals). Combining Eqs. (6-6) and (6-7) gives:

$$R_b = \frac{E}{2S_b}(OD) \tag{6-8}$$

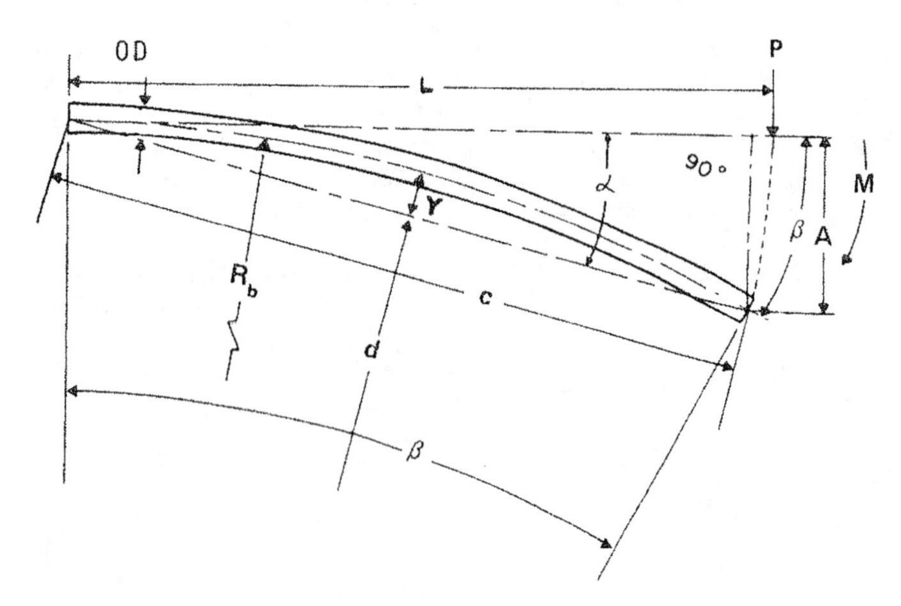

FIGURE 6-5. *Flexible pipe allowable bend.* Uni-Bel PVC Pipe Association. 2001. *Handbook of PVC pipe: Design and construction,* 4th ed. Dallas, Tex. Courtesy of the Uni-Bel PVC Pipe Association.

The central angle, β, subtended by the length of pipe is:

$$\beta = \frac{360}{2\pi R_b}(L) = \frac{57.3}{R_b}(L) \tag{6-9}$$

where L and R_b are both in the same units, and the angle of lateral deflection α [in degrees (radians)] of the curved pipe from a tangent to the circle is:

$$\alpha = \frac{\beta}{2} \tag{6-10}$$

with L = pipe length, in ft (m). The offset at the end of the pipe from the tangent to the circle, A, in ft (m), is:

$$A = 2R_b \sin\left(\frac{\beta}{2}\right)^2 = 2R_b(\sin\alpha)^2 \tag{6-11}$$

Because of the characteristics of a particular joint design, it is possible that a manufacturer's recommended bending radius may be greater or lesser than those calculated.

When change of direction in a flexible pipeline exceeds the permissible bending deflection angle for a given length of pipe, a manhole should be provided or, where allowable, an elbow should be used. If these options are not available, and after consultation with the manufacturer, longitudinal deflection may be a potential option.

6.6. TYPE OF CONDUIT

A variety of materials are available to the design engineer for use in sanitary sewers. (A discussion of properties and applications of the different sewer pipe materials is presented in Chapter 8.) The type of materials a design engineer chooses to specify is dictated by several factors, along with sound engineering judgment. These factors include the type of wastewater to be transported (residential, industrial, or a combination of both); the type of soil; trench load conditions; and the bedding and initial backfill material available. (Information on structural design for both rigid and flexible conduits is presented in Chapter 9.) There may also be regional differences in design criteria and regulatory requirements, and these should be considered in selecting a piping material.

6.7. VENTILATION

Forced ventilation of a sanitary sewer is generally considered a special application used to solve a specific problem. In most cases, natural ventilation from manholes, building vents, and flow variations of the wastewater is adequate to provide oxygen in the sanitary sewer atmosphere.

When forced ventilation is required, special airtight or pressure manhole covers must be used and the air exhausted to a high stack or to a deodorizing process. The design requirements for forced ventilation are specific to a particular sanitary sewer layout and design. They should not be interpreted as minimizing in any way the need for gas detection and ventilation before and during maintenance operations to eliminate any possible danger to sanitary sewer workers.

6.8. DEPTH OF SANITARY SEWER

Sanitary sewers should be installed at such depths that they can receive contributed flows from the tributary area by gravity flow. Deep basements and buildings on land substantially below street level may require individual pumping facilities. Sufficient sanitary sewer depth must be provided to prevent freezing and backflow of wastewater through connections.

No single method prevails for determining the minimum depth for a sanitary sewer. One suggestion is that the top of the sanitary sewer should not be less than 3 ft (1 m) below the basement floor of the building to be served. Another rule places the invert of the sanitary sewer not less than 6 ft (1.8 m) below the top of the house foundation. The latter assumes that it is not necessary for a sanitary sewer to serve basement drains. It also has the advantage of preventing the connection of exterior basement wall footing drains to the sanitary sewer. This, however, is acceptable only where basements are uncommon, where few basements have sanitary facilities, or if basement sump pumps are utilized. Where houses have no basements, sewers may be built at shallower depths, perhaps as little as 4 ft (1.2 m) below the house foundation, resulting in significant cost reductions. In business or commercial districts, however, it may be necessary to lay sewers as deep as 12 ft (3.6 m) or more to accommodate the underground facilities in such areas.

It is typical to lay house connections at a slope of 2%, with a minimum slope of 1%. In some developments in which houses are set well back from the street, the length and slope of the house connection may determine minimum sewer depths. In some cases, it may not be economically justifiable to lower the whole sewer system to provide service for only a few houses.

Because sewers usually are laid in public streets or easements, consideration must be given in design to the prevention of undue interference with other underground structures and utilities. The depth of the sanitary sewer is usually located such that it can pass under all other utilities, with the possible exception of storm sewers. Also, it is necessary to ensure that there is adequate clearance between both sewer mains and sewer service laterals and water mains. Unless local environmental regulations state otherwise, the Ten-States Standards require a minimum vertical separation of 18 inches (450 mm) between the outside of the water main and the outside of the sewer line. This may necessitate lowering of the sanitary sewer.

When sewers are to be laid at shallow depths, consideration should be given to live and impact loads since special requirements may be necessary in the selection and installation of the pipe. (Structural requirements, including these special cases, are discussed further in Chapter 9.)

6.9. FLOW VELOCITIES AND DESIGN DEPTHS OF FLOW

A sanitary sewer has two main functions: to convey the designed peak discharge and to transport solids so that deposits are kept to a minimum. It is essential, therefore, that the sanitary sewer has adequate capacity for the peak flow and that it function at minimum flows without excessive maintenance and generation of odors.

To meet these goals it has been customary to design sanitary sewers with some reserve capacity. When capacity requirements determine the sewer slope, often sanitary sewers through 15 inches (375 mm) in diameter are designed to flow half-full. Larger sanitary sewers have been typically designed to flow three-quarters full. However, where allowed, engineers are encouraged to strive for accurate peak flow estimation and design for a full pipe when capacity determines the pipe slope.

Velocity determination is based on calculated peak flow. Chapter 3 provides guidance regarding calculating the average and peak flows for sewer systems, both for the beginning of service and on the design day. It should be noted that the Ten-States Standards provide a peaking factor (PF) curve which calculates peaking factors as a function of population. The equation utilized in the Ten-States Standards is:

$$PF = \frac{18 + \sqrt{P}}{4 + \sqrt{P}} \tag{6-12}$$

where P is the population in thousands. The use of this equation may be mandated by certain agencies, although peaking factors calculated from this equation are considerably higher than those observed in modern systems (see Chapter 3). In addition, many communities have established specific values for design peak flows.

Traditional standards dictate that the minimum full-pipe velocity should not be less than 2 fps (0.60 mps) or generally greater than 10 fps (3.5 mps) at peak flow. A velocity in excess of 10 fps (3.5 mps) can be tolerated with proper consideration of pipe material, abrasive characteristics of the wastewater, turbulence, and thrust at changes of direction. The minimum velocity requirement is necessary to prevent the deposition of solids. The design engineer is urged to consider utilizing tractive force design, as outlined in Chapter 5, to ensure that self-cleansing occurs. Special attention must be given to conditions early in the life of each sewer reach, as initial flows are normally substantially lower than design flows and an adequate assessment of self-cleansing is crucial. When flat slopes and very small minimum design flows are encountered, the design engineer may elect to use a greater slope or the owner may desire to initiate a more intensive, well-planned line cleaning maintenance program for these reaches. Information on flow and velocities as they relate to hydrogen sulfide generation is presented in Chapter 4.

6.10. INFILTRATION/INFLOW

In many sanitary sewers, extraneous flow consisting of infiltration and inflow (I/I) is a major cause of hydraulic overloading of both the collection system and treatment plant (Santry 1964; "Municipal Requirements for Sewer Infiltration" 1965; Brown and Caldwell 1957). To handle this excess flow it may become necessary to construct relief sanitary sewers and expand existing treatment facilities. These measures add to the cost of the system. Other expenses incurred because of this unwanted flow include:

- Higher pumping costs.
- Replacement costs for failed sanitary sewers and surface pavements resulting from soil flushing into the sewer.
- Higher maintenance costs resulting from soil deposits in sanitary sewers and root penetration into leaky joints.

I/I can contribute substantially to sewer flows. Inflow is often the result of deliberately planned or expediently devised connections of stormwater sources into sanitary sewer systems; unintentional stormwater sources that result from structure design; or location or deterioration of sanitary sewers. The inflow from such sources can be prevented or corrected by regulation and inspection procedures aimed at enforcing regulations relating to sanitary sewer connections. It is also important to design the system to minimize potential inflow points. This can be accomplished through careful location of manholes to minimize the potential of flooding; use of waterproof covers or provision of manhole "diapers" to prevent inflow through covers; and provisions for locking of both manhole and cleanout covers to prevent their use as stormwater inlets.

Infiltration results from the age of the structure, soil conditions, materials, and methods of construction. It must be taken into consideration in the design, construction, and inspection of sanitary sewer systems. A more detailed discussion of infiltration and related matters is found in Chapter 3. Potential design solutions to minimize infiltration include the use of proper piping and gasket materials (see Chapter 8); the proper sealing of pipe entries into manholes; and the proper waterproofing of manholes. Chapter 7 discusses manholes and appurtenances and the steps that can be taken to improve their performance.

6.11. INFILTRATION/EXFILTRATION AND LOW-PRESSURE AIR TESTING

6.11.1. Infiltration/Exfiltration Test Allowance

The most effective way to control infiltration, and at the same time to ensure structural quality and proper installation of the new sanitary sewer, is to establish and enforce a maximum leakage limit as a condition of job acceptance. Limits may be stated in terms of water leakage quantities and should include both a maximum allowable test section rate and a maximum allowable system average rate. Current information indicates that a maximum allowable infiltration rate of 50 to 100 gal/inch-diameter/mi (5 to 10 L/m-diameter/km) of sewer pipe per day can be achieved without additional construction costs. Some model state laws ("Model Sewer Use Law" undated) limit infiltration to 25 gal/inch-diameter/mi (2.5 L/m-diameter/km) of sewer pipe per day.

Manholes may be tested separately and independently. If this is done, an allowance for manholes of 0.1 gal/hr/ft-diameter/ft-head (4 L/hr/m-diameter/m-head) would be appropriate.

6.11.2. Infiltration/Exfiltration Testing

When groundwater is observed to be at least 4 ft (1.2 m) above the top of the sewer pipe, the infiltration test will determine the integrity of the sewer line. Any leakage can be measured with a V-notch weir or similar flow measuring device. If no leakage is observed, it can be assumed that the line passes the test.

If the groundwater level is not at least 4 ft (1.2 m) above the top of the sewer pipe, then an exfiltration test is required. This is performed by plugging the manhole at the lower end of the test section and filling the line with water. In order to properly test a line, the elevation of water in the line should be at least 2 ft (0.6 m) above the groundwater level. If leakage does not exceed the limits specified, then the section tested is accepted. If leakage exceeds the limits specified, the leak must be located and repaired.

6.11.3. Low-Pressure Air Testing

A low-pressure air test may also be used to detect leaks in sewer pipe where hydrostatic testing is not practical (Uni-Bell 1998). Because of the physical difference between air and water and the difference in behavior under pressure conditions, the air test cannot be directly related to the water test, although either test can be used with confidence to prove the integrity of the sewer line. The air test depends on porosity, moisture content, and wall thickness of the sewer pipe. A well-constructed sanitary sewer that is impervious to water may still have some air loss through the sewer pipe wall. In applying low-pressure air testing to sanitary sewers designed to carry fluid under gravity conditions, it is necessary to distinguish between air losses inherent in the type of sewer pipe material used and those caused by damaged or defective pipe joints.

When testing polyvinyl chloride (PVC) pipe, the air test method recommended by Uni-Bell is the *time pressure* method. The line to be tested is plugged and the pressure raised to a minimum of 4 psig (28 kPa) greater pressure than the average back pressure of any groundwater. The pressure is then allowed to stabilize for a minimum of two minutes. The air supply is disconnected and the time required for the pressure to drop from 3.5 psig (24 kPa) to 2.5 psig (17 kPa) is determined. Test procedures and calculations are available from Uni-Bell. Ramseier's work (Ramseier and Rick 1964; Ramseier 1972) also should be considered in air testing.

For clay pipe, ASTM Standard C828 specifies a low-pressure test procedure that is designed specifically for vitrified clay pipe. The line to be tested is plugged and the pressure raised to a minimum of 4 psig (28 kPa). The pressure is then allowed to stabilize for two to five minutes. After the pressure stabilizes, the air pressure should be reduced to 3.5 psig (24 kPa) before starting the test. The air supply is disconnected and a timer started. If the air pressure does not drop more than 1 psig (7 kPa) during the test time (see Table 6-1), the line has passed.

In applying the low-pressure air test, the following factors should be understood and precautions followed during the test. The air test is intended to detect defects in the sewer line and establish the integrity of the line under sewer conditions. Since the pipe will be in a moist environment when in service, testing the pipe in wet conditions is appropriate. Plugs should be securely braced. Plugs should not be removed until all air pressure in the test section has been reduced to ambient pressure.

For safety reasons, no one should be allowed in the trench or manhole while the test is being conducted. The testing apparatus should be equipped with a pressure relief device to prevent the possibility of loading the test section with full compressor capacity.

Pipe that is large enough to permit personnel to conduct interior inspections can be accepted on the basis of such inspection, plus air testing of individual joints, if required. When individual joints are tested, allowable leakage is usually the computed rate per foot (meter) of pipe times the distance between joints. In practice, however, it is a go/no-go test.

TABLE 6-1. Minimum Test Time for Various Pipe Diameters

Nominal Pipe Size, inch (mm)	T (time), Min/100 ft	Nominal Pipe Size, inch (mm)	T (time), Min/100 ft
4 (100)	0.3	24 (600)	3.6
6 (150)	0.7	27 (675)	4.2
8 (200)	1.2	30 (750)	4.8
10 (250)	1.5	33 (825)	5.4
12 (300)	1.8	36 (900)	6.0
15 (375)	2.1	42 (1,050)	7.3
18 (450)	2.4	48 (1,200)	8.5
21 (525)	3.0		

6.12. DESIGN FOR VARIOUS CONDITIONS

6.12.1. Open Cut

The load on a sanitary sewer built in open cut is a function of the bedding, trench width, backfill material, and any superimposed loads. Consideration must be given to all of these elements. Backfill material placed 12 to 24 inches (0.3 to 0.6 m) over the top of the sewer pipe should be free of rocks or stones larger than 2 inches (50 mm) in diameter to avoid damage to the sewer pipe. Chapter 9, devoted to loads on sewer pipe, presents details of this phase of design.

6.12.2. Microtunneling

Microtunneling uses a steerable head and a jacking system to directly install lines as small as 6 inches (150 mm). The advantage of microtunneling is that it brings the benefits of trenchless construction while providing the accuracy necessary for construction of gravity sanitary sewers. This construction method should be considered for installations in areas where surface disruptions are either unacceptable or very costly, as the construction cost for microtunneling is significantly higher than open-cut methods. This construction method is described in greater detail in Chapter 12.

6.12.3. Directional Drilling

In directional drilling, a guided bore is made and a carrier pipe is pulled into location. Directional bores have been widely used for the installation of pressure pipes where grade is not critical. However, until very recently it was impossible to make an installation with sufficient accuracy to meet gravity sewer requirements. With the refinement of guidance methods, this has become feasible. Directional drilling is a trenchless method and should be considered where surface disruptions would be unacceptable. Although it is more costly than open-cut installation methods, it is more cost-effective than microtunneling in smaller sizes. Additional information on this construction method is contained in Chapter 12.

6.12.4. Conventional Tunnel

A thorough knowledge of tunnel construction methods is required before designing sanitary sewers for tunnel placement. This is especially necessary to effect economy of construction in this costly type of work. Tunneling methods are covered in Chapter 12.

6.12.5. Sanitary Sewers Built in Rock

Where sanitary sewers are built in rock trenches, special attention should be given to the method of bedding to avoid damage due to contact with the rock. Adequate clearances should be provided between the bottom and sides of the sanitary sewer and the adjacent rock trench. Granular bedding or a concrete cradle is normally provided. If blasting is anticipated in the area, the concrete cradle should be separated from any rock by a granular cushion.

6.12.6. Exposed Sanitary Sewers

Sometimes sanitary sewers have to be built above the ground surface. In these cases, the sewer pipe will be carried on supports or be designed as a self-supporting span. In these cases, the pipe should be rigid pipe and special consideration will need to be given to joint specification. In climates where freezing conditions exist, a method of freeze protection should be employed. For most climates in the United States, this can consist of polyurethane foam insulation with a protective outer jacket to prevent damage to the foam. In colder climates, heating elements may need to be provided to prevent freezing.

6.12.7. Foundations

Knowledge of foundation conditions should be obtained by borings, soundings, or test pits along the route of a sanitary sewer prior to design. Unstable foundation soils encountered in the form of silt, peat bog, saturated sand, or other soft or flowing materials require special bedding and must be considered in the design of a sanitary sewer under these conditions. If these soils are encountered during construction, costs usually will be higher than if anticipated beforehand.

Where the conditions are not severe, it may be possible to stabilize the trench bottom by placing a layer of crushed rock below the sewer pipe. The rock must be fine enough, or contain fines, so that settlement will not result from unstable bottom material flowing into the voids. The engineer may want to consider using filter fabric to encase the rock bedding, or even to envelop the pipe to prevent migration of bottom material into the pipe bedding. Concrete or wooden cradles often will suffice to spread the load in wet or moderately soft foundations. In some cases, underdrains laid beneath the sanitary sewer or well points will remove water held in the soil and permit dry construction, and they may eliminate the need for special foundations. Pipe joints that are tight, yet flexible, are particularly important when sanitary sewers are installed in areas with unstable soils.

6.12.8. Sanitary Sewers on Steep Slopes

Erosion control devices or methods may be required on steep slopes. It may also be necessary to provide anchorage or cutoff dams to prevent the sewer pipe from creeping downhill, or to prevent water from flowing along the pipe and causing the trench to wash out. If drop manholes are used and the flow is heavy, special energy dissipators may be required in the form of a special manhole bottom or a water cushion.

6.13. RELIEF SEWERS

An overloaded existing sanitary sewer may require relief, with the relief sewer constructed parallel to the existing line to divert flows to alternate outlets. In the design of a relief sanitary sewer, it must be decided whether (1) the proposed sewer is to share all rates of flow with the existing sanitary sewer; or (2) it is to take all flows in excess of some predetermined quantity; or (3) it is to divert a predetermined flow from the upper end of the system. The topography, available outlets, and available head may dictate which alternate is selected. If flows are to be divided according to some ratio, the inlet structure to the relief sanitary sewer must be designed to divide the flow. If it is to take all flows in excess of a predetermined quantity, the excess flow may be discharged over a side-overflow weir or through a regulator to the relief sanitary sewer. If flow is to be diverted in the upper reaches of a sewer system, the entire flow at the point of diversion may be sent to the relief sanitary sewer or the flow may be divided in a diversion structure.

An examination of flow velocities in the existing and relief sanitary sewers may determine the method of relief to use. If self-cleansing velocities cannot be maintained in either or both sanitary sewers when a division of flows is used, excessive maintenance and sulfide generation may result. If, on the other hand, the relief sanitary sewer is designed to take flows in excess of a fixed quantity, the relief sanitary sewer itself will stand idle much of the time and deposits in it may cause similar problems. Engineering judgment is required in deciding which method of relief to use. In some cases it might be better to design the new sanitary sewer with sufficient capacity to carry the total flow and to abandon the old one. Pipe bursting, which is covered in other manuals dealing with pipeline rehabilitation, allows increasing the size of an existing sewer, providing an alternative to relief sewers.

6.14. ORGANIZATION OF COMPUTATIONS

The first step in the hydraulic design of a sewer system is to prepare a map showing the locations of all the required sanitary sewers and from

TABLE 6-2. Typical Computation Form for Design of Sanitary Sewers

							Max Flow				Min Flow				
		Manhole No.			Area				Sewage and Infiltration				Sewage and Infiltration		
Line No. (1)	Location (2)	From (3)	To (4)	Length (ft) (5)	Increment (acre) (6)	Total (acre) (7)	Infiltration (mgd) (8)	Sewage (mgd) (9)	(mgd) (10)	(cfs) (11)	Infiltration (mgd) (12)	Sewage (mgd) (13)	(mgd) (14)	(cfs) (15)	Slope of Sewer (16)

Note: Ft × 0.305 = m; acre × 0.0405 = ha; mgd × 3.785 = m^3/day; cfs × 1.7 = m^3/min; cfs/acre × 4.2 = m^3/min/ha; in. × 2.54 = cm

which the area tributary to each point can be measured. Preliminary profiles of the ground surface along each line also are needed. These should show the critical elevations that will establish the sewer pipe grades, such as the basements of low-lying houses and other buildings and existing sanitary sewers which must be intercepted. Topographic maps and, where available, Geographic Information System (GIS) data are useful at this stage of the design.

											Sewer Invert Elevation		Elevation Ground Surface	
Diam (in.) (17)	Capacity Full (cfs) (18)	Velocity Full (fps) (19)	Min Velocity (fps) (20)	Max Velocity (fps) (21)	Max Depth (ft) (22)	Max Velocity Head (ft) (23)	Max Energy Head (ft) (24)	Manhole Loss: Transition + Curve + Junction (ft) (25)	Manhole Invert Drop (ft) (26)	Fall in Sewer (ft) (27)	Upper End (28)	Lower End (29)	Upper End (30)	Lower End (31)

Several trial designs may be required to determine which one will properly distribute the available hydraulic head. Time may be saved if grades are established tentatively by graphical means on profile paper before selecting final grades and computing the sewer pipe invert elevations.

Design computations, being repetitious, may best be done either on an electronic spreadsheet or on tabular forms permitting both wastewater quantity and the sanitary sewer design calculations to be placed on the same form. The form shown in Table 6-2 is fairly comprehensive and can be adapted to the particular needs of the designer. It is convenient in using this form to record the data for the sewer reaches on alternate lines, reserving intervening lines for the data on transition losses and invert drops.

In using forms of this type, it is assumed that uniform flow exists in all reaches. The form is therefore not recommended where a detailed analysis of the wastewater surface profile is to be based on nonuniform flow.

The use of Table 6-2 for sanitary sewer design requires supplementary charts, graphs, or tables for calculating wastewater flows and hydraulic data. As an alternative, computer programs and/or spreadsheet macros designed to calculate open-channel flow parameters can be used to calculate the necessary data.

Methods for computing the quantities of wastewater flow listed under Columns 8 through 15 in Table 6-2 are described in Chapter 3. Methods of calculating the hydraulic data in Columns 16 through 25 are set forth in Chapter 5. If the value of Column 26 is positive, an invert elevation drop is indicated. If it is negative, an invert elevation rise is indicated but would not be installed because of the adverse effect on solids deposition during low flows; thus, a value of zero should then be recorded in the column.

Another option is the use of proprietary sewer design programs. These programs, although intended for modeling and analysis, also include powerful features for design of new collection systems. They permit the selection of criteria for the design of new pipes, including such factors as limits on flow in each pipe, allowable pipe sizes, maximum flow velocity, and maximum allowable depth. The programs can automatically optimize the design based on specific criteria and calculate any needed data, such as:

- Invert elevations.
- Pipe diameter.
- Slope.
- Elevations to match crowns, inverts, or ⅔ points for transitions in pipe sizes.
- Needs for parallel pipes.

REFERENCES

Brown, K. W., and Caldwell, D. H. (1957). "New techniques for the detection of defective sewers." *Sewage and Industrial Wastes, 29,* 963.

Federal Housing Administration (FHA). (Undated). *Feasibility of curved alignment for residential sanitary sewers.* FHA Rep. No. 704. U.S. Government Printing Office, Washington, D.C.

Great Lakes Upper Mississippi River Basin Board of State Public Health and Environmental Managers and Sanitary Engineers (GLUMRB). (2004). *Recommended standards for wastewater facilities,* GLUMRB Health Education Services, Albany, N.Y.

"Model sewer use law." (Undated). New York State Department of Environmental Conservation, Albany, N.Y.

"Municipal requirements for sewer infiltration." (1965). *Pub. Works, 96*(6), 158.

National Technical Information Service (NTIS). (1977). *Sewer system evaluation, rehabilitation and new construction—A manual of practice.* U.S. Environmental Protection Agency Pub. 600/2-77-017d, NTIS, Springfield, Va.

U.S. Public Health Service (USPHS). (1964). *Pollution effects of stormwater and overflows from combined sewer systems.* USPHA Pub. No.1246, U.S. Government Printing Office, Washington, D.C.

U.S. Environmental Protection Agency (EPA). (1970). *Urban storm runoff and combined sewer overflow pollution.* EPA Pub. No.11023, U.S. Government Printing Office, Washington, D.C.

EPA. (1994). *Combined sewer overflow control policy.* EPA 59 Fed. Reg. 18688, U.S. Government Printing Office, Washington, D.C.

Ramseier, R. E., and Rick, G. C. (1964). "Low pressure air test for sanitary sewers." J. San. *Engrg. Div., Proc. ASCE., 90*(SA2), 1.

Ramseier, R. E. (1972). "Testing new sewer pipe installations." *J. Water Poll. Control Fed., 44*(4), 557–564.

Reinhart, F. W. (1961). "Long-term working stress of thermoplastic pipe." *Soc. Petroleum Engrs. J., 17*(8), 75.

Reissner, E. (1959). "On finite bending of pressurized tubes." *J. Appl. Mech., Trans. ASME*, 386–392.

Santry, I. W., Jr. (1964). "Infiltration in sanitary sewers." *J. Water Poll. Control Fed., 36*, 1256.

Timoshenko, S. (1948). *Strength of materials*, D. Van Nostrand Co., New York.

"Uni B-6, recommended practice for low-pressure air testing of installed sanitary sewer pipe." (1998). Uni-Bell PVC Pipe Association, Dallas, Tex.

CHAPTER 7

APPURTENANCES AND SPECIAL STRUCTURES

7.1. INTRODUCTION

Certain appurtenances are essential to the proper functioning of sanitary sewer systems. They may include manholes, terminal cleanouts, service connections, inverted siphons, junction chambers, and other structures or devices of special design.

Many states have established criteria through their regulatory agencies which govern, to some extent, the design and construction of appurtenances to sanitary sewer systems. In addition, each municipal engineering office or private office acting for a municipality has its own design standards. It is to be expected, therefore, that many variations will be found in the design of even the simplest structures. The discussion to follow is limited to a general description of each of the various appurtenances, with special emphasis on the features considered essential to good design.

7.2. MANHOLES

7.2.1. Objectives

A manhole design should pass at least these major tests. It should:

- Provide convenient access to the sewer for observations and maintenance operations.
- Cause a minimum of interference with the hydraulics of the sewer.
- Be durable and generally a watertight structure.
- Be strong enough to support applied loads.

7.2.2. Manhole Spacing and Location

When designing sewer systems which parallel or cross streams, rivers, or other waterways, the engineer should consider evaluating evidence of stream channel movement to determine whether proposed sewer appurtenance locations are subject to future erosion. Should the evidence suggest the channel and/or banks are migrating, it may be beneficial to further evaluate potential future scenarios of this movement. Additional considerations to be made at that juncture include bank stabilization and protective measures.

For additional discussion, see Section 6.4. of Chapter 6.

7.2.3. General Shape and Dimensions

Most manholes are essentially cylindrical in shape, with the inside dimensions sufficient to perform inspecting and cleaning operations without difficulty. On small sewers, a minimum inside diameter of 4 ft (1.2 m) at the bottom has been widely adopted in the United States. A diameter of 3 ft (1 m) is more common in some other countries. The diameter is generally constant up to a cone at the top, where the diameter is reduced to receive the frame and cover (Fig. 7-1A). In some areas where brick manholes are used, the 4-ft- (1.2-m)-diameter cylinder is tall enough to provide an adequate working space and, above that, a 3-ft (1-m) shaft is constricted up to the cone (Fig. 7-1C). It has become common practice in recent years to use eccentric cones, especially in precast concrete manholes, thus providing a vertical side for the steps (Fig. 7-1B). Most often the orientation places the steps over the bench, but some designs place the steps opposite the outlet pipe, thus preserving the maximum working space on the bench.

Another design used under special conditions, especially where a larger working space is needed, specifies a reinforced concrete slab instead of a cone, as shown in Fig. 7-1D. This is applicable whether the working space is circular or rectangular. The slab must be suitably reinforced to withstand traffic and earth loads.

7.2.4. Shallow Manholes

Irregular topography sometimes results in shallow manholes. A manhole of standard design does not provide a space in which a maintenance worker can work effectively if the depth is only 3 to 4.5 ft (1 to 1.5 m). An extra-large cover with a 30- to 36-inch- (0.75- to 0.9-m)-diameter opening helps improve this condition. A manhole that is cylindrical up to a flat slab at the surface is suitable if the head room is 4 ft (1.2 m). Usually, the best option is to plan on maintenance work being done from the surface. Sometimes slots have been provided, no wider than the diameter of the sewer. A special foundation is needed if a slot access of this type must support traffic loads.

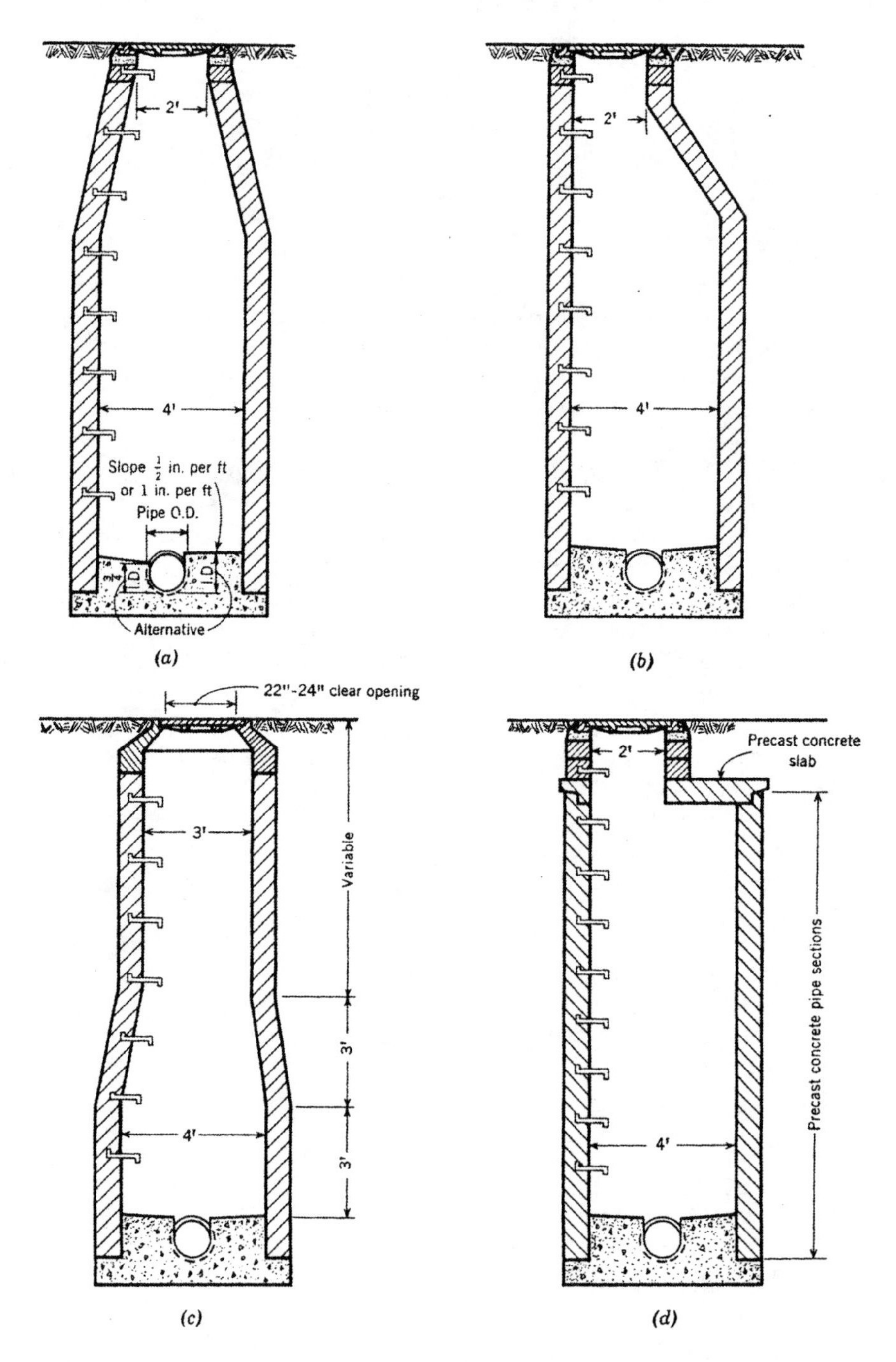

FIGURE 7-1. Typical manholes for small sewers (ft × 0.3 = m; inch × 2.54 = cm).

7.2.5. Construction Material

The materials most commonly used for manhole walls include precast concrete sections and cast-in-place concrete. Manholes built before the middle of the twentieth century were usually made of brick, followed by concrete block with occasional cast-in-place designs. Since then, precast concrete manhole sections have become dominant, at least in the United States, and preformed fiberglass-reinforced manholes or high-density polyethylene (HDPE) manholes are being used in some places. The sections are available in various heights and include properly spaced steps. When these manholes are used, the design should consider provision of antiflotation collars or other means of anchorage to prevent flotation of the manhole. Precast manhole bottoms also have been used in some places and are frequently preferred for new installations.

Transition sections are furnished to reduce the diameter of the manhole at the top to accommodate the frame and cover. It is common practice to allow three or four courses of brick or concrete rings just below the rim casting to permit easier future adjustment of the top elevation.

Brick or concrete block manhole walls normally are built 8 inches (200 mm) thick at the shallower depths, and may increase to 12 inches (300 mm) below 8 to 12 ft (2.5 to 3.5 m) from the surface. Joints should be filled completely with cement mortar. The outside walls of brick or block manholes should be plastered with cement mortar not less than 0.5 inch (13 mm) thick. In wet areas, a bituminous damp-proofing compound is often applied to the exterior of the cement mortar.

It is difficult to make brick manholes watertight, and precast manholes may leak because of imperfect sealing of the joints. Preferably the sections of precast manholes are joined in the same manner as in a pipeline, using elastomeric gaskets or a joint filler of proven effectiveness.

In some cases it may be beneficial to build sewers out of newer materials, such as fiberglass-reinforced plastic (FRP) or HDPE, and in those cases the engineer should consider utilizing pipe sections as manhole stacks. With HDPE, the vertical and horizontal pipe sections are fused together to form the sewer line with a vertical manhole stack, thereby reducing potential sources of groundwater infiltration.

7.2.6. Frame and Cover

The manhole frame and cover normally are made of cast or ductile iron. The cover is designed with these objectives:

- *Provision of an adequate aperture for access to the sewer.* The most common practice in the United States is to require a 24-inch (600-mm) clear opening. Sometimes 22-inch (550-mm) openings have been used, but current standards have been moving toward the larger dimension. It is difficult for a large worker, properly equipped, to enter through a 22-inch (550-mm) opening. For sewers up to 3 ft (900 mm) in diameter, covers sometimes are used that have a clear opening of 3 ft (900 mm) to accommodate cleaning equipment equaling the sewer diameter. However, collapsible cleaning equipment is the rule for large sewers. Many authorities do not use openings larger than 2 ft (600 mm). Openings larger than 3 ft (900 mm) are used in special cases.

- *Adequate strength to support superimposed loads.* A typical traffic-weight cover for a 24-inch (600-mm) clear opening weighs about 160 lb (75 kg) and the frame about the same amount or somewhat more. Weights up to 440 lb (200 kg) are specified in some places as the total for cover and frame. Lighter weights may be used where there is no danger that they will be subjected to heavy loads.
- *A good fit between cover and frame,* so there will be no rattling in traffic. It is usually specified that the seat in the frame on which the cover rests and the matching face of the cover be machined to ensure good fit.
- *Provision for opening.* Most commonly, this takes the form of a notch at the side where a pick or bar can be used to pry the cover loose, often supplemented by a pick hole a short distance in from the edge.
- *Prevent earth and gravel from falling into the sewer when the manhole must be opened.* To intercept sticks or earth inserted or falling through the pick hole, a dustpan is sometimes placed under the cover. Usually this is an iron disc, slightly smaller than the manhole opening, resting on lugs at the base of the frame. A polyethylene bowl-shaped diaphragm is now on the market that will retain dirt, and it is gasketed so that it is supposed to prevent the inflow of water to the sewer. Rubber gaskets sometimes are laid on the seat under the cover to maintain tightness in low areas subject to flooding.
- *Resistance to unauthorized entry.* The principal defense against a cover being lifted by children is its weight, but more persistent and competent vandals bent on throwing debris into a manhole are not deterred easily. An emergency measure in an area particularly plagued by mischief of this sort or by illegal disposals is a covering of planks over the channel. Sometimes covers are bolted in place and occasionally a lock is provided. The theft of manhole covers is a problem in some places.
- *Seal between frame and manhole.* The seal between the manhole frame and the manhole stack is a common source of debris and water influx. This is especially true in areas with significant seasonal temperature fluctuations and those subject to vibration (manholes in the wheel path of vehicles). This can be reduced by providing internal or external flexible seals or coatings.

7.2.7. Connection between Manhole and Sewer

Differential settling of the manhole and the sewer sometimes breaks the sewer pipe. A pipe joint just outside the manhole permits flexibility and lessens this danger. If the soil conditions are quite unstable, a second joint within 3 ft (1 m) of the first may be necessary. To accomplish this purpose, the joints must not be rigid. Elastomeric gaskets and couplings are available to form flexible, watertight connections between the manhole and the sewer pipe, thus allowing not only flexure but also a minor amount of differential settlement that otherwise would break the pipe (Fig. 7-2). These connectors are specified in ASTM C923.

7.2.8. Steps

Manhole steps should be wide enough for a worker to place both feet on one step, with the design to prevent lateral slippage off the step and far enough from the wall to be easy to stand on. They are generally spaced at 12- to 16-inch (0.3- to 0.4-m) intervals. Attention must be given to prevailing safety regulations, such as those issued by the Occupational Safety and Health Administration in the United States.

Types that have been most commonly used are made of ductile iron or shaped from 0.75- or 1-inch (20- to 25-mm) galvanized steel or wrought iron bars. These metals corrode in the moist atmosphere prevailing in most manholes, but under normal conditions corrosion is not rapid. If the wastewater contains much hydrogen sulfide, and especially if there also is much turbulence, the lower steps will fail in a few years. Steps formed of ⅝-inch (15-mm) AISI Type 304 or 316 stainless steel

FIGURE 7-2. A and B, flexible joint connection between pipe and manhole. Courtesy of NPC, Inc., Milford, N. H.

rods have been used in a few places. They should have a long life under most conditions. In recent years, steps made of plastic or steel armored with plastic have been used. Aluminum has also been successfully used in many sanitary sewers. However, it is a very corrodible metal under conditions sometimes encountered in sewers. In some areas, steps are omitted entirely. This is sometimes done to reduce the danger that a corroded step might break under a person's weight. Eliminating steps has also become common practice to prevent manhole entry without following proper safety procedures requiring harnesses, tripods, and multiworker crews.

7.2.9. Channel and Bench

A channel of good hydraulic properties is an important objective that frequently is not realized because of careless construction. The channel should be, insofar as possible, a smooth continuation of the pipe. In fact, the pipe sometimes is laid through the manhole with the top half removed to provide the channel. If this is done merely by breaking out the upper half, it is difficult to make a satisfactory channel. The best practice seems to be to lay a neatly cut half-pipe then build up the sides with concrete, or to use steel forms.

The completed channel cross section should be U-shaped. Some engineers specify a channel constructed only as high as the centerline of the pipe on small sizes. Others require that the height be three-quarters of the diameter or the full diameter. For sizes 15-inch (375-mm) and larger, the required channel height rarely is less than three-quarters of the diameter.

Loss of energy caused by expansion and contraction of the stream from pipe to U-shaped channel to pipe, with the pipe running full, should be less than 3% of the velocity head if the U-shaped channel is the full depth of the pipe. It will be much greater with a channel that allows the water to swirl out over the bench. Close attention to detail is required to secure well-constructed U-shaped channels.

The bench should provide good footing for a workman and a place where minor tools and equipment can be laid. It must have enough pitch to drain to the channel, but not too much. A slope of 1 in 12 is common, but 1 in 24 is specified in some areas to provide a safer footing.

In the past it was ordinary practice to allow an arbitrary drop of 0.1 ft (30 mm) in the invert across the manhole, or a slope of 0.025 in a standard manhole, regardless of the slope of the adjacent pipeline. If the channel is constructed properly, this drop is unnecessary. The drop is, in fact, objectionable for it causes excessive turbulence just where it is less desirable and sacrifices head that might better be used toward the attainment of good slopes along the entire sewer. The usual practice calls for a continuation of the pipe slope through the manhole.

7.2.10. Manholes on Large Sewers

The operations and methods of maintenance in large sewers are not the same as in small ones, and manhole designs are specified accordingly. Sometimes a platform is provided at one side, or the manhole is simply a vertical shaft over the center of the sewer. In the latter case, a block of reinforced concrete is cast around the pipe and designed to form an adequate foundation to support the shaft plus transmitted traffic loads. Such manholes usually are built without steps since a worker cannot step off into the water anyway. When entry is necessary, a worker is lowered in a chair hoist or cage. Large, factory-made T-sections also can be used if adequate support is provided for the shaft and transmitted loads.

Where a sewer is larger than 2 ft (600 mm) and the small-sewer type of manhole is used, the diameter of the manhole should be increased sufficiently to maintain an adequate width of bench, preferably 1 ft (0.3 m) or more on each side. In sewers that a worker cannot straddle, maintenance workers frequently lay planks to bridge the channel. Hence, there must be adequate and well-formed benches on each side. Sometimes the entering pipe is extended 2 to 3 ft (0.6 to 1 m) into the manhole and mortared over to form a smooth platform, as shown for the larger manhole in Fig. 7-3.

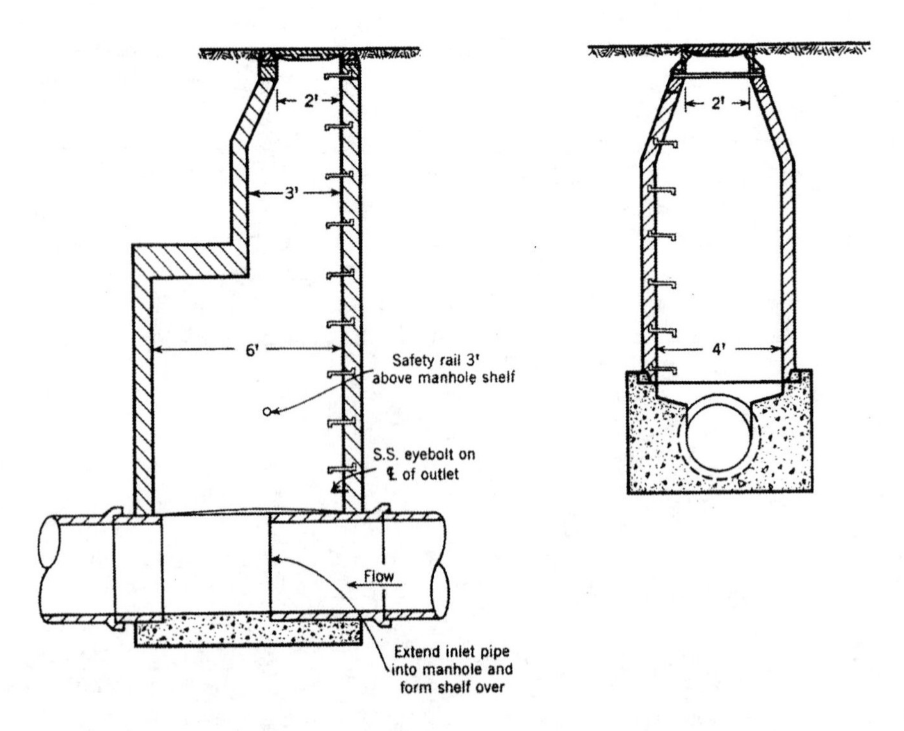

FIGURE 7-3. Two manholes for intermediate-sized sewers (ft × 0.3 = m).

7.3. BENDS

Particular care must be used in the construction of curved channels to accommodate bends. The highest workmanship is necessary to produce channels that are smooth, with uniform sections, radius, and slope.

A curve of very short radius causes energy-wasting turbulence. Some authorities recommend that for optimum performance the radius of the centerline be three times the pipe diameter or channel width. Reasonably satisfactory conditions usually can be obtained if the radius is not less than 1.5 times the diameter. If the velocity is supercritical, surface turbulence and energy losses arise even with long-radius bends.

The radius of curvature of a bend within a manhole is maximized if the points of tangency of the outer curve of the channel with the walls of the pipes are at the ends of the manhole diameter, as shown in Fig. 7-4 for the 12- and 18-inch (300- and 450-mm) pipes. This is true regardless of

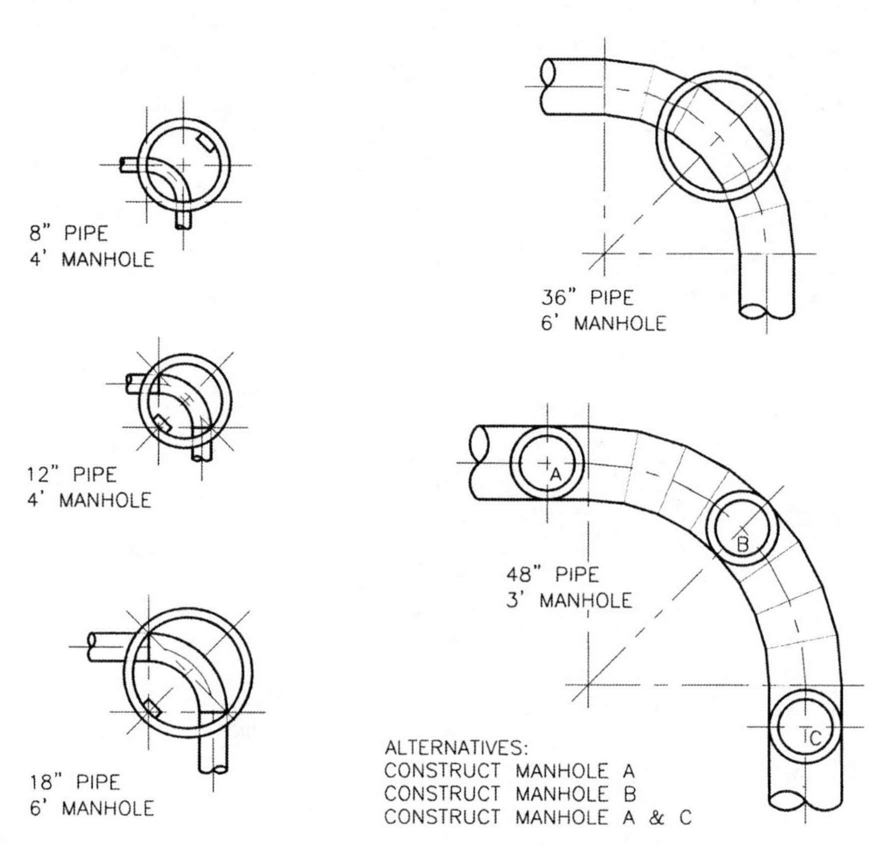

FIGURE 7-4. Manhole placement for various types of bends (1 ft. = 300 mm; 1 in. = 25.4 mm)

the angle. Bends of less than 90 degrees can, of course, be accommodated more easily. For angles substantially less than 90 degrees on sewers larger than 12 inches (300 mm) in diameter, the manhole is usually centered over the pipe.

Completion of a bend within a manhole is not necessary and becomes impossible as the pipe approaches the size of the manhole. Furthermore, when the size of the sewer is such that the manhole is only a chimney over the sewer, a manhole may be placed over the center of the curve or on the downstream tangent, or perhaps two may be used—one upstream and one downstream. Figure 7-4 shows some of the possible designs for manholes on bends.

Frequently, extra fall in the channel invert is provided on bends to compensate for bend energy losses. When this is done, the extra fall should reflect only the expected losses. Although experimental data for large conduits are scarce, it would appear that for a well-made 90-degree bend with a centerline radius of curvature not less than one pipe diameter, the loss in an open channel should not exceed 0.4 of the velocity head. Thus, for a velocity of 3 fps (1 mps), the loss would probably be not more than 0.06 ft (20 mm) for subcritical flow. Energy loss will be greater with supercritical flow but conservation of the energy of the stream is not likely to be important under that condition. In sewers where flows are small or velocities moderate, the energy losses at bends are usually ignored. The slight backing-up of the water due to the energy loss is usually inconsequential. (A more complete discussion of energy losses in bends, including those associated with supercritical flow, is found in Chapter 5.)

7.4. JUNCTIONS AND DIVERSIONS

On small sewers, junctions are made in ordinary manholes, with the branch line curved into the main channel. Excessive widening of the main channel at the junction should be avoided. Eddying flows and accumulations of sludge and rags are the result of poor flow patterns prevailing in many junction manholes. To minimize these objectionable conditions, the invert of the branch lines may be brought in somewhat higher than the invert of the main channel where the two join. Channels are generally constructed in the manhole bottom for all lines. Sometimes right-angle junctions are used in small sewers. This usually causes less energy loss than the attempt to conserve some of the energy of the side stream by use of a curved channel.

Large junctions generally are constructed in cast-in-place reinforced concrete chambers entered through manhole shafts (their hydraulic design is discussed in Chapter 5).

Diversion structures are specialty features that commonly occur when retrofitting an existing sewer with a parallel sewer to provide extra capacity. In those cases it is advantageous to consider using a diversion structure at a location (or multiple locations) that allows for future isolation of all or portions of the downstream system. In ideal situations, the structures can be designed to allow for automatic flow equalization using overflow weirs between the sewer lines, and can be supplemented with gates to provide for manual flow equalization or diversion. The true cost benefits of these structures are realized during maintenance operations when the cost of bypass pumping is considerable.

7.5. DROP MANHOLES

If a sewer enters a manhole at an elevation considerably higher than the outgoing pipe, it is generally not satisfactory to let the stream merely pour into the manhole because the structure then does not provide an acceptable working space. Drop manholes are usually provided in these cases. Figure 7-5 shows common types. These structures are not trouble-free. Sticks may bridge the drop pipe, starting a stoppage. Because of such a stoppage or merely because of high flow, wastewater may spill out of the end of the pipe, making the manhole a dangerous and objectionable place to work. Cleaning equipment also may lodge in the drop pipe. Sometimes the drop pipe is placed inside the manhole or the drop pipe is made of a larger diameter to minimize stoppages. Another arrangement sometimes used is to provide a cross instead of a tee outside the manhole, with the vertical pipe extended to the surface of the ground and with a suitable cover so that it is accessible for cleaning.

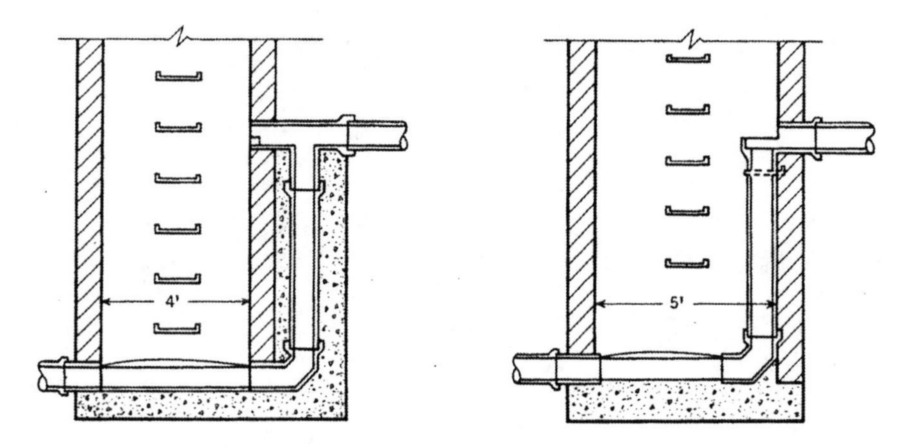

FIGURE 7-5. Drop manholes (ft × 0.3 = m).

Drop manholes should be used sparingly and, generally, only when it is not economically feasible to steepen the incoming sewer. Some engineers eliminate drops by using vertical curves. It would appear that this should be a general rule for elevation differences of less than about 3 ft (1 m) and often for larger drops as well.

7.6. TERMINAL CLEANOUTS

Terminal cleanouts sometimes are used at the upstream ends of sewers, although most engineers now specify manholes. Their purpose is to provide a means for inserting cleaning tools, flushing, or inserting an inspection light into the sewer.

A terminal cleanout consists of an upturned pipe coming to the surface of the ground. The turn should be made with bends so that flexible cleaning rods can be passed through it. The diameter should be the same as that of the sewer. The cleanout is capped with a cast-iron frame and cover (Fig. 7-6).

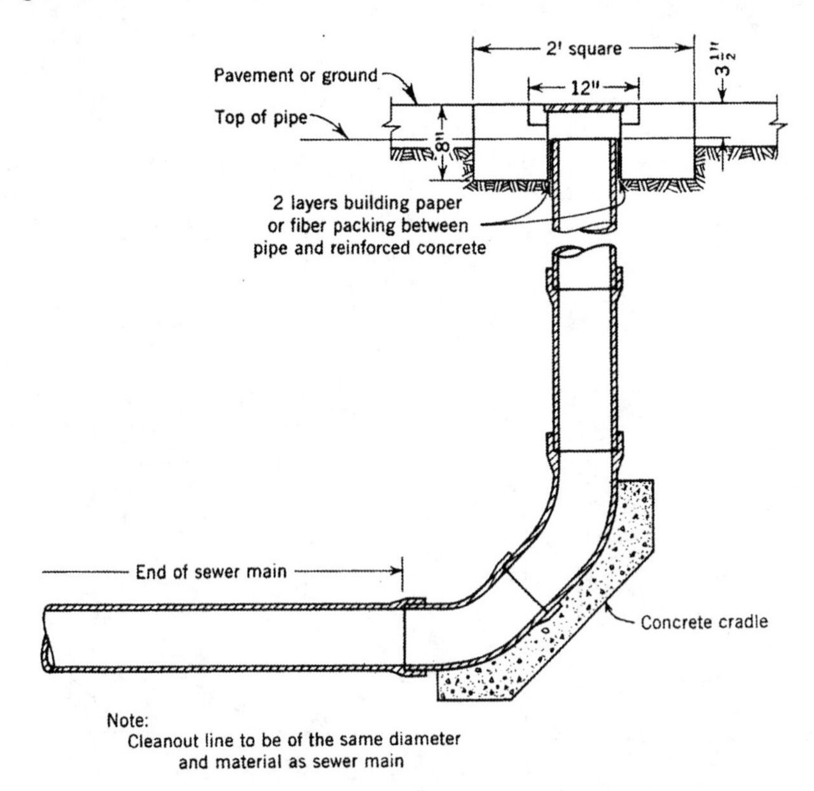

FIGURE 7-6. Terminal cleanout (ft × 0.3 = m; inch × 2.54 = cm).

In the past, tees often were used instead of pipe bends and the structures were called lampholes. Since cleaning equipment cannot be passed into the sewer through a tee, their use is no longer considered to be good design practice.

Regulations in most areas allow terminal cleanouts (if at all) only within 150 to 200 ft (45 to 60 m) of a manhole.

7.7. SERVICE LATERALS

Service laterals, also called house connections or service connections, are the branches between the street sewer and the property or curb lines, serving individual properties. They usually are required to be 4, 5, or 6 inches (100, 125, or 150 mm) in diameter, preferably with a slope of 1 in 48, or 2%. Sometimes 1% slopes are allowed and this seems to serve just as well. If a stoppage occurs in a service lateral, it may be due to root penetration, grease, or sometimes corrosion (in the case of iron pipes). Steeper slopes are of no benefit in coping with those problems.

Materials, joints, and workmanship for service laterals should be equal to those of the street sewer to minimize infiltration and root penetration. Particular attention should be paid to the construction of service laterals, especially compaction of bedding and backfill material, and jointing techniques, since these sewers frequently represent the major source of infiltration/inflow in a sanitary sewer system (see Chapter 3).

Often building sewers are constructed to the property or curb line at the time the street sewer is constructed. To meet the future needs of unsubdivided properties, wyes or tees sometimes are installed at what are presumed to be convenient intervals. Laterals or stubs not placed in use should be plugged tightly. Figure 7-7 shows typical connections. Typical connections to a deep sewer are shown in Fig. 7-8.

If wyes or tees are not installed when the sewer is constructed, the sewer must be tapped later, and deplorably poor connections often have resulted. This is especially true for those connections made by breaking into the sewer and grouting-in a stub. Either a length of pipe should be removed and replaced with a wye or tee fitting or, better, a clean opening should be cut with proper equipment and a tee-saddle or tee-insert attached. Any connection other than to existing fittings must be made by experienced workers under close supervision.

In some places, test tees are required on the service lateral, which permit the outlet to the street sewer to be plugged. This makes it possible to test the service lateral.

When a connection is to a concrete trunk sewer, a bell may be installed at the outside of the pipe. Three designs are shown in Fig. 7-9. Preferably the bell is provided by the manufacturer, but it can be installed in the field

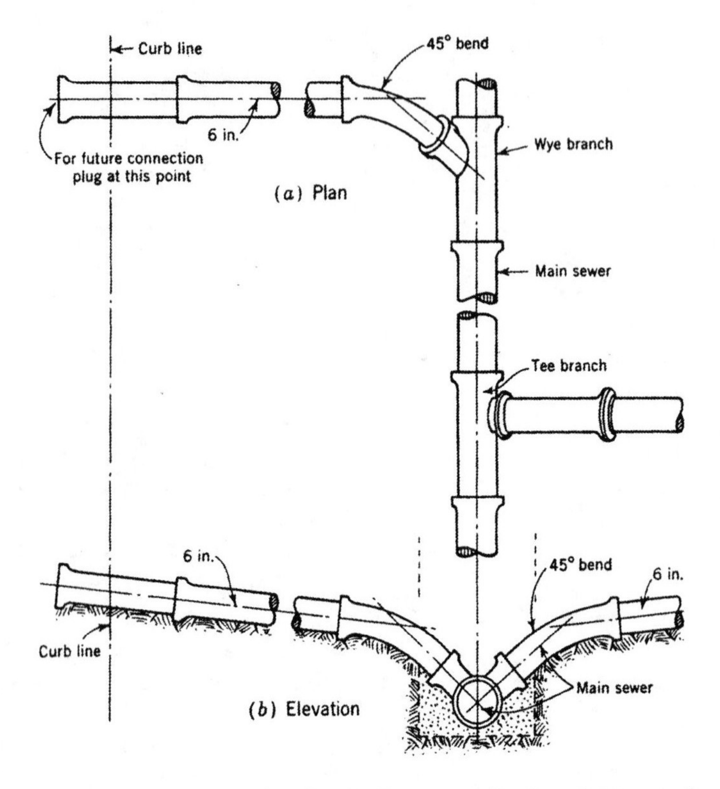

FIGURE 7-7. Service connection for shallow sewer (inch × 2.54 = cm).

if necessary. It must be high enough on the pipe so that the lateral will not be flooded by high flows in the sewer.

Large trunks are not ordinarily used as collecting sewers. When they are more than 3 or 4 ft (900 to 1,200 mm) in diameter, they frequently are paralleled by smaller collecting sewers that enter the trunks at manholes.

7.8. CHECK VALVES AND RELIEF OVERFLOWS

Where the floor of a building is at an elevation lower than the top of the next upstream manhole on the sewer system, a stoppage in the main sewer can lead to overflow of wastewater into the building. Devices that sometimes are used to guard against such occurrences include backflow preventers or check valves and relief overflows.

Backflow preventers or check valves may be installed where the house plumbing discharges to the house sewers. Usually a double check valve is specified. Even so, such devices frequently do not remain effective over long periods of time.

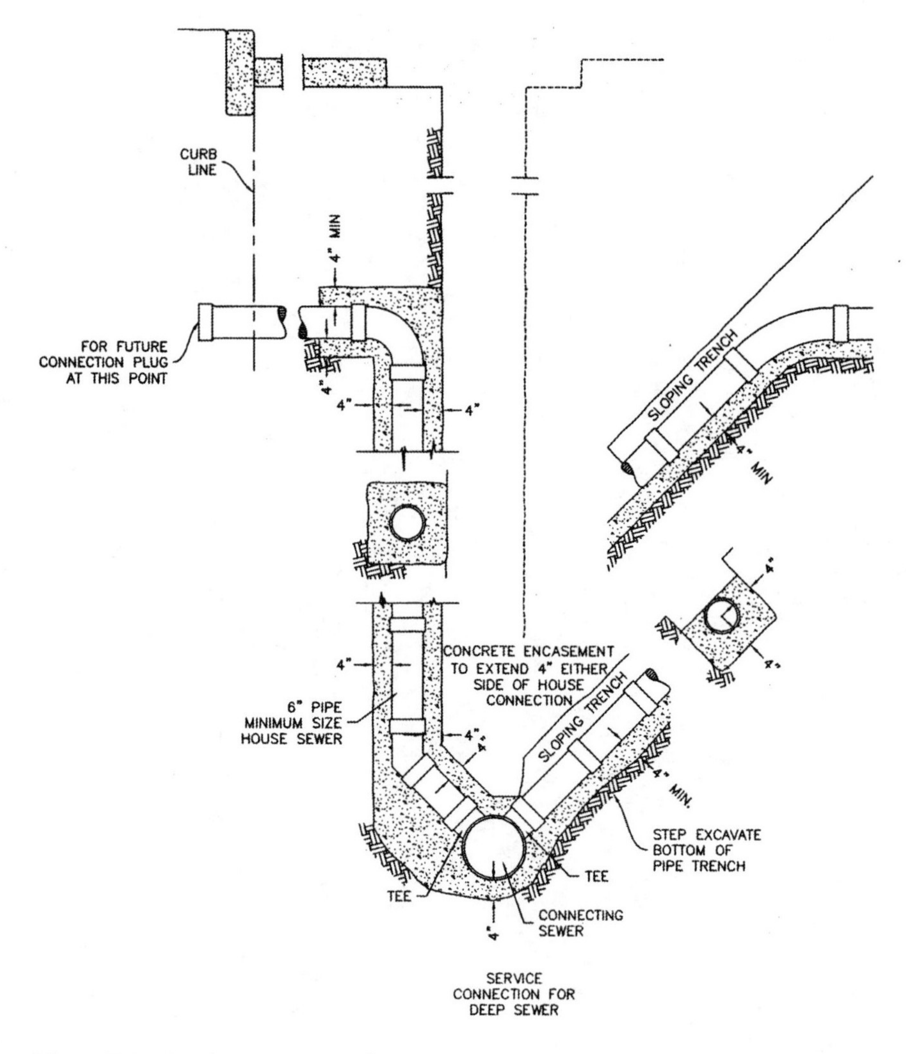

Figure 7-8. Service connection for deep sewer (inch × 2.54 = cm).

Any overflow of wastewater is undesirable but if a stoppage occurs in a street sewer, overflow may result. It will be from a manhole in the street, into a building, or on occasion at a designated overflow point. For the latter to be effective, it must be at an elevation lower than the floor level being protected. At this point, a relief device may be installed that encases a ball resting on a seat to close the end of a vertical riser and prevent flow into the sewer. The relief device must be constructed so that the ball will rise and allow overflowing wastewater to escape and thus provide relief. This practice is not encouraged except in the most extreme conditions and

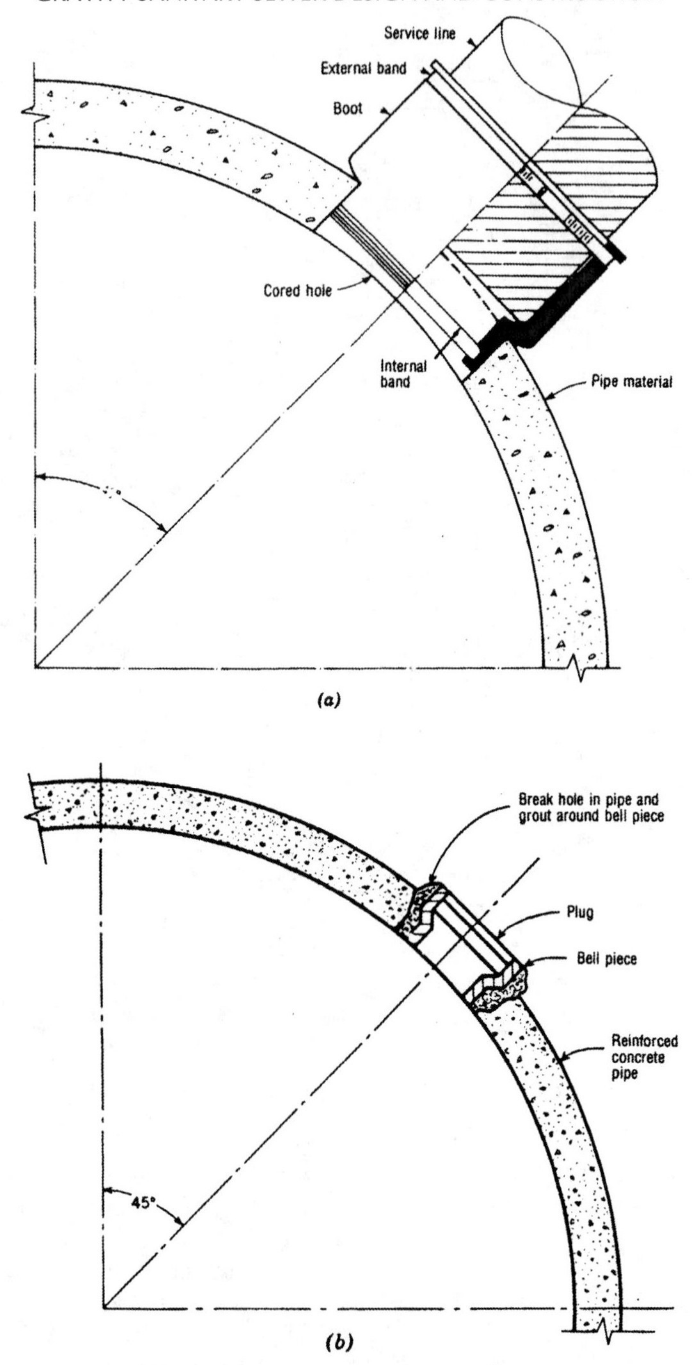

FIGURE 7-9. A–C Various connections to large sewers. A and C courtesy NPC, Inc., Milford, N. H.

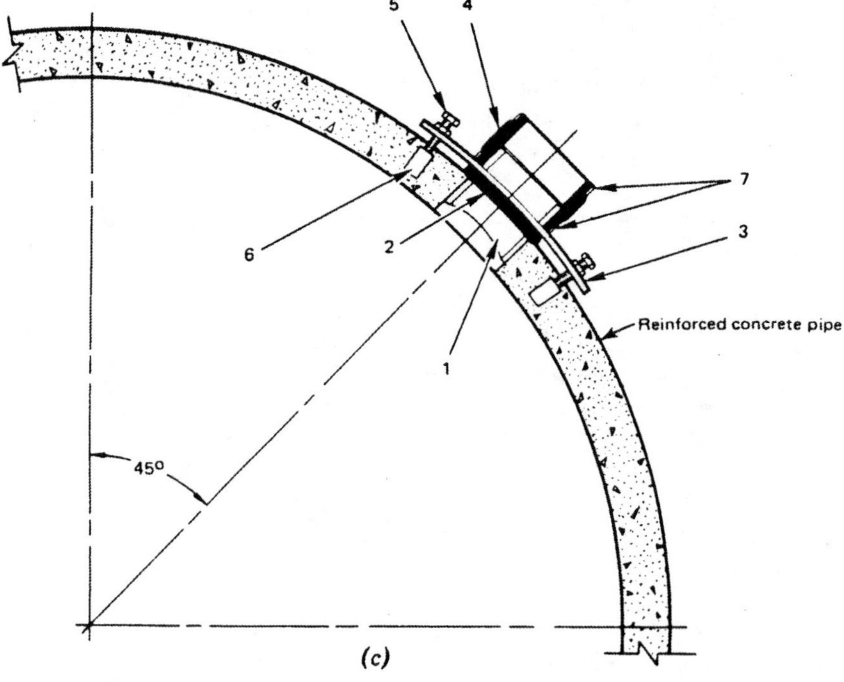

1—Acrylonitrile-butadiene-styrene (ABS), PVC, or cast iron (CI) nipple;
2—Large neoprene "donut";
3—Cast ductile flange;
4—Neoprene coupling;
5—Stainless steel nuts, bolts, and washers;
6—Lead lag anchors;
7—Stainless steel clamps.

FIGURE 7-9. (Continued).

usually with regulatory approvals. Building owners who have valuable property in basements that might be flooded usually protect themselves, insofar as possible, by check valves.

7.9. SIPHONS

"Siphon" in sewerage practice almost always refers to an inverted siphon or depressed sewer which would stand full even with no flow. Its purpose is to carry the flow under an obstruction, such as a stream or depressed highway, and to regain as much elevation as possible after the obstruction has been passed.

7.9.1. Single- and Multiple-Barrel Siphons

Siphons are often constructed with multiple barrels. The objective is to provide adequate self-cleansing velocities and maintenance flexibility under widely varying flow conditions. The primary barrel is designed so that a velocity of 2 to 3 fps (0.6 to 1.0 mps) will be reached at least once each day, even during the early years of operation. Additional pipes, regulated by lateral overflow weirs, assist progressively in carrying flows of greater magnitude (i.e., maximum dry-weather flow to maximum storm flow). The overflow weirs may be considered as submerged obstacles, causing loss in head as flow passes over them. The weir losses may be assumed equal to the head necessary to produce critical velocity across the crest. Weir crest elevations are dependent on the depths of flow in the upstream sewer for the design quantity increments. Sample crest length calculations are presented in textbooks (Fair and Geyer 1954).

Many engineers maintain that for sanitary sewers there is usually no need for multiple barrels. They reason that solids which settle out at low flows will flush out when higher flows are obtained, except for those heavy solids that would accumulate even at high flows. Single-barrel siphons generate less sulfide and cause less loss of hydraulic head than do multiple barrels. Single-barrel siphons have been built with diameters ranging from 6 to 90 inches (150 to 2,300 mm) or more. Engineers holding to this concept generally favor a small barrel if initial flows are to be much lower than in later years, with a larger barrel constructed at a later date or constructed at the outset and blocked off so that it can always operate as a single-barrel siphon. In some situations a spare barrel may be desirable purely for emergency or maintenance use.

7.9.2. Profile

Two considerations that govern the profile of a siphon are provision for hydraulic losses and ease of cleaning. The friction loss through the barrel will be determined by the design velocity. For calculating this head loss, it is advisable to use a conservative Hazen-Williams friction coefficient of 100 (Manning n from 0.014 for small sizes to 0.015 for the largest). In the case of multiple-barrel siphons, additional losses due to side-overflow weirs must be considered.

Siphons may need cleaning more often than gravity sewers. For easy cleaning by modern methods, the siphon should not have any sharp bends, either vertical or horizontal. Only smooth curves of adequate radius should be used. The rising leg should not be so steep as to make it difficult to remove heavy solids by cleaning tools that operate hydraulically. Some agencies limit the rising slope to 15%, but slopes as great as 50% (30 degrees) are used in some places. There should be no change of pipe diameter within the length of a barrel since this would hamper cleaning operations.

The engineer should also incorporate the use of gates and/or stop logs at the opening of each siphon pipe. This facilitates easy isolation of one of more of the pipes for future cleaning operations.

7.9.3. Air Jumpers

Positive pressure develops in the sewer atmosphere upstream from a siphon because of the downstream movement of air induced by the sewage flow. In extreme cases, this pressure may equal several inches (centimeters) of water. Air therefore tends to exhaust from the manhole at the siphon inlet, escaping in large amounts even from a pick hole. Under all except maximum flow conditions, there is a drop in water surface elevation into a siphon, with consequent turbulence and release of odors. The exiting air can thus be the cause of serious odor problems. Conversely, air is drawn in at the siphon outlet.

Attempts to close the inlet structure tightly will usually force the air out of plumbing vents or manholes farther upstream. Insofar as the attempt to close the sewer tightly is successful, oxygen depletion in the sewer atmosphere occurs, aggravating sulfide generation where this is a problem.

To overcome this difficulty, a number of siphons built in recent years have used air jumpers (i.e., pipes that take the air off the top of the inlet structure and return it at the end of the siphon). Usually, the jumper pipe is one-third to one-half the diameter of the siphon. Sometimes the pipe can be suspended above the hydraulic grade line of the sewer, but in other cases it must run more or less parallel to the siphon. In these cases, provision must be made for dewatering the jumper; otherwise it will fill with condensate. In some cases a drain has been installed to a percolation pit. One large air jumper in use consists of 48-inch- (1,200-mm)-diameter pipe paralleling a 90-inch- (2,300-mm) siphon, 2,000 ft (610 m) long, utilizing a sump pump for dewatering.

7.9.4. Sulfide Generation

Sulfide may be produced in a long siphon. There is nearly always a hydraulic jump or turbulence at a siphon inlet, which causes absorption of oxygen and delays the onset of sulfide buildup in comparison with pressure mains that lack this initial aeration. Thus, sulfide buildup may be delayed for as much as an hour if the wastewater is of low temperature or low biochemical oxygen demand (BOD). When higher temperatures prevail, and especially if oxygen absorption at the inlet is minimal, sulfide buildup may be underway in 20 minutes. For any given flow and wastewater characteristics, the sulfide concentration produced in a filled pipe is roughly proportional to the pipe diameter. (For further details, see Chapter 4.)

7.10. FLAP GATES OR DUCKBILL VALVES

Flap gates or duckbill valves are installed at or near sewer outlets to prevent back-flooding of the sewer system by high tides or high stages in the receiving stream. These are common only in combined or storm sewers.

Duckbill valves are made of elastomers which are not susceptible to corrosion. They are commercially available in sizes up to 96-inch (2,438-mm) diameter. The valves have been found to be self-cleansing.

Flap gates may be made of wood, cast iron, or steel. They are commercially available in sizes up to 8 ft (2.4 m) in diameter. Larger gates can be fabricated from plates and structural shapes. They should be hinged by a link-type arrangement which allows the gate shutter to seat more securely. Hinge pins, linkages, and seats should be corrosion-resistant.

The maintenance of flap gates requires regular inspection and removal of debris from the pipe and outlet chamber, lubrication of hinge pins, and cleaning of seating surfaces.

7.11. SEWERS ABOVE GROUND

Occasionally, in rolling or hilly terrain, it is desirable and economical to build sewers above the surface of the ground or across gullies and stream valleys. Such sewers often are constructed in carefully compacted fill. Sometimes it is better to suspend a sewer over a waterway or a highway than to go under it by means of an inverted siphon. Sewer crossings in such cases have been constructed by installing or hanging the pipes on bridges, by fastening them to structural supports which rest on piers, by supporting them with suspension spans and cables, and by means of sewer pipe beams.

Structural design of suspended sewers is similar to that of comparable structures with supporting members of timber, steel, or reinforced concrete. Foundation piers or abutments should be designed to prevent overturning and settlement. The impact of flood waters and debris should be considered.

If the sewer is exposed, as on a trestle, steel pipe may be used with coating and lining for corrosion protection. Sometimes sewers of other materials are carried inside steel pipes. The steel pipe may be supported by simple piers at suitable intervals.

In recent years, prestressed concrete pipe beams have been used to span waterways and other obstacles. Generally, they have been of three types:

1. A rectangular section with a circular void extending the full length of member, either pretensioned or post-tensioned, and similar to a hollow box highway girder. This section is normally used for smaller sewers.

2. A pretensioned circular pipe section which may be produced in most any diameter. This type is economical for long crossings.
3. Reinforced concrete pipe sections assembled and post-tensioned to form the required sewer pipe beam. These beams may be fabricated economically using standard pipe forms and prestressing equipment. The pipes are cast with longitudinal cable ducts in the walls. After curing, the pipes are aligned and post-tensioning cables are inserted and jacked to the design tension, and anchored. Pressure grouting of the ducts completes the manufacture and the sewer beam is then shipped to the job site for installation.

Protection against freezing and prevention of leakage are important design and construction considerations for aboveground sewers. It has been found necessary in some designs to employ expansion jointing between aboveground and belowground sewers. Special couplings are available for such purposes. Anchorage provision also must be made to prevent permanent creep. Expansion joints in sewers supported on bridges or buildings should match the expansion joints in the structures to which the sewer is attached.

7.12. UNDERWATER SEWERS AND OUTFALLS

7.12.1. Ocean Outfalls

Communities adjacent to a seacoast may discharge their treated wastewater into the ocean. Disinfected secondary effluents generally are discharged relatively close to shore but usually beyond the distance designated for body contact. Primary effluents are carried far enough to sea to avoid any undesirable effects. In the United States, with very few exceptions, all ocean discharges must have secondary treatment, but this is not the usual policy in other countries where the discharge of primary effluent is considered satisfactory if suitable depths and distances are reached.

For proper design, it is essential to obtain detailed data on the following:

- Bathymetric profiles of possible outfall routes.
- Nature of the ocean bottom.
- Water density stratification or thermoclines, by seasons.
- Patterns of water movement at point of discharge and travel time to shore.

Since seawater is 2.5% denser than sanitary wastewater, the discharged wastewater rises rapidly, normally producing a boil at the surface. The rising plume mixes with a quantity of seawater, which is generally from

10 to 100 or more times the wastewater flow. Dilution increases rapidly as the wastewater field moves away from the boil. The required length and depth of the outfall is related to the degree of treatment of the wastewater. The length must be calculated so that time and dilution will adequately protect the beneficial uses of the adjacent waters.

Much research has been done regarding the dilution of the wastewater and the die-away of the bacteria (Rawn and Palmer 1930; Brooks 1960; Abraham 1963; Pomeroy 1960; Pearson 1956; Gunnerson 1961). A full treatment of that subject is beyond the scope of this Manual.

Where the outfall is deep and there is good density stratification (thermocline), the rising plume may pick up enough cold bottom water so that the mixture is heavier than the surface water. The rising plume therefore stops beneath the surface or reaches the surface and then resubmerges.

Diffusers may be used to gain maximum benefit of density stratification. If, however, they merely divide the flow into many small streams in a small area (a gas burner type of diffuser), they do little good. The flow must be dispersed widely so that huge flows of dilution water can be utilized at low velocity.

The diffuser must be approximately level if it is to accomplish reasonably uniform distribution. For design of the diffuser, the rule of thumb may be used that the total cross-sectional area of the ports should not be more than half the cross-sectional area of the pipe. In large diffusers, often exceeding 0.6 miles (1 km) in length, the diffuser diameter may be stepped down in size toward the end (Rawn et al. 1960). Computerized calculations are used in the design of these large diffusers.

These principles are well-illustrated by the Los Angeles City outfall in Santa Monica Bay, California. The effluent is carried by a 12-ft- (3.7-m)-diameter pipe to a point 5 miles (8 km) from shore at a depth of 190 ft (58 m), then dispersed through a Y-shaped diffuser with the two arms totaling 8,000 ft (2,400 m) in length. Except for certain periods in winter when the thermocline is practically nonexistent, no wastewater can be seen rising to the surface. The flow of effluent, which has been upgraded to full secondary standards, exceeds 340 mgd (14 m^3/sec), yet the bathing waters of the highly popular beaches on the Bay show no bacterial evidence of the wastewater discharge.

Ocean outfalls require specialized design expertise; however, a general overview is provided here. Outfalls into the open ocean generally are buried to a point where the water is deep enough to protect them from wave action, hydrodynamic forces, and shifting bottom sands—usually about 30 ft (10 m). Trenches in rock are formed by blasting. Beyond the buried portion, the outfall rests on the bottom, with a flanking of rock to prevent currents from undercutting it where the bottom is soft.

Ocean outfalls in the smaller sizes are now usually made of steel pipe, mortar-lined and coated. Steel pipes are welded and usually can be dragged into place from the shore. Relatively short pipelines are sometimes floated into place. Reinforced concrete pipe is used for the larger sizes. The joints for concrete outfalls usually are made with rubber gaskets similar to those used for construction on land. Special bolted restraints are used to secure the joints in small outfalls. Where the depth exceeds 30 ft (10 m), large pipes are simply laid on the bottom, sometimes with a rock cradle. They must be adequately flanked with rock; in this condition they are stable. Many such outfalls are operating successfully.

7.12.2. Other Outlets

If effluent is discharged into an estuary or landlocked bay, special studies are needed to explore tidal currents, upstream flow of salt water at the bottom, available dilution, etc., in order to determine which discharge locations are compatible with various degrees of treatment.

Sewers discharging into streams with high-velocity flood flows require thought in design to prevent undermining of the outlet structure and the pipe itself. Large outlets into the Mississippi River, for example, have been built with massive headwalls and wingwalls with deep foundations.

7.13. MEASURING WASTEWATER FLOWS

The design of a sewer system often requires measurements of flows in sewers as well as the design of permanent monitoring facilities to measure and record future flows at one or more points in the system. There are many reasons why such measurements are made, two of which are to collect information needed in the administration of contracts between cooperating parties, and to aid engineers in the planning of future expansions.

Two classes of metering devices may be distinguished: those that are adapted to filled pipes ("closed-channel flow"), and those that make measurements in streams that have a surface exposed to the air, as in a partly filled gravity sewer (the "open-channel flow" condition).

A force main from a pump station is a closed channel in which a Venturi-type meter, a sonic meter depending upon the Doppler effect, or a magnetic meter may be used. In rare cases the stream in a gravity sewer is caused to flow through a depressed reach as a pressure conduit where one of these devices has been placed. This section of the chapter, however, deals only with measurements of the open-channel type, since most sewer measurements are made by these methods. Generally, flow measurement in open channels is accomplished using a calculation or spot check, a portable flow logger, or a permanent metering station.

There are several spot check methods for determining flows. The most popular technique is the Manning equation, which relates level to flow. For the Manning equation to be effective, there must be at least 250 ft

(60 m) of a straight course or channel, with 1,000 ft (300 m) preferred. Other factors in the equation are the slope of the pipe and the coefficient of roughness for the channel material (the coefficient of roughness is available in Chapter 5 and in other technical manuals). Another crucial requirement for the Manning equation is that the downstream flow be unobstructed. If there is a downstream control device or other obstruction, the relationship between level and flow is destroyed [see p. 129 of the Isco Flow Handbook (Grant and Dawson 1997)].

Another spot check is to measure the depth of flow with a ruler or level probe and determine the velocity of the stream using a dye or other indicator. Such a measurement is valid only if a reliable determination of average flow depth and velocity can be made. In large sewers, and especially if depths can be measured at several manholes, quite precise results can be obtained. At a flow of less than about 1 cfs (30 L/sec) in a sewer, the results are likely to be uncertain.

Still another spot check method is the use of a dye (usually Rhodamine) or a radioactive element, with subsequent sampling at a downstream point and analyses to determine the dilution of the tracer. It is possible to secure very precise results by this method but only if the amount of tracer added is accurately measured and the downstream concentration is determined with a high degree of precision. It is a good method of calibrating any type of meter but it can be costly due to the sophisticated equipment required and the skill required to perform the study.

Flow determinations can often be made at a pump station by timing the filling of a wet well. If an average filling time is determined as well as an average time of pumping out, calculations can be made of the pumping rate. A time meter on the pump will then give a measure of total flow, but the pumping rate must be checked from time to time since it may not remain constant. In those types of pump time analyses, consideration must be given to the discharge head pressures since some systems are designed to operate on a force main discharge in place of a gravity flow discharge. The residual pressure in the discharging force main significantly impacts the pumping rate, thus adversely affecting the accuracy of the timing method. It may be best to incorporate the use of noninvasive external flow meters. The same wet well can be used for a simple drawdown calculation where the volume of the wet well is calculated and the differences in volume are noted over time. Some companies make special-purpose data loggers that will calculate inflow and outflow from a pump station using the wet well volume at various alarm (level switch) points and pump run time. Importantly, these do not depend on pump performance curves for their flow calculation.

Since the mid-1980s, the area/velocity meter (Fig. 7-10) has become increasingly popular for making temporary measurements in open channels. Most area/velocity flow meters are designed to operate on battery

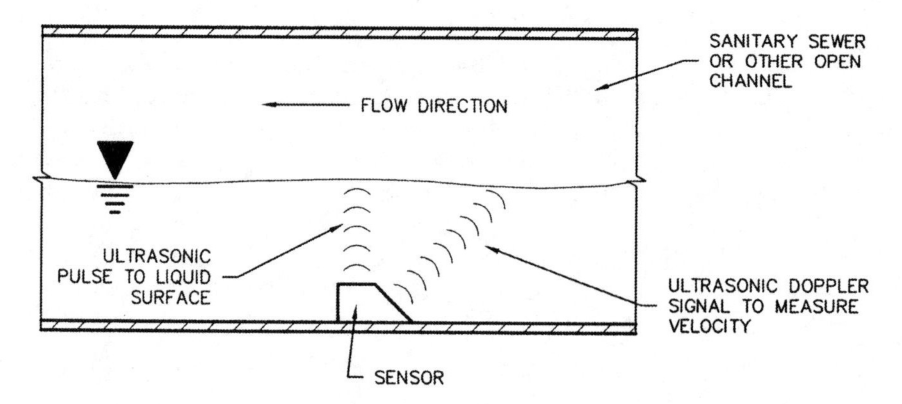

FIGURE 7-10. Area/velocity flow.

power and hang in the sewer. They can be programmed for any channel that can be defined mathematically. They are easy to install and their accuracy can be quite good with the right site conditions. The same meter can be moved and used at different sites as often as desired. Area/velocity meters directly solve the continuity flow equation:

$$Q = V \times A \text{ (flow rate = velocity} \times \text{area).}$$

These devices use Doppler ultrasonic frequency shift, electromagnetic induction, or radar to measure fluid velocity. By using differential pressure or ultrasonic pulses to measure the level, they can calculate what percentage of the total available area is being used by the fluid at the time of measurement. The user installs a single probe (although some manufacturers offer dual probes) in the flow stream and then programs the electronics with the channel dimensions. Many types of channels are available as menu choices. For example, the user selects Round Pipe and then types in the diameter. The most popular version of the area/velocity meter is a flow logger, which will take readings on a timed basis, log these readings to an onboard data logger, then go back to sleep. The user will download the data at selected intervals via direct connection to a computer, via phone (landline or cellular), or via spread-spectrum radio. The data is in digital form so it is easily displayed as charts, graphs, or tables in computer software programs. Area/velocity meters are especially useful in surcharge conditions and reverse-flow conditions. Because they measure level and velocity, events when the level is high but the channel is not flowing will be sensed by the low, zero, or negative velocity measurement.

For a permanent metering station, the most traditional device for obtaining a continuous flow record in a gravity sewer is a weir, using that word in its general meaning of "an obstruction over which the water flows." In engineering practice in the United States, the word has come to be associated mostly with the sharp-crested weir used for measuring flow quantities, but a sharp crest is not universally implied. More common today is the compound weir. One of the more popular compound weirs for sewer flow will use a V-notch portion for lower flows, then one or more rectangular weirs for increasing flows. Many of these weirs are designed to minimize obstruction.

In flowing over a weir, the stream passes through a control section as shown in Fig. 7-11. For any given flow, the elevation of the water surface

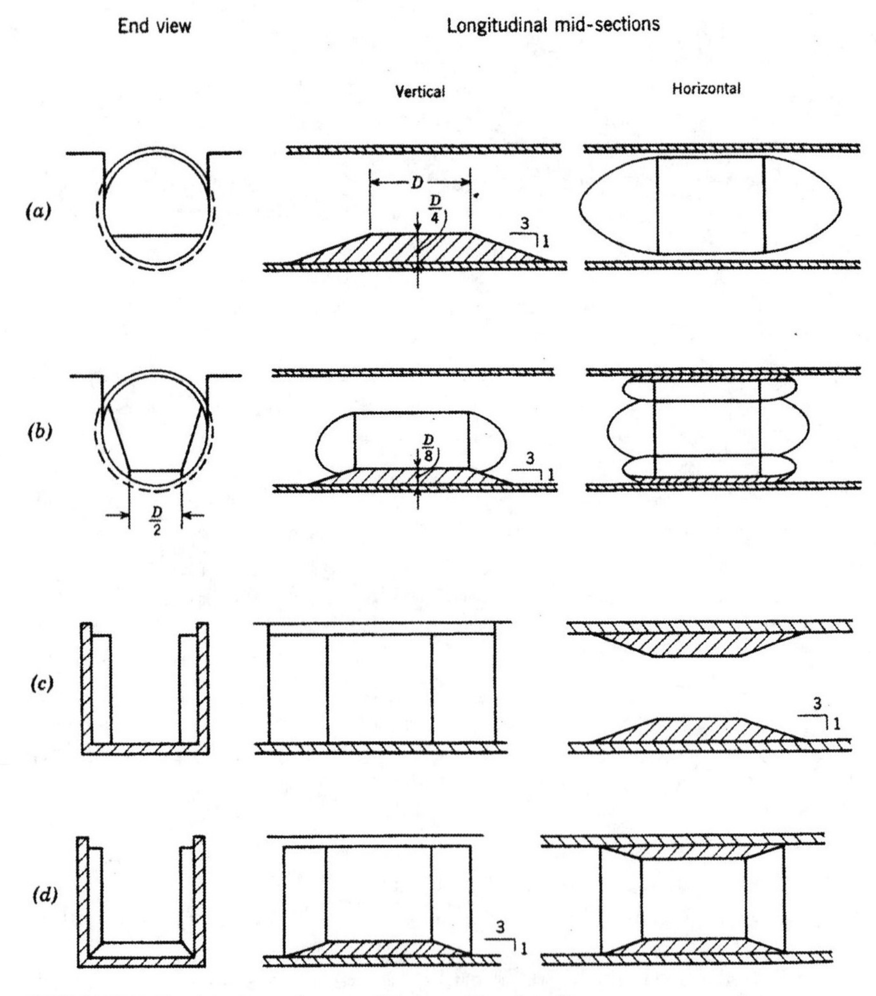

FIGURE 7-11. *Various shapes of Palmer-Bowlus flume.*

at the control section is such that the total energy (elevation plus kinetic energy) is minimal. This section controls the upstream water elevation. The upstream elevation near the weir is used as a measure of the discharge over the weir, but the elevation reading needs to be increased by the amount of the kinetic energy of the velocity of the water at the place of measurement so as to obtain a figure for the total energy. Inherent in the calculations is the assumption that there is no energy loss between the point of measurement and the control section. Under extreme conditions this assumption may entail a significant error.

In 1936 H. K. Palmer and Fred Bowlus, both engineers at the Los Angeles County Sanitation Districts, showed the advantage of a streamlined weir (Palmer and Bowlus 1936). They devised a trapezoidal form which came to be known as a Palmer-Bowlus flume (Fig. 7-11B). (They described it as a Venturi flume, but it has little in common with the Venturi meters used for closed-channel measurements.) Palmer-Bowlus flumes are available commercially in various sizes. Later, Bowlus devised a simple slab form (unpublished work). Figure 7-11A shows a Bowlus weir constructed for insertion into a 18-inch (450-mm) pipe. It is easily placed and easily retrieved by means of a chain attached to the upstream toe. Portable weirs of this type are used in sewers up to 27 inches (675 mm) in diameter, or larger if there is access by an opening larger than a standard manhole.

Since this type of weir can be placed in a sewer without altering the invert, it is particularly well-suited where it is necessary to construct a metering station on an existing sewer. A Parshall flume, which is a widely used form of a streamlined weir with dimensions generally in arbitrarily fixed ratios, specifies a drop in the invert.

Ludwig and Ludwig (1951) and Wells and Gotaas (1958) experimented with both the trapezoidal and slab-type streamlined weirs. They found that these devices installed in sewers can meter flows up to 90% of the sewer capacity and that the differences between actual flows and the flows shown by the theoretically calculated rating curves were less than 3%.

Streamlined weirs have often been built in channels of rectangular sections, and in some of these the control section has been produced by merely constricting the sides, leaving the invert as a clear continuation of the invert of the rectangular channel (Fig. 7-11C). Calculation of the rating curve follows the same form as that for other rectangular streamlined weirs.

The theoretical equation for a rectangular streamlined weir is:

$$Q = \frac{2}{3} B \sqrt{\frac{2}{3} g H^3} \quad \text{(any fully consistent units)} \qquad (7\text{-}1)$$

in which Q is the flow, B is the width at the control section, g is the gravitation constant, and H is the total energy of the stream (i.e., the elevation relative to the elevation of the crest of the weir plus the kinetic energy at the same location). For a rectangular streamlined weir, the formula reduces to:

$$Q = 3.09\, BH^{3/2} \text{ (cfs)} \tag{7-2a}$$

or

$$Q = 1.705\, BH^{3/2} \text{ (m}^3\text{/sec)} \tag{7-2b}$$

The discharge over a sharp-crested weir is a few percent greater than that shown by the calculations in the preceding equations because the control section over the sharp crest is slanted slightly upstream. The Parshall flume has a discharge about 7% greater than that for a simple rectangular shape. Empirically determined rating curves have been published for Parshall flumes of various sizes.

Because of the wide utility of the Bowlus weir, a rating curve for it is shown in Table 7-1 for a l-ft- (305-mm)-diameter pipe. The height of the weir is one-quarter of the pipe diameter. It is assumed that the weir is placed so that the critical section is in the outlet pipe from a manhole and the depth measurement is made in a U-shaped channel in the manhole.

If the flow is very small it may be desirable to use a weir with a height of less than one-quarter the pipe diameter. The Palmer-Bowlus trapezoidal flume has a height equal to one-eighth the diameter of the flume. If the velocity of approach is too great, a weir height greater than one-quarter-diameter may be used. A computer can easily be programmed to provide a rating table for any chosen design.

The attainment of a critical flow condition over any type of weir is essential if a correct measurement is to be obtained. Usually, this condition is ensured if a hydraulic jump is observed downstream. If the slope of the pipe is such that the normal flow of the water is supercritical, a hydraulic jump will be seen upstream from the weir instead of downstream as a result of the retardation of the stream approaching the weir.

Any kind of weir will cause some retardation of the upstream velocity. Therefore, there may be some stranding of solids upstream. Usually this is of no consequence, but if much grit accumulates it may materially increase the upstream velocity and reduce the apparent discharge. This effect is probably greater with a sharp-crested weir and may be minimal with the Parshall flume and other weirs that do not have a raised bottom. A wooden shovel or paddle may be used to move excessive accumulations of sand over or through the streamlined weir.

Permanent metering stations are designed with vaults that provide access to the hydraulic device for maintenance and calibration. The meas-

TABLE 7-1. Rating Curve for Streamlined Bowlus Weir Placed in Outlet Pipe of Manhole*

h or h/D (1)	Q, in cubic feet per second (2)	h, in feet (3)	Q, in cubic feet per second (4)	h, in feet (5)	Q, in cubic feet per second (6)
0.05	0.031	0.26	0.43	0.47	1.13
0.06	0.041	0.27	0.45	0.48	1.17
0.07	0.052	0.28	0.48	0.49	1.21
0.08	0.064	0.29	0.51	0.50	1.25
0.09	0.077	0.30	0.54	0.51	1.29
0.10	0.091	0.31	0.57	0.52	1.33
0.11	0.106	0.32	0.60	0.53	1.37
0.12	0.121	0.33	0.63	0.54	1.41
0.13	0.138	0.34	0.66	0.55	1.45
0.14	0.155	0.35	0.70	0.56	1.49
0.15	0.174	0.36	0.73	0.57	1.53
0.16	0.193	0.37	0.76	0.58	1.58
0.17	0.213	0.38	0.80	0.59	1.62
0.18	0.233	0.39	0.83	0.60	1.66
0.19	0.255	0.40	0.87	0.61	1.70
0.20	0.277	0.41	0.90	0.62	1.74
0.21	0.300	0.42	0.94	0.63	1.79
0.22	0.324	0.43	0.98	0.64	1.83
0.23	0.348	0.44	1.01	0.65	1.87
0.24	0.373	0.45	1.05	0.66	1.91
0.25	0.399	0.46	1.09		

*Height of the weir is $D/4$, in which D is the pipe diameter in feet. The calculated flows, Q, are for a pipe 1 ft (305 mm) in diameter. For any other pipe size, use h/D in place of h, and multiply the flows by $D^{5/2}$.
Note: 1 ft + 0.305 m; 1 cfs + 8.3 L/sec.

urement of water elevation upstream from the weir or flume may be made with a streamlined float, provided that an arm or cords are used to hold it in place. Quite commonly a float well is provided outside the channel, sometimes outside the vault that houses the channel. The float well is connected to the channel with a pipe generally in the size range of 0.5 to 1 inch (15 to 25 mm) in diameter, terminating in a smooth opening flush with the wall of the channel. A major problem with float wells is the accumulation of solids. Fresh water is provided for periodic or continuous flushing of the well and connecting pipe. A drain is also useful in the bottom of the float well. Safety of the fresh water supply is protected by an air gap located higher than the highest water level under flooding conditions, which generally means higher than the ground surface. A leveling drain also should be provided, so arranged that when the connection to the channel is shut off and the leveling valve is opened the water will discharge to a sump or other low point until the level in the well exactly equals the elevation of the crest of the weir.

An alternative to the float well is a bubble tube. The pressure required to discharge air from the end of a pipe dipping into the water serves as a measure of the depth of water at that point. The end of the pipe is usually cut cleanly at a right angle to the axis of the pipe. The pipe may be perpendicular to the flow or it may be angled downstream so as to reduce the accumulation of paper and stringy material. Erroneous results will be obtained if the pipe angles upstream. Only a small air stream is needed; a bubble every 5 to 10 seconds will suffice. Usually the air is supplied from a pressure tank recharged from time to time by a compressor. For temporary installations, a cylinder of air or carbon dioxide is a convenient source. Pressure sensors send signals that are converted to flow rates for indication and recording. Usually the recorder is outside the vault to escape the humid, corrosive atmosphere in the vault.

REFERENCES

Abraham, G. (1963). *Jet diffusion in stagnant ambient fluid.* Pub. No. 29, Delft Hydraulics Laboratory, Delft, The Netherlands.

ASTM International (ASTM). (2007). "Standard specification for resilient connectors between reinforced concrete manhole structures, pipes and laterals." ASTM Standard C 923-02 (active as of January, 2007), ASTM International, West Conshohocken, Penn.

Brooks, N. H. (1960). "Diffusion of sewage effluent in an ocean current." *Proc. 1st Conf. on Waste Disposal in Marine Environment,* University of California, Berkeley, Pergamon Press Ltd., London.

Fair, G. M., and Geyer, J. C. (1954). *Water supply and wastewater disposal,* first ed., John Wiley & Sons, Inc., New York, N.Y.

Grant, D. M., and Dawson, B. D. (1997). *Isco open channel flow measurement handbook,* fifth ed., Teledyne Isco, Lincoln, Neb.

Gunnerson, C. G. (1961). "Marine disposal of wastes." *J. San. Engrg. Div. Proc. ASCE, 87*(SaAl), 23.

Ludwig, J. H., and Ludwig, R. G. (1951). "Design of Palmer-Bowlus flumes." *Sewage and Industrial Wastes, 23,* 1096.

Palmer, H. K., and Bowlus, F. D. (1936). "Adaption of Venturi flumes to flow measurements in conduits." *Trans. ASCE, 101,* 1195.

Pearson, E. A. (1956). *An investigation of the efficacy of submarine outfall disposal of sewage and sludge.* Pub. No. 14, California Water Pollution Control Board, Sacramento, Calif.

Pomeroy, R. D. (1960). "The empirical approach for determining the required length of an ocean outfall." *Proc. 1st Conf. on Waste Disposal in Marine Environment*, University of California, Berkeley, Pergamon Press Ltd., London.

Rawn, A. M., and Palmer, H. K. (1930). "Predetermining the extent of a sewage field in sea water." *Trans. ASCE, 94*, 1036.

Rawn, A. M., Bowerman, F. R., and Brooks, N. H. (1960). "Diffusers for disposal of sewage in sea water." *J. San. Engrg. Div. Proc. ASCE, 86*(SA2), 65.

Wells, E. A., Jr., and Gotaas, N. (1958). "Design of Venturi flumes in circular conduits." *Trans. ASCE, 123*, 749.

CHAPTER 8

MATERIALS FOR SEWER CONSTRUCTION

8.1. INTRODUCTION

Pipe for sanitary sewer construction generally is manufactured from various basic materials in accordance with nationally recognized product specifications. Each type of sanitary sewer pipe, and its advantages and limitations, should be evaluated carefully in the selection of pipe materials for given applications.

Various factors are involved in the evaluation and selection of materials for sewer construction and are dependent on the anticipated conditions of service. Factors that may be involved and should be considered are:

- Intended use—type of wastewater.
- Scour or abrasion conditions.
- Installation requirements—pipe characteristics and sensitivities.
- Corrosion conditions—chemical, biological.
- Flow requirements—pipe size, velocity, slope, and friction coefficient.
- Infiltration/exfiltration requirements.
- Product characteristics—pipe size, fitting and connection requirements, laying length.
- Cost-effectiveness—materials, installation, maintenance, life expectancy.
- Physical properties—crush strength for rigid pipe, pipe stiffness or stiffness factor for flexible pipe, soil conditions, pipe beam loading strength, hoop strength for force main pipe, pipe shear loading strength, pipe flexural strength.
- Handling requirements—weight, impact resistance.

No single pipe product will provide optimum capability in every characteristic for all sanitary sewer design conditions. Specific application requirements should be evaluated prior to selecting or specifying pipe materials.

With the advancement of technology, new pipe materials are periodically being offered for use in sanitary sewer construction. Discussion of pipe materials provided in this chapter has been limited to the commonly accepted pipe materials currently available today for sanitary sewer applications in new construction. These products are listed below, alphabetically within the two commonly accepted classifications of rigid pipe and flexible pipe:

1. Rigid pipe
   a. Concrete pipe
   b. Vitrified clay pipe (VCP)
2. Flexible pipe
   a. Ductile iron pipe (DIP)
   b. Thermoplastic pipe
      i. Acrylonitrile-butadiene-styrene (ABS)
      ii. ABS composite
      iii. Polyethylene (PE)
      iv. Polyvinyl chloride (PVC)
   c. Thermoset plastic pipe
      i. Reinforced plastic mortar (RPM)
      ii. Reinforced thermosetting resin (RTR)
   d. Steel pipe

Materials utilized in rehabilitation of sanitary sewer lines include many of the above as well as some newer materials. These materials are discussed further in Chapter 12.

8.2. SEWER PIPE MATERIALS

8.2.1. Rigid Pipe

Sanitary sewer pipe materials in this classification derive a substantial part of their basic earth load carrying capacity from the structural strength inherent in the rigid pipe wall. Commonly specified rigid sanitary sewer pipe materials are discussed in the following subsections.

8.2.1.1. Concrete Pipe

Reinforced and nonreinforced concrete pipe and polymer concrete pipe are used for gravity sanitary sewers. Reinforced concrete pressure pipe and prestressed concrete pressure pipes are used for pressure sewers and force mains as well as sanitary gravity sewers. Nonreinforced concrete pipe is available in nominal diameters from 4 through 36 inches (100 through 900 mm). Reinforced concrete pipe is available in nominal diameters from 12 through 200 inches (300 mm through 5 m). Pressure pipe is available in diameters from 12 through 120 inches (300 mm through 3 m). Polymer concrete pipe used for gravity sewers is available in diameters from 6 through 144 inches (150 mm through 3.6 m). Polymer concrete pipe is also used for jacking and direct bury design. Concrete fittings and appurtenances such as wyes, tees, and manhole sections are generally available. A number of jointing methods are available depending on the tightness required and the operating pressure. Various linings and coatings are available.

A number of mechanical processes are used in the manufacture of concrete pipe. These processes use various techniques, including centrifugation, vibration, and packing and tamping for consolidating the concrete in forms. Gravity and pressure concrete pipe may be manufactured to any reasonable strength requirement by varying the wall thickness, concrete strength, and quantity and configuration of reinforcing steel or prestressing elements (ACPA 1994).

Potential advantages of concrete pipe include:

- Wide range of structural and pressure strengths.
- Wide range of nominal diameters.
- Wide range of laying lengths [generally 4 to 24 ft (1.2 to 7.4 m)].
- Allows direct installation for microtunnelling (without casing).

Potential disadvantages of concrete pipe include:

- High weight.
- Susceptible to corrosion where acids are present; linings are generally required (except for polymer concrete pipe).
- Susceptible to shear and beam breakage when improperly bedded.

Concrete pipe is normally specified by nominal diameter, class or D-load strength, and type of joint. The product should be manufactured in accordance with one or more of the following standard specifications:

- Concrete Sewer, Storm Drain, and Culvert Pipe, ANSI/ASTM C14
- Reinforced Concrete Culvert, Storm Drain, and Sewer Pipe, ANSI/ASTM C76
- Reinforced Concrete Arch Culvert, Storm Drain, and Sewer Pipe, ANSI/ASTM C506
- Reinforced Concrete D-Load Culvert, Storm Drain, and Sewer Pipe, ANSI/ASTM C655
- Reinforced Concrete Elliptical Culvert, Storm Drain, and Sewer Pipe, ANSI/ASTM C507

- Reinforced Concrete Low-Head Pressure Pipe, ANSI/ASTM C361
- Joints for Circular Concrete Sewer and Culvert Pipe, Using Rubber Gaskets, ANSI/ASTM C443
- External Sealing Bands for Non-Circular Concrete Sewer, Storm Drain, and Culvert Pipe, ANSI/ASTM C877
- Standard Specification for Manufacture of Reinforced Concrete Sewer, Storm Drain and Culvert Pipe for Direct Design, ASTM C1417

Additional information relative to the selection and design of concrete pipe may be obtained from the American Concrete Pipe Association (ACPA) Concrete Pipe Design Manual (ACPA 1994) and Concrete Pipe Handbook (ACPA 2000).

8.2.1.2. Vitrified Clay Pipe (VCP)

VCP is used for gravity sanitary sewers. The product is manufactured from clay and shale. Clay pipe is vitrified at a temperature at which the clay mineral particles become fused. The product is available in diameters from 3 through 36 inches (75 through 900 mm) and in some areas up to 42 inches (1.05 m). Clay fittings are available to meet most requirements, with special fittings manufactured on request. A number of jointing methods are available (NCPI 1995).

VCP is manufactured in standard and extra-strength classifications, although in some areas the manufacture of standard-strength pipe is not common in sizes 12-inch (300-mm) and smaller. The strength of VCP varies with the diameter and strength classification. The pipe is manufactured in lengths up to 10 ft (3 m).

Potential advantages of VCP include:

- High resistance to chemical corrosion.
- High resistance to abrasion.
- Wide range of fittings available.

Potential disadvantages of VCP include:

- Limited range of sizes available.
- High weight.
- Susceptible to shear and beam breakage when improperly handled or bedded.

VCP is specified by nominal pipe diameter, strength, and type of joint. The product should be manufactured and tested in accordance with one or more of the following standard specifications:

- Standard Specification for Vitrified Clay Pipe, Extra Strength, Standard Strength and Perforated, ANSI/ASTM C700

- Compression Joints for Vitrified Clay Pipe and Fittings, ASTM C425
- Pipe, Clay, Sewer, Federal Specification SS-P-361d
- Standard Methods of Testing Vitrified Clay Pipe, ANSI/ASTM 301

Additional information relative to the selection and design of VCP may be obtained from the National Clay Pipe Institute (NCPI) Clay Pipe Engineering Manual (NCPI 1995).

8.2.2. Flexible Pipe

Sanitary sewer pipe materials in this classification derive load-carrying capacity from the interaction of the flexible pipe and the embedment soils affected by the deflection of the pipe to the point of equilibrium under load. Commonly specified flexible sanitary sewer pipe materials are discussed below.

8.2.2.1. Ductile Iron Pipe (DIP)

DIP is used for gravity and pressure sanitary sewers and force mains. DIP is manufactured by adding cerium or magnesium to cast (gray) iron just prior to the pipe casting process. The product is available in nominal diameters from 4 through 64 inches (75 mm through 1350 mm) and in lengths to 20 ft (6.1 m). Cast iron (gray iron) or ductile iron fittings are used with DIP. Various jointing methods for the product are available.

DIP is manufactured in various thicknesses, classes, and strengths. Linings for the interior of the pipe (e.g., cement mortar lining with asphaltic coating, coal tar epoxies, epoxies, fusion-bonded coatings) may be specified. An exterior asphaltic coating and polyethylene exterior wrapping are also commonly specified (DIPRA 2003).

Potential advantages of DIP include:

- Long laying lengths (in some situations).
- High pressure and load-bearing capacity.
- High impact strength.
- High beam strength.

Potential disadvantages of DIP include:

- Susceptible to corrosion where acids are present.
- Susceptible to chemical attack in corrosive soils.
- High weight (however, additional sacrificial material is available to resist corrosion since the walls are thicker than necessary).

DIP is specified by nominal diameter, class, lining, and type of joint. DIP should be manufactured in accordance with one or more of the following standard specifications:

- Polyethylene Encasement for Gray and Ductile Cast-Iron Piping for Water and Other Liquids, ANSI A21.5 (AWWA C105)
- Ductile Iron Gravity Sewer Pipe, ASTM A746
- Gray-Iron and Ductile Iron Fittings, 3 Inch through 48 Inch, for Water and Other Liquids, ANSI/AWWA C110
- American National Standard for Rubber-Gasket Joints for Ductile-Iron Pressure Pipe and Fittings, ANSI/AWWA C111/A21.11
- American National Standard for Protective Fusion-Bonded Epoxy Coatings for the Interior and Exterior Surfaces of Ductile-Iron and Gray-Iron Fittings for Water Supply Service, ANSI/AWWA C116/A21.16
- American National Standard for the Thickness Design of Ductile-Iron Pipe, ANSI/AWWA C150/A21.50
- Cement Mortar Lining for Cast-Iron and Ductile-Iron Pipe and Fittings for Water, ANSI A21.4 (AWWA C104)

Additional information relative to the selection and design of DIP may be obtained from the Ductile Iron Pipe Research Association (DIPRA) at www.dipra.org.

8.2.2.2. Thermoplastic Pipe

Thermoplastic materials include a broad variety of plastics that can be repeatedly softened by heating and hardened by cooling through a temperature range characteristic for each specific plastic. Thermoplastic pipe product design should be based on long-term data. Generally, thermoplastic materials used in sanitary sewers are limited to acrylonitrile-butadiene-styrene (ABS), polyethylene (PE), and polyvinyl chloride (PVC).

8.2.2.2.1. Acrylonitrile-Butadiene-Styrene (ABS) Pipe

ABS pipe is used for both gravity and pressure sanitary sewers. Non-pressure-rated ABS sewer pipe is available in nominal diameters from 3 through 12 inches (75 through 300 mm) and in lengths up to 35 ft (11.2 m). A variety of ABS fittings and several jointing systems are available.

ABS pipe is manufactured by extrusion of ABS plastic material. ABS gravity sanitary sewer pipe is available in three dimension ratio (DR) classifications (23.5, 35, and 42) depending on nominal diameter selected. The classifications relate to three pipe stiffness values, PS 150, 45 and 20 psi (PS 1000, 300, and 150 kPa), respectively. The DR is the ratio of the average outside diameter to the minimum wall thickness of the pipe.

Potential advantages of ABS pipe include:

- Light weight.
- Long laying lengths (in some situations).
- High impact strength.
- Ease in field cutting and tapping.

Potential disadvantages of ABS pipe include:

- Limited range of sizes available.
- Subject to environmental stress cracking.
- Subject to excessive deflection when improperly bedded and haunched.
- Subject to attack by certain organic chemicals.
- Subject to surface change effected by long-term ultraviolet exposure.

ABS pipe is specified by nominal diameter, dimension ratio, pipe stiffness, and type of joint. ABS pipe should be manufactured in accordance with one or more of the following specifications:

- Acrylonitrile-Butadiene-Styrene (ABS) Sewer Pipe and Fittings, ANSI/ASTM D2751
- Solvent Cement for Acrylonitrile-Butadiene-Styrene (ABS) Plastic Pipe and Fittings, ANSI/ASTM D2235
- Elastomeric Seals (Gaskets) for Joining Plastic Pipe, ANSI/ASTM F477
- Joints for Drain and Sewer Plastic Pipes Using Flexible Elastomeric Seals, ANSI/ASTM D3212
- PVC and ABS Injected Solvent Cemented Plastic Pipe Joints, ANSI/ASTM F545

8.2.2.2.2. Acrylonitrile-Butadiene-Styrene (ABS) Composite Pipe

ABS composite pipe is used for gravity sanitary sewers. The product is available in nominal diameters from 8 through 15 inches (200 through 375 mm) and in lengths from 6.25 to 12.5 ft (2 to 4 m). ABS fittings are available for the product. The jointing systems available include elastomeric gasket joints and solvent cemented joints.

ABS composite pipe is manufactured by extrusion of ABS plastic material with a series of truss annuli that are filled with filler material such as lightweight Portland cement concrete.

Potential advantages of ABS composite pipe include:

- Light weight.
- Long laying lengths (in some situations).
- Ease in field cutting.

Potential disadvantages of ABS composite pipe include:

- Limited range of sizes available.
- Susceptible to environmental stress cracking.
- Susceptible to rupture when improperly bedded.
- Susceptible to attack by certain organic chemicals.
- Susceptible to surface change effected by long-term ultraviolet exposure.

ABS composite pipe is specified by nominal diameter and type of joint. ABS composite pipe should be manufactured in accordance with one or more of the following standard specifications:

- Acrylonitrile-Butadiene-Styrene (ABS) Composite Sewer Piping, ANSI/ASTM D2680
- Solvent Cement for Acrylonitrile-Butadiene-Styrene (ABS) Plastic Pipe and Fittings, ANSI/ASTM D2235
- Joints for Drain and Sewer Plastic Pipes Using Flexible Elastomeric Seals, ANSI/ASTM D3212
- Elastomeric Seals (Gaskets) for Joining Plastic Pipe, ANSI/ASTM F477.

8.2.2.2.3. Polyethylene (PE) Pipe

PE pipe is used for gravity and pressure sanitary sewers and force mains. Nonpressure PE pipe, primarily used for sewer relining, is available in nominal diameters from 4 through 48 inches (100 mm through 1200 mm). PE fittings are available. Jointing is primarily accomplished by butt-fusion or flanged adapters, with bell and spigot sometimes used.

PE pipe is manufactured by extrusion of PE plastic material. Nonpressure PE pipe is produced at this time in accordance with individual manufacturers' product standards.

Potential advantages of PE pipe include:

- Long laying lengths (in some situations).
- Light weight.
- High impact strength.
- Ease in field cutting.

Potential disadvantages of PE pipe include:

- Relatively low tensile strength and pipe stiffness.
- Limited range of diameters available.
- Susceptible to environmental stress cracking.
- Susceptible to excessive deflection when improperly bedded and haunched.

- Susceptible to attack by certain organic chemicals.
- Susceptible to surface change effected by long-term ultraviolet exposure.
- Special tooling required for fusing joints.

PE pipe is specified by material designation, nominal diameter (inside or outside), standard dimension ratios, and type of joint. PE pipe should be manufactured in accordance with one or more of the following specifications:

- Butt Heat Fusion Polyethylene (PE) Plastic Fittings for Polyethylene (PE) Plastic Fittings for Polyethylene (PE) Pipe and Tubing, ANSI/ASTM D3261
- Polyethylene (PE) Plastic Pipe (SDR-PR), ANSI/ASTM D2239
- Polyethylene (PE) Plastic Pipe (SDR-PR) Based on Controlled Outside Diameter, ASTM D3035
- Large Diameter Profile Wall Sewer Pipe, ASTM F894

8.2.2.2.4. Polyvinyl Chloride (PVC) Pipe

PVC pipe is used for both gravity and pressure sanitary sewers. Nonpressure PVC sewer pipe is available in nominal diameters from 4 through 48 inches (100 through 1,220 mm). PVC pressure and nonpressure fittings are available. PVC pipe is generally available in lengths up to 20 ft (6.1 m). Jointing is primarily accomplished with elastomeric seal gasket joints, although solvent cement joints for special applications are available (Uni-Bell 2001).

PVC pipe is manufactured by extrusion of the plastic material. Nonpressure PVC sanitary sewer pipe is provided in two dimension ratios (DR 35 and 26) that relate to two pipe stiffness values, PS 46 and 115 psi (PS 320 and 800 kPa), respectively.

Potential advantages of PVC pipe include:

- Light weight.
- Long laying lengths (in some situations).
- High impact strength.
- Ease in field cutting and tapping.

Potential disadvantages of PVC pipe include:

- Subject to attack by certain organic chemicals.
- Subject to excessive deflection when improperly bedded and haunched.
- Subject to surface changes effected by long-term ultraviolet exposure.

PVC pipe is specified by nominal diameter, dimension ratio, pipe stiffness, and type of joint. PVC pipe should be manufactured in accordance with one or more of the following standard specifications:

- Type PSM Poly(Vinyl Chloride) (PVC) Sewer Pipe and Fittings, ANSI/ASTM D3034
- Type PSP Poly(Vinyl Chloride) (PVC) Sewer Pipe and Fittings, ANSI/ASTM D3033
- Elastomeric Seals (Gaskets) for Joining Plastic Pipe, ANSI/ASTM F477
- Joints for Drain and Sewer Plastic Pipes Using Flexible Elastomeric Seals, ANSI/ASTM D3212
- PVC and ABS Injected Solvent Cemented Plastic Pipe Joints, ANSI/ASTM F545
- Solvent Cements for Poly(Vinyl Chloride) (PVC) Plastic Pipe and Fittings, ANSI/ASTM D2564
- Standard Specification for Polyvinyl Chloride (PVC) Large Diameter Plastic Gravity Sewer Pipe and Fittings, ASTM F679
- Polyvinyl Chloride (PVC) Profile Gravity Sewer Pipe & Fittings Based on Controlled Inside Diameter, ASTM F794
- Polyvinyl Chloride (PVC) Closed Profile Gravity Sewer Pipe & Fittings Based on Controlled Inside Diameter, ASTM F1803

Additional information relative to the selection and design of PVC pipe may be obtained from the Uni-Bell Handbook of PVC Pipe—Design and Construction (Uni-Bell 2001).

8.2.2.3. *Thermoset Plastic Pipe*

Thermoset plastic materials include a broad variety of plastics. These plastics, after having been cured by heat or other means, are substantially infusible and insoluble. Thermoset plastic pipe product design should be based on long-term data. Generally, thermoset plastic materials used in sanitary sewers are provided in two categories—reinforced plastic mortar (RPM) and reinforced thermosetting resin (RTR).

8.2.2.3.1. Reinforced Plastic Mortar (RPM) Pipe

RPM pipe is used for both gravity and pressure sewers. RPM pipe is available in nominal diameters from 8 through 144 inches (200 mm through 3600 mm). In smaller diameters, RPM fittings are generally available; in larger diameters, RPM fittings are manufactured as required. A number of jointing methods are available. Various methods of interior protection (e.g., thermoplastic or thermosetting liners or coatings) are available.

RPM pipe is manufactured containing fibrous reinforcements, such as fiberglass, and aggregates, such as sand embedded in or surrounded by cured thermosetting resin.

Potential advantages of RPM pipe include:

- Light weight.
- Long laying lengths (in some situations).

Potential disadvantages of RPM pipe include:

- Susceptible to strain corrosion in some environments.
- Susceptible to excessive deflection when improperly bedded and haunched.
- Susceptible to attack by certain organic chemicals.
- Susceptible to surface change effected by long-term ultraviolet exposure.

RPM pipe is specified by nominal diameter, pipe stiffness, stiffness factor, beam strength, hoop tensile strength, lining or coating, thermosetting plastic material, and type of joint. RPM pipe should be manufactured in accordance with one or more of the following standard specifications:

- Reinforced Plastic Mortar Sewer Pipe, ANSI/ASTM D3262
- Reinforced Plastic Mortar Sewer and Industrial Pressure Pipe, ASTM D3754

8.2.2.3.2. Reinforced Thermosetting Resin (RTR) Pipe

RTR pipe is used for both gravity and pressure sanitary sewers. RTR pipe is generally available in nominal diameters from 1 through 12 inches (25 through 300 mm), manufactured in accordance with ASTM standard specifications. The product is also available in nominal diameters from 12 through 144 inches (300 mm through 3600 mm), manufactured in accordance with individual manufacturers' specifications. In small diameters, RTR fittings are available; in larger diameters, RTR fittings are manufactured as required. A number of jointing methods are available, including flush bell and spigot. Various methods of interior protection (e.g., thermoplastic or thermosetting liners or coatings) are available.

RTR pipe is manufactured using a number of methods, including centrifugal casting, pressure laminating, and filament winding. In general, the product contains fibrous reinforcement materials, such as fiberglass, embedded in or surrounded by cured thermosetting resin.

Potential advantages of RTR pipe include:

- Light weight.
- Long laying lengths (in some situations).

Potential disadvantages of RTR pipe include:

- Susceptible to strain corrosion in some environments.
- Susceptible to excessive deflection when improperly bedded and haunched.
- Susceptible to attack by certain organic chemicals.
- Susceptible to surface change effected by long-term ultraviolet exposure.
- Higher cost than PVC.

RTR pipe is specified by nominal diameter, pipe stiffness, lining and coating, method of manufacture, thermosetting plastic material, and type of joints. RTR pipe should be manufactured in accordance with one or more of the following standard specifications:

- Filament-Wound Reinforced Thermosetting Resin Pipe, ASTM D2996
- Centrifugally Cast Reinforced Thermosetting Resin Pipe, ANSI/ASTM D2997
- Machine-Made Reinforced Thermosetting Resin Pipe, ASTM D2310

8.2.2.4. Corrugated Steel Pipe

Steel pipe is rarely used for sanitary sewers. When used, it usually is specified with interior protective coatings or linings (polymeric, bituminous, asbestos, etc.). It is not recommended for use in sanitary sewer systems. Steel pipe is fabricated in diameters from 8 through 120 inches (200 mm through 3 m). Appurtenances include tees, wyes, elbows, and manholes fabricated from steel. Various linings and coatings are available (AISI 1994). Steel is generally manufactured in lengths up to 40 ft (12.2 m).

Potential advantages of steel pipe include:

- Light weight.
- Long laying lengths (in some situations).

Potential disadvantages of steel pipe include:

- Susceptible to corrosion where acids are present.
- Susceptible to chemical attack in corrosive soils.
- Difficulty in making lateral connections.
- Poor hydraulic coefficient (unlined corrugated steel pipe).
- Susceptible to excessive deflection when improperly bedded or haunched.
- Susceptible to turbulence abrasion.

Steel pipe is specified by size, shape, wall profile, gage or wall thickness, and protective coating or lining.

8.3. PIPE JOINTS

8.3.1. General Information

The requirement for the control of ground water infiltration and wastewater exfiltration in sanitary sewer systems renders the specification of pipe joint design essential to proper sanitary sewer design. A substantial variety of pipe joints are available for the different pipe materials used in sanitary sewer construction. A common requirement that must be imposed on the design of all sanitary sewer systems, regardless of the type of sewer pipe specified, is the use of reliable, tight pipe joints. A good pipe joint must be watertight, root-resistant, flexible, durable, and easily assembled in the field. In current practice, various forms of gasket (elastomeric seal) pipe joints are used in sanitary sewer construction. They generally can be assembled by unskilled labor in a broad range of weather conditions and environments with good assurance of a reliable, tight seals.

Water infiltration/exfiltration testing or air exfiltration testing is commonly specified for typical nonpressure sanitary sewer system construction to demonstrate that infiltration/exfiltration is within acceptable limits. (The subject of infiltration/exfiltration testing is discussed in Chapter 6.)

8.3.2. Types of Pipe Joints

Commonly specified sanitary sewer pipe joints are described in the following subsections.

8.3.2.1. Gasket Pipe Joints

Gasket joints effect a seal against leakage through compression of an elastomeric seal or ring. Gasket pipe joint design is generally divided into two types: push-on pipe joint and mechanical compression pipe joint.

8.3.2.1.1. Push-on Pipe Joint

This type of pipe joint uses a continuous elastomeric ring gasket that is compressed into an annular space formed by the pipe, fitting or coupler socket, and the spigot end of the pipe, thereby providing a positive seal when the pipe spigot is pushed into the socket. When using this type of pipe joint in pressure sanitary sewers, thrust restraint may be required to prevent joint separation under pressure. Push-on pipe joints (fittings, couplers, or integral bells) are available on nearly all pipe products mentioned.

8.3.2.1.2. Mechanical Compression Pipe Joint

This type of pipe joint uses a continuous elastomeric ring gasket that provides a positive seal when the gasket is compressed by means of a mechanical device. When using this type of pipe joint in pressure sanitary sewers, thrust restraint may be required to prevent joint separation under pressure. This type of pipe joint may be provided as an integral part of DIP. When incorporated into a coupler, this type of pipe joint may be used to join two similarly sized plain spigot ends of any commonly used sewer pipe materials.

8.3.2.2. Cement Mortar Pipe Joint

This type of pipe joint involves use of shrink-compensating cement mortar placed into a bell-and-spigot pipe joint to provide a seal. The use of this joint is discouraged in that reliable, watertight joints are not assured. Cement mortar joints are not flexible and may crack due to any pipe movement. This type of joint is used on concrete gravity and pressure pipe to protect otherwise exposed steel.

8.3.2.3. Elastomeric Sealing Compound Pipe Joints

Elastomeric sealing compound may be used in jointing properly prepared concrete gravity sanitary sewer pipe. Pipe ends must be sandblasted and primed for elastomeric sealant application. The sealant—a thixotropic, two-compound elastomer—is mixed on the job site and applied with a caulking gun and spatula. The pipe joint, when assembled with proper materials and procedures, provides a positive seal against leakage in gravity sewer pipe.

8.3.2.4. Solvent Cement Pipe Joints

Solvent cement pipe joints may be used in jointing thermoplastic pipe materials such as ABS, ABS composite, and PVC pipe. This type of pipe joint involves bonding a sewer pipe spigot into a sewer pipe bell or coupler using solvent cement. Solvent cement joints can provide a positive seal provided the proper cement is applied under proper ambient conditions with proper techniques. Reference should be made to ASTM F402 for safe handling procedures. Required precautions should be taken to ensure adequate trench ventilation and protection for workers installing the pipe. Solvent cement pipe joints may be desired in special situations and with some plastic fittings.

8.3.2.5. Heat Fusion Pipe Joints

Heat fusion pipe joints are commonly specified for PE sanitary sewer pipe and are now being used on some PVC pipe. The general method of jointing PE sanitary sewer pipe involves butt fusion of the pipe lengths, end-to-end. After the ends of two lengths of PE pipe are trimmed and softened to a melted state with heated metal plates, the pipe ends are forced together to the point of butt fusion, providing a positive seal (Plastics Pipe Institute, undated). The pipe joint does not require thrust restraint in pressure applications. Trained technicians with special apparatus are required to achieve reliable, watertight pipe joints.

8.3.2.6. *Mastic Pipe Joints*

Mastic pipe joints are frequently used for special, nonround shapes of concrete gravity sewer pipe that are not adaptable for use with gasket pipe joints. The mastic material is placed into the annular space to provide a positive seal. Application may be by troweling, caulking, or by the use of preformed segments of mastic material m a manner similar to gaskets. Satisfactory performance of the pipe joints depends upon the proper selection of primer, mastic material, and good workmanship in application.

8.3.2.7. *Sealing Band Joints*

External sealing bands of rubber made in conformance with ASTM C877 are also used on noncircular concrete sewer pipe. These elastomeric bands are wrapped tightly around the exterior of the pipe at the joint and extend several inches (centimeters) on each side of the joint. Sealing against the concrete is achieved by mastic applied to one side of the band.

8.4. SUMMARY

With consideration of appropriate factors essential to the proper design of sanitary sewer systems, it is apparent that the broad variety of materials for sanitary sewer construction available is a source of significant benefit. Each sewer pipe product and each type of pipe joint offers distinct advantages that can prove to be specifically beneficial in given applications. Each sewer pipe product and each type of pipe joint also presents limitations that must be understood for proper system design for given applications.

REFERENCES

American Iron and Steel Institute (AISI). (1994). *Handbook of steel drainage and highway construction products*, AISI, Washington, D.C.

American Concrete Pipe Association (ACPA). (2000). *Concrete pipe handbook*, ACPA, Irving, Tex.

ACPA. (1994). *Concrete pipe design manual*, ACPA, Irving, Tex. Available on the ACPA web site www.concrete-pipe.org/designmanual.htm, accessed November 7, 2006.

ASTM International (ASTM). (2005). "Safe handling of solvent cements used for joining thermoplastic pipe and fittings," *ASTM F402*, ASTM, Philadelphia, Penn.

Committee of the Great Lakes-Upper Mississippi River Board of State Sanitary Engineers (GLUMRB). (2006). *Recommended standards for sewage works*, GLUMRB Health Education Services, Albany, N.Y.

Ductile Iron Pipe Research Association (DIPRA). (2003). *Handbook—Ductile iron pipe*, DIPRA, Birmingham, Ala. Available on the DIPRA web site www.dipra.org, accessed November 7, 2006.

Institute for Hydromechanic and Hydraulic Structures (IHHS). (1973). *Wear data of different pipe materials at sewer pipelines*, IHHS, Technical University of Darmstadt, Darmstadt, West Germany.

National Clay Pipe Institute (NCPI). (1995). *Clay pipe engineering manual*, NCPI, Washington, D.C. Available on the NCPI web site www.ncpi.org/engineer.htm, accessed November 7, 2006.

Plastics Pipe Institute. (Undated). "Underground installation of polyethylene piping," New York, N.Y.

Uni-Bell PVC Pipe Association (Uni-Bell). (2001). *Handbook of PVC pipe—Design and construction*, Uni-Bell, Dallas, Tex. Available on the Uni-Bell web site www.uni-bell.org/pubs/handbook.pdf, accessed November 7, 2006.

CHAPTER 9
STRUCTURAL REQUIREMENTS

9.1. GENERAL

The structural design and load analysis of buried sewers is essentially a problem of soil–structure interaction. The influence of each factor is dependent on the stiffness of the structure relative to the soil, and two major categories of pipe have been defined:

- Rigid pipe
- Flexible pipe

Pipe materials in the rigid category may include concrete, reinforced concrete, clay, fiber cement, and cast-in-place pipe. Pipe materials in the flexible category may include thermoplastic plastic, thermosetting plastic, corrugated steel, corrugated aluminum, and ductile iron. Flexible pipe has historically been designed for the installed condition, whereas rigid pipe has been designed for a test load condition and then correlated with an assumed earth pressure distribution for the installed condition. However, advances in analytical methods and computer technology enable both categories of pipe to be designed for the installed condition.

The theory of loads and supporting strength of buried pipe was developed more than 70 years ago and refined over the years as additional data became available. The theory referred to as the Marston-Spangler Theory is based on a detailed earth load analysis for specifically defined installations and a semi-empirical, simplified interpretation of earth pressure distribution around the pipe for each type of installation. For rigid pipe, the load-carrying capacity of the pipe is established by an in-plant, three-edge-bearing (T.E.B.) test load. Under this test load condition, the pipe is subjected to a concentrated line load with high stresses induced within the pipe wall. This test load condition is then correlated with the more favorable and less severe installed load condition by means of a bedding factor. The bedding factor values developed under the Marston-Spangler Theory do not reflect current construction practices and methods. Designing directly for the installed condition enables evaluation of the actual stresses induced in the pipe for a known external pressure distribution; then, by applying the principles of engineering mechanics, the pipe can be structurally designed based on stress analysis similar to the design of all other structures.

The designs of rigid and flexible pipes are treated separately in this Manual. There are no specific design procedures given for flexible pipes of intermediate stiffness. For such cases, design procedures (such as computer analysis based on soil–structure interaction or the designs for rigid or flexible pipe) may be used (not interchangeably) for conservative results.

Because installation conditions have such an important effect on both load and supporting strength, a satisfactory sewer construction project requires accurate assumed design conditions for the job site. Therefore, this chapter also includes recommendations for construction and field observations to obtain this goal.

This chapter does not include information of design of cast-in-place reinforced concrete sewer pipe sections. Reference should be made to standard textbooks and to American Concrete Institute (ACI)/ASTM specifications or industry handbooks for such design data.

9.2. LOADS ON SEWERS CAUSED BY GRAVITY EARTH FORCES

9.2.1. Earth Loads

Sanitary sewers are normally installed in a relatively narrow trench excavated in undisturbed soil and then covered with backfill to the proposed ground level. In this type of installation, the backfill material will tend to settle downward relative to the undisturbed soil in which the trench is excavated. This relative settlement of backfill generates upward friction forces, causing an arching action of the backfill which relieves the load on the pipe, such that the load is equal to the weight of the prism of backfill within the trench minus the upward friction forces. For narrow trenches the backfill load is less than the weight of the prism of backfill over the pipe. As the trench width is increased for any given pipe size, backfill material, and trench depth, a limiting width is reached beyond which the trench width no longer affects the backfill load. The trench width at which this condition occurs is defined as the transition width. Once the transition width is realized, the backfill load is a maximum and remains constant regardless of any further increase in the trench width.

For the Direct Design procedure described below, a detailed earth load analysis is not conducted but, rather, the backfill load is related to the weight of the prism of backfill over the pipe multiplied by a vertical arching factor, varying from 1.35 to 1.45.

The backfill or fill load on buried pipe depends on the installation, and standard installations have been defined for rigid pipe and flexible pipe. The earth load on flexible pipe is determined as equal to the weight of the prism of earth over the pipe and is given by the equation:

$$W_e = wHD_0 \tag{9-1}$$

where

W_e = the vertical earth load, lb/ft (N/m)
w = unit weight of soil, lb/ft^3 (kg/m^3)
H = design height of earth above top of pipe, ft (m)
D_0 = outside diameter of pipe, ft (m).

The earth load on rigid pipe is determined as equal to the weight of the prism of earth over the pipe increased by a vertical arching factor (VAF) from 1.35 to 1.45, depending on the standard installation type, and is given by the equation:

$$W_e = VAF \times [wD_0] \times [H + (0.107D_0)] \tag{9-2}$$

The variables are the same as for Eq. (9-1). Figure 9-1 shows the difference between these two loading conditions.

9.2.2. Marston-Spangler Load Analysis

9.2.2.1. General

The loads given above may be used with the various design methods outlined by the pipe manufacturing associations and by ASCE to calculate required pipe strengths. Many times these calculations are integrated into

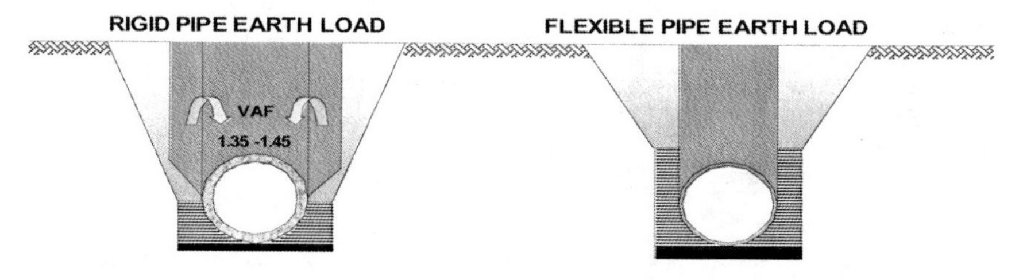

FIGURE 9-1. Earth loads on pipe.

a computer program designed for a particular piping material. These methods, which are specific to various piping organizations, are outlined in greater detail below.

Although these Direct Design methods are typically preferred, there may be times when the older Marston-Spangler load analysis may be preferred. Marston developed methods for determining the vertical load on buried conduits caused by soil forces in all of the most commonly encountered construction conditions (Marston and Anderson 1913; Marston 1930). These methods are historically based on both theory and experiment and have generally achieved acceptance as being useful and reliable, although perhaps overly conservative.

In general, the theory states that the load on a buried pipe is equal to the weight of the prism of soil directly over it, called the interior prism, plus or minus the frictional shearing forces transferred to that prism by the adjacent prisms of soil—the magnitude and direction of the frictional forces being a function of the relative settlement between the interior and adjacent soil prisms. The theory makes the following assumptions:

- The calculated load is the load that will develop when ultimate settlement has taken place.
- The magnitude of the lateral pressures that induce the shearing forces between the interior and adjacent soil prisms is computed in accordance with Rankin's theory.

The general form of Marston's equation is:

$$W = C\omega B^2 \tag{9-3}$$

in which W is the vertical load per unit length acting on the sewer pipe because of gravity soil loads, ω is the unit weight of soil; B is the trench width or sewer pipe width, depending on installation conditions; and C is a dimensionless coefficient that marries the effect of the following variables:

- The ratio of the height of fill to width of trench or sewer pipe.
- The shearing forces between interior and adjacent soil prisms.
- The direction and amount of relative settlement between interior and adjacent soil prisms for embankment conditions.

9.2.2.2. *Types of Loading Conditions*

Although the general form of Marston's equation includes all the factors necessary to analyze all types of installation conditions, it is convenient to classify these conditions, write a specialized form of the equation, and prepare separate graphs and tables of coefficients for each. In performing a Marston-Spangler load analysis, there are three types of open-cut installation methods. These are:

- Trench.
- Negative-projecting embankment.
- Positive-projecting embankment.

Figure 9-2 shows the various types of installation.

Trench conditions are defined as those in which a sewer pipe is installed in a relatively narrow trench cut in undisturbed ground and covered with soil backfill to the original ground surface.

Embankment conditions are defined as those in which the sewer pipe is covered above the original ground surface or when a trench in undisturbed soil is so wide that trench wall friction does not affect the load on the sewer pipe. The embankment classification is further subdivided into two major subclassifications—positive-projecting and negative-projecting. Sewer pipes are defined as positive-projecting when the top of the sewer pipe is above the original adjacent ground surface. Negative-projecting sewer pipe is that installed with the top of the sewer pipe below the adjacent original ground surface in a trench which is narrow with respect to the size of pipe and the depth of cover, as shown on Fig. 9-2, and when the

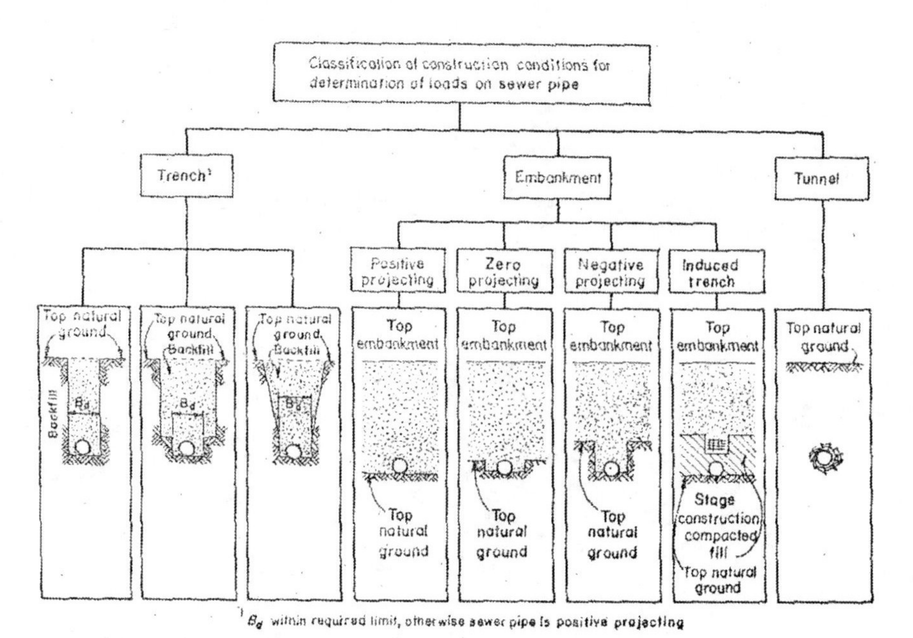

FIGURE 9-2. *Classification of construction conditions.*

native material is of sufficient strength that the trench shape can be maintained dependably during placement of the embankment.

A special case, called the induced trench condition, may be employed to minimize the load on a conduit under an embankment of unusual height.

9.2.2.3. *Loads for Trench Conditions*

This type of installation is normally used in the construction of sewers, drains, and water mains. The pipe is installed in a relatively narrow trench excavated in undisturbed soil and then covered with backfill extending to the ground surface. The vertical soil load to which a sewer pipe in a trench is subjected is the resultant of two major forces. The first is produced by the mass of the prism of soil within the trench. The second is the friction or shearing forces generated between the undisturbed soil outside the trench and the backfill placed within the trench.

The backfill soil has a tendency to settle in relation to the undisturbed soil in which the trench is excavated. This downward movement or tendency for movement induces upward shearing forces which support a part of the weight of the backfill. Thus, the resultant load on the horizontal plane at the top of the sewer pipe within the trench is equal to the weight of the backfill minus these upward shearing forces, as illustrated in Fig. 9-3.

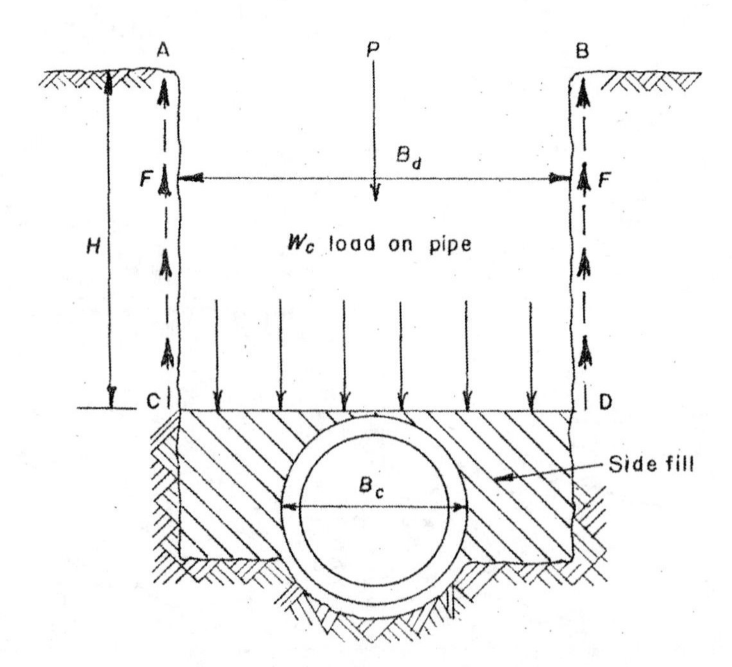

FIGURE 9-3. Load-producing forces. P, *Weight of backfill ABCD;* F, *Upward shearing forces on AC and BD; and* $W_c = P - 2F$.

Unusual conditions may be encountered in which poor natural soils may effect a change from trench to embankment conditions with considerably increased load on the sewer pipe. This is covered in the next subsection.

In general, the weight of backfill is given by:

$$W_c = C_d w B_d^2 \tag{9-4}$$

where

W_c = load, lb/ft (N/m)
w = density of backfill material, lb/ft^3 (kg/m^3)
B_d = width of trench at top of pipe, ft (m)

and

$$C_d = \frac{1 - e^{-2K\mu' \frac{H}{B_d}}}{2K\mu'} \tag{9-5}$$

For this equation, H is the height of backfill material above the pipe (feet or meters); K = Rankine's ratio of active lateral unit pressure to vertical unit pressure; and $\mu' = \tan \phi'$, the coefficient of friction between fill material and the sides of the trench. Rankine's ratio is defined as:

$$K = \frac{\sqrt{\mu^2 + 1} - \mu}{\sqrt{\mu^2 + 1} + \mu} = \frac{1 - \sin \phi}{1 + \sin \phi} \tag{9-6}$$

where $\mu = \tan \phi$ = the coefficient of internal friction of backfill material (μ' may be equal to or less than μ, but never greater than μ.) The value of C_d for various ratios of H/B_d and various types of soil backfill may be obtained from Fig. 9-4.

Typical values of $K\mu'$ are:

$K\mu'$ = 0.1924 Max. for granular materials without cohesion
$K\mu'$ = 0.165 Max. for sand and gravel
$K\mu'$ = 0.150 Max. for saturated top soil
$K\mu'$ = 0.130 Max. for ordinary clay
$K\mu'$ = 0.110 Max. for saturated clay.

The trench load formula, Eq. (9-4), gives the total vertical load on a horizontal plane at the top of the sewer pipe. If the sewer pipe is rigid, it will carry practically all of this load. If the sewer pipe is flexible and the soil at the sides is compacted to the extent that it will deform under vertical load

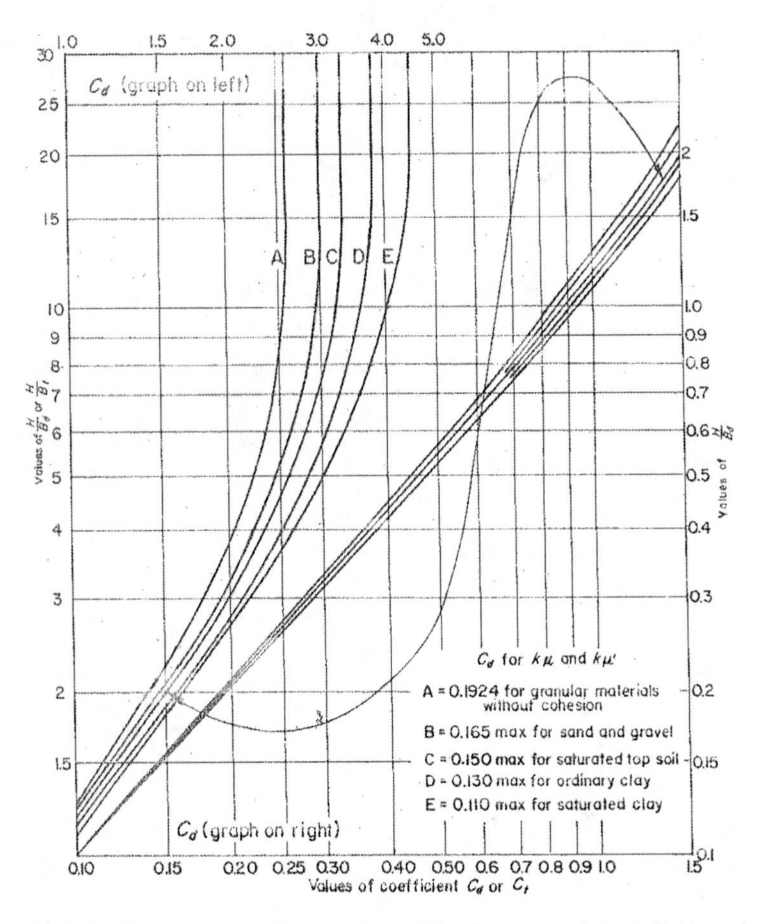

FIGURE 9-4. Computation diagram for soil loads on trench installations (sewer pipe completely buried in trench).

less than the sewer pipe itself will deform, the side fills may be expected to carry their proportional share of the total load. Under these circumstances, the trench load formula may be modified to:

$$W_c = C_d w B_c B_d \tag{9-7}$$

in which B_c is the outside width of the pipe, in feet or meters. It must be emphasized that Eq. (9-7) is applicable only if the backfill is compacted as described above. The equation should not be used merely because the pipe is a flexible type.

The term "side fill" refers to the soil backfill that is placed between the sides of sewer pipe and the sides of the trench. The character of this material and the manner of its placement have two important influences on the structural behavior of a sewer pipe.

First, the side fill may carry a part of the total vertical load on the horizontal plane at the elevation of the top of the sewer pipe. Second, the side fill plays an important role in helping the sewer pipe carry vertical load. Every pound (newton) of force that can be brought to bear against the sides of an elastic ring increases the ability of the ring to carry the vertical load by nearly the same amount.

Examination of Eq. (9-4) indicates the important influence the width of the trench exerts on the load as long as the trench condition formula applies. This influence has been verified by extensive experimental evidence. These experiments have also indicated that the width of the trench at the top of the sewer pipe is the controlling factor.

The width of the trench below the top of the sewer pipe is also important. It must not be permitted to exceed the safe limit for the strength of the sewer pipe and class of bedding used. The minimum width must be consistent with the provision of sufficient working space at the sides of the sewer pipe to assemble joints properly, to insert and strip forms, and to compact backfill. The engineer must establish and allow reasonable tolerances in width for variations in field conditions and accepted construction practice.

As defined previously, the transition width occurs at a certain limiting value at which the backfill load reaches a maximum and remains constant, regardless of any increase in the width of the trench. There are sufficient experimental data at hand to show that it is safe to calculate the imposed load by means of the trench-conduit formula [Eq. (9-4)] for all widths of trench less than that which gives a load equal to the load calculated by the projecting-conduit formula (discussed in Section 9.2.2.4. on embankment conditions). In other words, as the width of the trench increases (other factors remaining constant), the load on a rigid sewer pipe increases in accordance with the theory for a trench sewer pipe until it equals the load determined by the theory for a projecting sewer pipe. The width of the trench at which this transition occurs may be determined from Fig. 9-5 (Schlick 1932). The curves in Fig. 9-5 are calculated for sand and gravel, where $K\mu' = 0.165$, but can be used for other types of soil because the change with varying values of $K\mu$ is small. In any event, the engineer can check by calculating loads for both trench and embankment conditions. There is little research on the appropriate value of $r_{sd}p$ (the projection ratio times the settlement ratio) to use in the application of the transition width concept. In the absence of specific information, a value of $+0.5$ is suggested as a reasonably good working value. These quantities are defined in the subsection on loads for embankment conditions.

It is advisable, in the structural design of sewers, to evaluate the effect of the transition width on both the design criteria and the construction

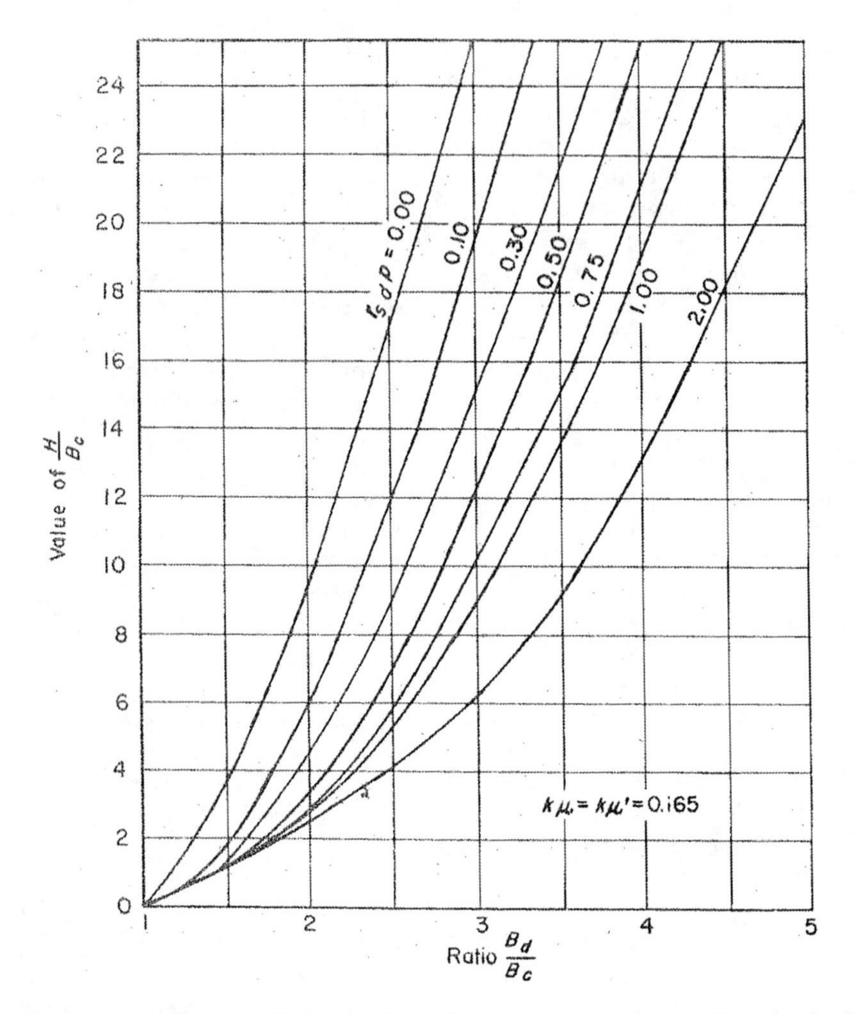

FIGURE 9-5. Values of B_d/B_c *at which pipes in trench and projecting pipe load formulas have equal loads.*
Schlick, W. J. (1932). "Loads on pipe in wide ditches." Bulletin 108, Iowa Engineering Experiment Station, Ames, Iowa, courtesy of American Concrete Pipe Association, Irving, Tex.

latitude. A contractor may wish, for example, to place well points for drainage in the trench. In another scenario, the trench walls may need to be sloped to meet Occupational Safety and Health Administration (OSHA) requirements for trench safety. If either of these scenarios requires a wider trench than usual, a stronger sewer pipe or higher class of bedding may be necessary.

To comply with OSHA requirements, it may be advantageous to excavate the trench with sloping sides in undeveloped areas where no inconvenience to the public or danger to property, buildings, subsurface structures, or pavements, will result. A subtrench, as shown in Fig. 9-6, may be used in such cases to minimize the load on the pipe. When sheeting of the subtrench at the pipe is necessary, it should extend about 1.5 ft (0.5 m) above the top of the pipe.

Loads on sewer pipe in sheeted trenches should be calculated from a trench width measured to the outside of the sheeting if the sheeting is pulled, or to the inside if it is left in place. Voids created by removal of the sheeting should be backfilled with a flowable material such as pea gravel or flowable concrete fill.

If a shield or trench box is used in sewer pipe-laying operations, the shield or box width controls the width of the trench at the top of the sewer pipe. This width, with a small addition for the space needed to advance the shield without a large friction loss, should be the width factor used in computing loads on the sewer pipe. Extreme care must be taken when advancing the shield in the trench to prevent pipe joints from pulling apart and disturbance of the pipe bedding.

Sanitary sewers that are to be constructed in sloping-side trenches with the slopes extending to the invert or to any plane above the invert, but below the top of the sewer, should be designed for loads computed by using the actual width of the trench at the top of the sewer pipe, or by the projecting sewer formula (covered in Section 9.2.2.4.)—whichever gives the least load on the sewer pipe.

If for any reason the trench becomes wider than that specified and for which the sewer pipe was designed, the load on the sewer pipe should be checked and a stronger sewer pipe or higher class of bedding should be used, if necessary.

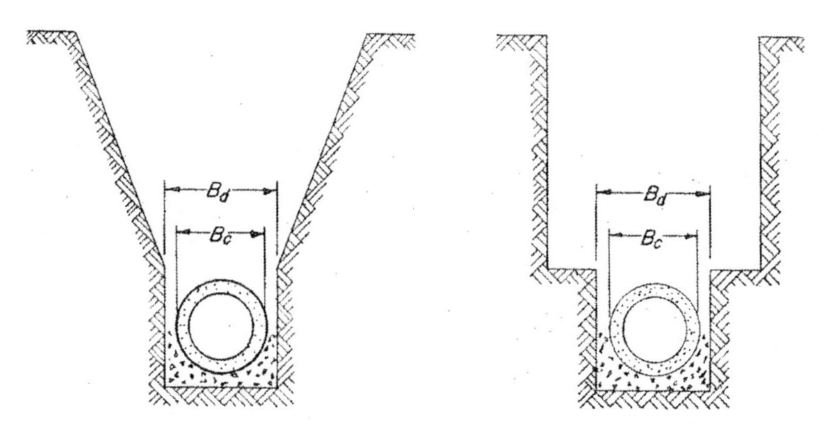

FIGURE 9-6. Examples of subtrenches.
Courtesy of American Concrete Pipe Association, Irving, Tex.

Sample Calculations

Example 9-1. Determine the load on a 24-inch-diameter rigid sewer pipe under 14 ft of cover in trench conditions.

Assume that the sewer pipe wall thickness is 2 inches; $B_c = 24 + 4 =$ 28 inches $= 2.33$ ft; $B_d = 2.33 + 2.00 = 4.33$ ft; and $w = 120$ lb/ft^3 for saturated top soil backfill. Then $H/B_d = 14/4.33 = 3.24$; C_d (from Fig. 9-4) $=$ 2.1; and $W_c = 2.1 \times 120 \times (4.33)^2 = 4{,}720$ lb/ft (68,880 N/m).

Example 9-2. Determine the load on the same-sized sewer laid on a concrete cradle and with trench sheeting to be removed.

Assume that the wall thickness is 2 inches; the cradle projection outside of the sewer pipe is 8 inches (4 inches on each side); and the maximum clearance between cradle and outside of sheeting is 14 inches. Then $B_d =$ $24 + (2 \times 2$ inches$) + 8 + (2 \times 14) = 64$ inches $= 5.33$ ft.

As this seems to be an extremely wide trench, a check should be made on the transition width of the trench; $B_c = 2.33 = 2.33$ ft; $H = 14$ ft; $r_{sd}p =$ 0.5; and $H/B_c = 14/2.33 = 6$.

From Fig. 9-5, $B_d/B_c = 2.39$ (the ratio of the width of the trench to the width of the sewer at which loads are equal by both trench sewer theory and projecting-sewer theory); $B_d = 2.33 \times 2.39 = 5.57 > 5.33$; $H/B_d =$ $14/5.33 = 2.63$; C_d (from Fig. 9-4) $= 1.85$; and $W_c = 1.85 \times 120 \times (5.33)^2 =$ 6,300 lb/ft (91,700 N/m).

Example 9-3. Determine the load on the same sewer if (rough) sheeting is left in place.

B_d becomes 4 inches less $= 5$ ft; $H/B_d = 14/5 = 2.8$; C_d (from Fig. 9-4) $=$ 1.92; and $W_c = 1.92 \times 120 \times (5)^2 = 5{,}750$ lb/ft (84,040 N/m).

Example 9-4. Determine the load on a 30-inch-diameter flexible sewer pipe installed in a trench 4 ft, 6 inches wide at a depth of 12 ft.

Assume the soil is clay weighing 120 lb/ft^3 and that it will be well-compacted at the sides of the sewer pipe. Then $H = 12$ ft; $B_d = 4.5$ ft; $B_c =$ 2.5 ft; $H/B_d = 2.67$; $C_d = 1.9$; and $W_c = 1.9 \times 120 \times 4.5 \times 2.5 = 2{,}565$ lb/ft (37,450 N/m).

For conservative design, the prism load should be determined. The prism load on flexible sewer pipe will be $W = 2.5 \times 12 \times 120 = 3{,}600$ lb/ft (52,460 N/m).

9.2.2.4. Loads for Positive-Projecting Embankment Conditions

This type of installation is normally used when the pipe is installed in a relatively flat stream bed or drainage path. The pipe is installed on the original ground or compacted fill, and then covered by an earth fill or embankment.

The load on a positive-projecting sewer pipe is equal to the weight of the prism of soil directly above the structure, plus (or minus) vertical shearing forces which act on vertical planes extending upward into the embankment from the sides of the sewer pipe. For an embankment installation of sufficient height, these vertical shearing forces may not extend to the top of the embankment, but terminate in a horizontal plane at some elevation above the top of the sewer pipe known as the "plane of equal settlement," as shown in Fig. 9-7.

The shear increment acts downward when $(s_m + s_g) > (s_f + d_c)$ and vice versa. In this expression, s_m is the compression of the columns of soil of height pB_c; s_g is the settlement of the natural ground adjacent to the sewer pipe; s_f is the settlement of the bottom of the sewer pipe; and d_c is the deflection of the sewer pipe.

The location of the plane of equal settlement is determined by equating the total strain in the soil above the pipe to that in the side fill plus the settlement of the critical plane. When the plane of equal settlement is an imaginary plane above the top of the embankment (i.e., shear forces extend to the top of the embankment), the installation is called either "complete trench condition" or "complete projection condition," depending on the direction of the shear forces. When the plane of equal settlement

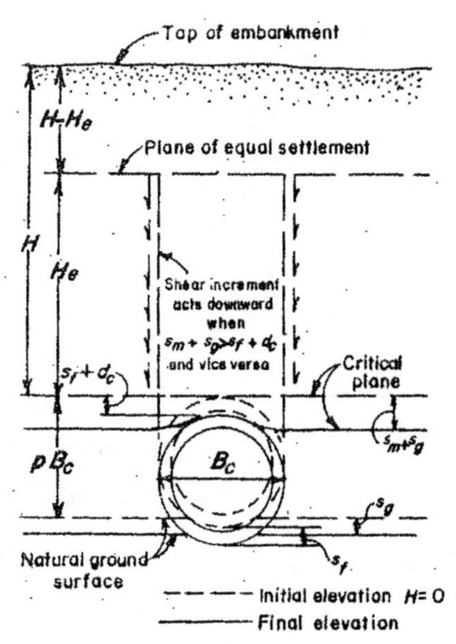

FIGURE 9-7. Settlements that influence loads on positive-projecting sewer pipe. s_g, *settlement of natural ground adjacent to sewer pipe;* s_m, *compression of columns of soil of height* pB_c; d_c, *deflection of sewer pipe; and* s_f, *settlement of bottom of sewer pipe.*

Courtesy of American Concrete Pipe Association, Irving, Tex.

is located within the embankment as shown in Fig. 9-7, the installation is called an "incomplete trench condition" or "incomplete projection condition," as shown in Fig. 9-8.

In computing the settlement values, the effect of differential settlement caused by any compressible layers below the natural ground surface also must be considered. An exceptional situation for a sewer pipe in a trench can be encountered where the natural soil settles more than the trench backfill, such as where the natural soils are organic or peat and the trench backfill is relatively incompressible compacted fill. A more common situation is where the sewer pipe is pile-supported in organic soils. In such cases, the load on the sewer pipe is greater than that of the prism above the pipe, and down-drag loads should be considered in the design of the piles.

9.2.2.4.1. Fill Loads

The fill load on a pipe installed in a positive-projecting embankment condition is computed by the equation:

$$W_c = C_c w B_c^2 \tag{9-8}$$

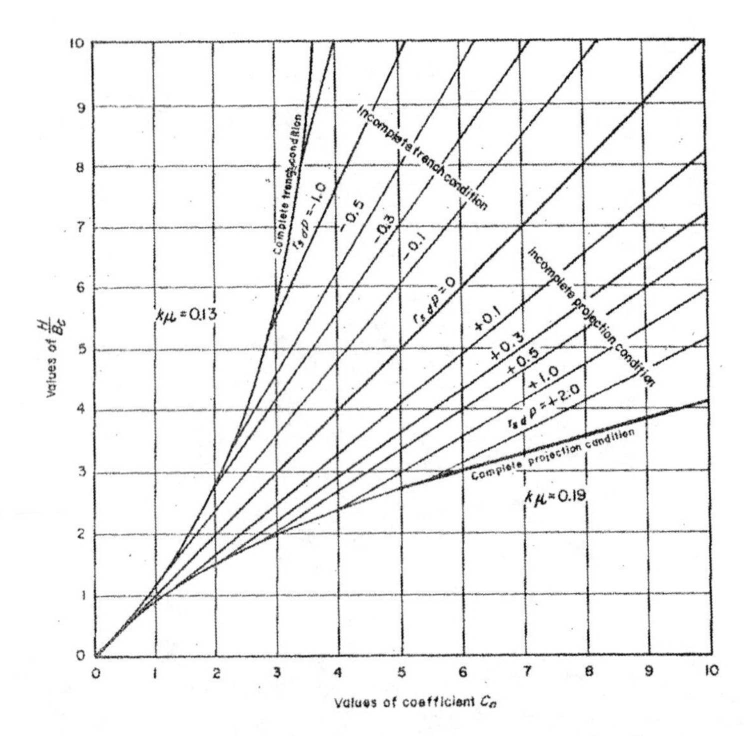

FIGURE 9-8. Diagram for coefficient C_c *for positive-projecting sewer pipes.* Courtesy of American Concrete Pipe Association, Irving, Tex.

where

W_c = load, lb/ft (N/m)
w = density of backfill material, lb/ft^3 (kg/m^3)
B_c = outside horizontal span of the pipe, ft (m)

and

$$C_c = \frac{e^{-2K\mu\frac{H}{B_c}} - 1}{2K\mu} \text{ when } H \leq H_e \tag{9-9}$$

and

$$C_c = \frac{e^{-2K\mu\frac{H}{B_c}} - 1}{2K\mu} + \left(\frac{H}{B_c} - \frac{H_e}{B_c}\right)e^{2K\mu\frac{H_e}{B_c}} \text{ when } H > H_e \tag{9-10}$$

The settlements that influence loads on positive-projecting embankment installations are shown in Fig. 9-7. To evaluate the H_e term in Eq. (9-9), it is necessary to determine, numerically, the relationship between the pipe deflection and the relative settlement between the prism of fill directly above the pipe and the adjacent soil. This relationship is defined as a settlement ratio, expressed as:

$$r_{sd} = \frac{(s_m + s_g) - (s_f + d_c)}{s_m} \tag{9-11}$$

where s_g is the settlement of the natural ground adjacent to the sewer pipe, s_m is the compression of the columns of soil of height pB_c, $(s_m + s_g)$ is the settlement of the critical plane, s_f is the settlement of the bottom of the sewer pipe, and d_c is the deflection of the sewer pipe.

The fill load on a pipe installed in a positive-projecting embankment condition is influenced by the product of the settlement ratio, r_{sd}, and the projection ratio, p. The projection ratio p is the vertical distance the pipe projects above the original ground surface, divided by the outside vertical height of the pipe (B'_c). Recommended settlement ratio design values are listed in Table 9-1.

Figure 9-8 is a graphical solution by Spangler that permits reasonable estimates of C_c for various conditions of $H/B_c r_{sd}$ and p. Since the effect of μ' is nominal, $K\mu'$ was assumed to be 0.19 for the projection condition and 0.13 for the trench condition. Figure 9-8 will provide an estimate of C_c, which is well within the accuracy of the theoretical assumptions.

In Fig. 9-8, the family of straight lines represents the incomplete conditions, whereas the curves represent the complete conditions. The straight

TABLE 9-1. Recommended Design Values of r_{sd}

Type of Sewer Pipe	Soil Conditions	Settlement Ratio, r_{sd}
Rigid	Rock or unyielding foundation	+1.0
Rigid	Ordinary foundation	+0.5 to +0.8
Rigid	Yielding foundation	0 to +0.5
Rigid	Negative-projecting installation	−0.3 to −0.5
Flexible	Poorly compacted side fills	−0.4 to 0
Flexible	Well-compacted side fills	0

lines intersect the curves where H_e equals H. These diagrams can be used to determine the minimum height of fill for which the plane of equal settlement will occur within the soil mass.

Where the $r_{sd}p$ product is zero, the load coefficients term C_c is equal to H/B_c. Substituting this value in Eq. (9-8) results in the load, W_c, being equal to the weight of fill above the pipe. For positive values of $r_{sd}p$, the load on the pipe will be greater than the weight of fill above the pipe, and for negative values the load will be less than the weight of fill above the pipe.

9.2.2.5. *Loads for Negative-Projecting Embankment and Induced Trench Conditions*

This type of installation is normally used when the pipe is installed in a relatively narrow or deep stream bed or drainage path. The pipe is installed in a shallow trench of such depth that the top of the pipe is below the natural ground surface or compacted fill, then covered with an earth fill or embankment which extends above the original ground level.

Sometimes straw, hay, cornstalks, sawdust, or similar materials may be added to the trench backfill to augment the settlement of the interior prism. The greater the value of the negative projection ratio, p', and the more compressible the trench backfill over the sewer pipe, the greater will be the settlement of the interior prism of soils in relation to the adjacent fill material. In using this technique, the plane of equal settlement must fall below the top of the finished embankment. This action generates upward shearing forces which relieve the load on the sewer pipe.

An induced trench sewer pipe, as illustrated in Fig. 9-9, is first installed as a positive-projecting sewer pipe. The embankment then is built up to some height above the top and is thoroughly compacted as it is placed. A trench of the same width as the sewer pipe is next excavated directly over the sewer pipe down to or near its top. This trench is refilled with loose, compressible material and the balance of the embankment is completed in a normal manner.

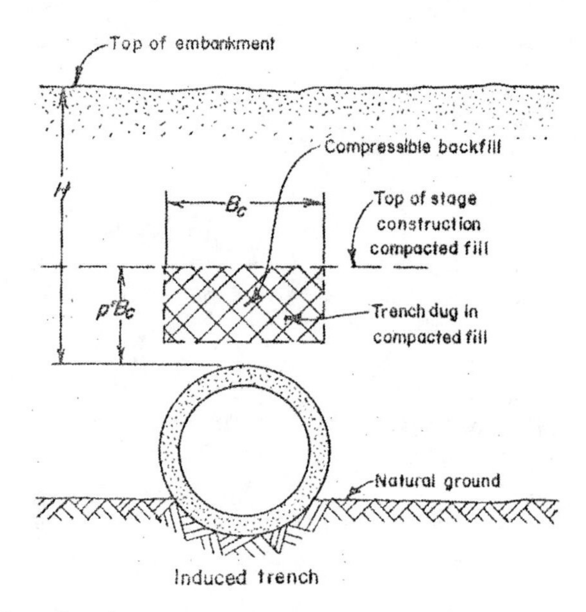

FIGURE 9-9. Induced trench pipe.

The fill load on a pipe installed in a negative-projecting embankment condition is computed by the equation:

$$W_c = C_n w B_d^2 \tag{9-12}$$

where
W_c = load, lb/ft (N/m)
w = density of backfill material, lb/ft^3 (kg/m^3)
B_d = width of the trench, ft (m)

and

$$C_n = \frac{e^{-2K\mu\frac{H}{B_d}} - 1}{-2K\mu} \text{ when } H \leq H_e \tag{9-13}$$

and

$$C_n = \frac{e^{-2K\mu\frac{H}{B_d}} - 1}{-2K\mu} + \left(\frac{H}{B_d} - \frac{H_e}{B_d}\right)e^{-2K\mu\frac{H_e}{B_d}} \text{ when } H > H_e \tag{9-14}$$

If the material within the subtrench is densely compacted, Eq. (9-12) can be expressed as:

$$W_c = C_n w B_d B'_d \tag{9-15}$$

where B'_d is the average of the trench width and the outside diameter of the pipe.

In the case of the induced trench sewer pipe, B_c is substituted for B_d in Eq. (9-12). B_c is the width of the sewer pipe in feet or meters, assuming the trench in the fill is no wider than the sewer pipe.

The settlements that influence loads on negative-projecting embankment installations are shown in Fig. 9-10. To evaluate the H_e term in Eqs. (9-13) and (9-14), it is necessary to determine, numerically, the relationship between the pipe deflection and the relative settlement between the prism of fill directly above the pipe and the adjacent soil. This relationship is defined as a settlement ratio, expressed as:

$$r_{sd} = \frac{s_g - (s_d + s_f + d_c)}{s_d} \tag{9-16}$$

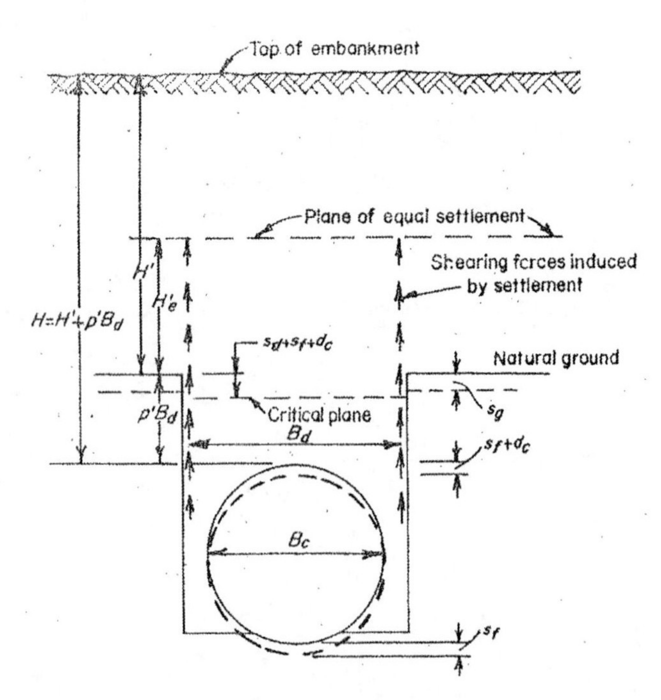

FIGURE 9-10. Settlements that influence loads on negative-projecting sewer pipes.

Recommended settlement ratio design values are listed in Table 9-1. For negative-projecting embankment installations, the projection ratio, p', is the vertical distance from the top of the pipe to the original ground surface or compacted fill, at the time of installation, divided by the width of the trench.

In general, the notation for calculation of loads on negative-projection conditions follows that given for positive projections. The depth of the top of pipe below the critical plane is defined by $p'B_d$, in which p' is defined as the negative projection ratio. If the natural ground surface is on a transverse slope, the vertical distance may be taken as the average distance from the top of the pipe to the top of the trench, at both sides of the trench. Furthermore, s_d is defined as the compression within the fill, for height $p'B_d$.

Present knowledge of the value of the settlement ratio for induced trench sewer pipe is meager. Research reported by Taylor (1971) of the Illinois Department of Highways indicated that the measured settlement ratio of 48-inch (1,200-mm) reinforced concrete pipe culvert installed under induced trench conditions under 30 ft (9 m) of fill, varied from -0.25 to -0.45.

Figure 9-11 provides values of C_n versus H/B_d for various values of r_{sd}, for values of p' equal to 0.5, 1, 1.5, and 2. For other values of p' between 0.5 and 2, values of C_n may be obtained by interpolation. As with the previous figures, only one value of $K\mu$ is used. The family of straight lines represents the incomplete conditions, whereas the curves represent the complete conditions. The straight lines intersect the curves at the point where the height of the plane of equal settlement, H_e, equals the height of the top of embankment, H. These diagrams can therefore be used to determine the height of the plane of equal settlement above the top of the pipe.

9.2.2.6. *Sewer Pipe under Sloping Embankment Surfaces*

Cases arise where the sewer pipe has different heights of fill on the two sides because of the sloping surface of the embankment or when an embankment exists on one side of the sewer pipe only. Design based on the larger fill height may not yield conservative values. When yielding ground may envelope the sewer pipe, a surcharge on one side of the sewer pipe may result in vertical displacement.

Sample Calculations

Example 9-5. Determine the load on a 48-inch-diameter reinforced concrete sewer pipe installed as a positive-projecting pipe under a fill 32 ft high above the top of the pipe. The wall thickness of the sewer pipe is 5 inches and the density of fill is 125 lb/ft^3.

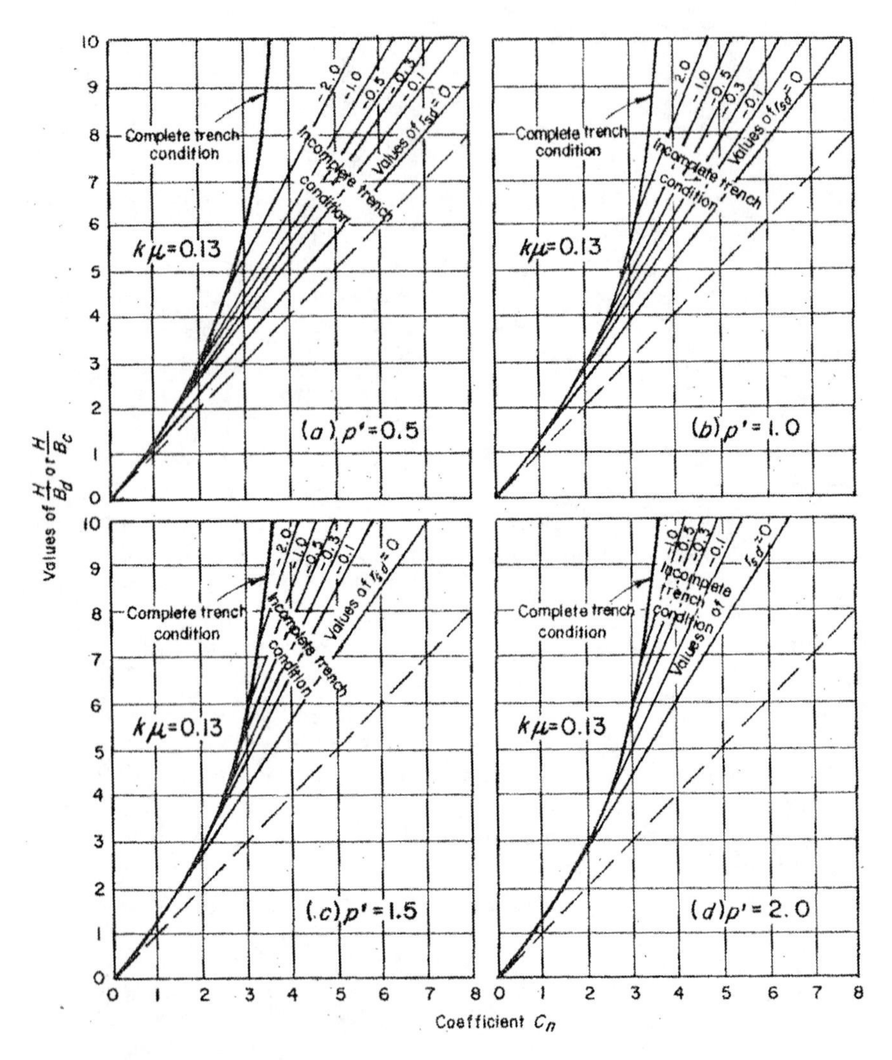

FIGURE 9-11. Diagrams for coefficient C_n *for negative-projecting and induced trench sewer pipes.*

Assume the projection ratio is +0.5 and the settlement ratio is +0.6. Then H = 32 ft; B_c = 4.83 ft; H/B_c = 6.63; $r_{sd}p$ = 0.5 × 0.6 = 0.3; C_c (from Fig. 9-8) = 9.2; and W_c = 9.2 × 125 × $(4.83)^2$ = 26,800 lb/ft (392,300 N/m).

Example 9-6. Determine the load on the sewer pipe of Example 9-5 when installed as a negative-projecting sewer pipe in a trench whose depth is such that the top of the sewer pipe is 7 ft below the surface of the natural ground in which the trench is dug. The width of the trench is 2 ft greater than the outside diameter of the sewer pipe.

Assume the settlement ratio = −0.3. Then H = 32 ft; B_d = 4.83 + 2 = 6.83 ft; H/B_d = 4.69; p' = 1; C_n (from Fig. 9-11) = 3; and W_c = 3 × 125 × $(6.83)^2$ = 17,500 lb/ft (255,950 N/m).

Example 9-7. Determine the load on the sewer pipe of Example 9-5 when installed as an induced trench sewer pipe with its top 2.5 ft below the elevation to which the soil is compacted thoroughly for a distance of 12 ft on each side of the sewer pipe.

Assume the settlement ratio = −0.3. Then H = 32 ft; B_c = 4.83; H/B_c = 6.63; p' = approximately 0.5; C_n (from Fig. 9-11) = 4.8; and W_c = 4.8 × 125 × $(4.83)^2$ = 14,000 lb/ft (204,300 N/m).

9.2.2.7. *Loads for Trenchless Installation Conditions*

This type of installation is used where surface or other conditions prohibit installation of the pipe by conventional open-cut methods or where it is necessary to install the pipe under an existing embankment. The theories set forth in this Manual will usually be appropriate for materials where jacking of the sewer pipe is possible and for tunnels in homogeneous soils of low plasticity. Where a tunnel is to be constructed through materials subject to unusually high internal pressures and stresses, such as some types of clay or shales which tend to squeeze or well, or through blocky and seamy rock, the loads on the sewer pipe cannot be determined from the factors discussed here. Reference should be made to the following section on tunnels.

Horizontal directional drilling (HDD) and microtunneling have become very common since the last edition of this Manual. Each of these methods imposes its own type of loads on the pipe being installed. In the past, HDD pipe engineering has focused primarily on installation techniques. Because HDD installations are not typically utilized for gravity sanitary sewers, the design guidance is not covered in this Manual. For specific projects that may utilize HDD, the reader is advised to consult ASTM F1962, Standard Guide for Use of Maxi-Horizontal Directional Drilling for Placement of Polyethylene Pipe or Conduit Under Obstacles, Including River Crossings, and the ASCE Manual of Practice 108, Pipeline Design for Installation by Directional Drilling. Pipe installed by microtunneling can be treated similarly to pipe installed by pipe jacking, utilizing the appropriate equations herein. (Jacked sewer pipe is assumed to carry the earth load as it is pushed into place ["Jacked-in-Place Pipe Drainage" 1960; ACPA 1960].)

The various trenchless methods of constructing sanitary sewers are described in Chapter 12.

9.2.2.7.1. *Load-Producing Forces*

For the materials considered in this Manual, the vertical load acting on the jacked sewer pipe or tunnel supports, and eventually the sewer pipe in the tunnel, is the resultant of two major forces. First is the weight of the overhead prism of earth within the width of the jacked sewer pipe or tunnel excavation. Second is the shearing forces generated between the interior prisms and the adjacent material caused by the internal friction and cohesion of soils.

During excavation of a tunnel, and varying somewhat with construction methods, the soil directly above the face of the tunnel tends to settle slightly in the period immediately after excavation and prior to placement of the tunnel support. Also, the tunnel supports and the sewer pipe must deflect and settle slightly when the vertical load comes on them. This downward movement or tendency for movement induces upward shearing forces which support a part of the weight of the prism of earth above the tunnel. In addition, the cohesion of the material provides further support for the weight of the prism of earth above the tunnel. The resultant load on the horizontal plane on the top of the tunnel and within the width of the tunnel excavation is equal to the weight of the prism of earth above the tunnel minus the upward friction forces and cohesion of the soil along the limits of the prism of soil over the tunnel.

Hence, the forces involved with gravity earth loads on jacked sewer pipe or tunnels in such soils are similar to those described for loads on sewer pipe in trenches, except for the cohesion of the material. Cohesion also exists in the case of loads in trenches and embankments, but is neglected because the cohesion of the disturbed soil is of minor consequence and may be absent altogether if the soil is saturated. However, in the case of jacked sewer pipe or in tunnels where the soil is undisturbed, cohesion can be an appreciable factor in the loads and may be considered safely if reasonable coefficients are assumed.

Jacking stresses must be investigated in pipe that is to be jacked into place. The critical section is at the pipe joint where the transfer of stress from one pipe to the adjacent pipe occurs. Jointing materials should be used which will provide uniform bearing around the pipe circumference (ACPA 1960). Thrust at the joint is usually transmitted through the tongue or groove, but not both. Concrete stress in the tongue or groove should be checked and additional reinforcements for both longitudinal and bending stresses provided if required.

9.2.2.7.2. Marston's Formula

The earth load on a pipe installed under these conditions is computed by the equation:

$$W_t = C_t w B_t^2 - 2cC_t B_t \qquad (9\text{-}17)$$

where

W_t = load, lb/ft (N/m)
w = density of soil material, lb/ft^3 (kg/m^3)
B_t = maximum width of the tunnel excavation, ft (m)
= B_c in the case of jacked pipe
c = cohesion coefficient, lb/ft^2 (N/m^2)

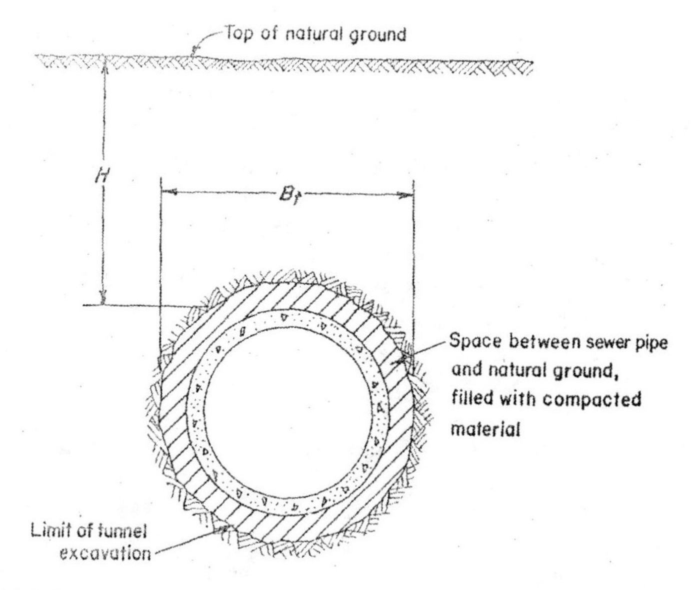

FIGURE 9-12. Sewer pipe in tunnel.

and

$$C_t = \frac{1 - e^{-2K\mu' \frac{H}{B_t}}}{2K\mu'} \tag{9-18}$$

In Eq. (9-17), the $C_t w B_t^2$ term is similar to the trench equation (9-4) for trench loads, except that H is the distance from the ground surface to the top of the tunnel, and B_t is substituted for B_d. The $2cC_tB_t$ term accounts for the cohesion of the undisturbed soil. Conservative design values of the coefficient of cohesion for various soils are listed in Table 9-2. To obtain

TABLE 9-2. Recommended Safe Values of Cohesion

	Values of c	
Material	kPa	lb/ft^2
Clay, very soft	2	40
Clay, medium	12	250
Clay, hard	50	1,000
Sand, loose dry	0	0
Sand, silty	5	100
Sand, dense	15	300

the total earth load for any given height of cover, width of bore or tunnel, and type of soil, the value of the cohesion term is subtracted from the value of the trench load term.

The values of the coefficient for C_t for various ratios of H/B_t and various materials can be obtained from Fig. 9-13 or Fig. 9-4. Values of $K\mu$ and $K\mu'$ are the same as those noted in Fig. 9-4. An analysis of the formula for computing C_t indicates that for very high values of H/B_t, the coefficient C_t approaches the limiting value of $1/(2K\mu')$. Hence, when the tunnel is very deep, the load on the tunnel can be calculated readily by using the limiting value of C_t.

In addition to the earth loads, axial loads encountered during installation must be considered. In estimating axial loads, it is necessary to provide for a uniform distribution of the jacking force around the periphery of the pipe. This can be accomplished either through maintaining close tolerances for parallelism of the ends of the pipe; by providing a spacer of plywood or rubber; or by ensuring the jacking forces are distributed through a jacking frame to the pipe parallel to the axis of the pipe. Under most circumstances, the cross-sectional area of the pipe wall will be adequate to resist the pressures encountered.

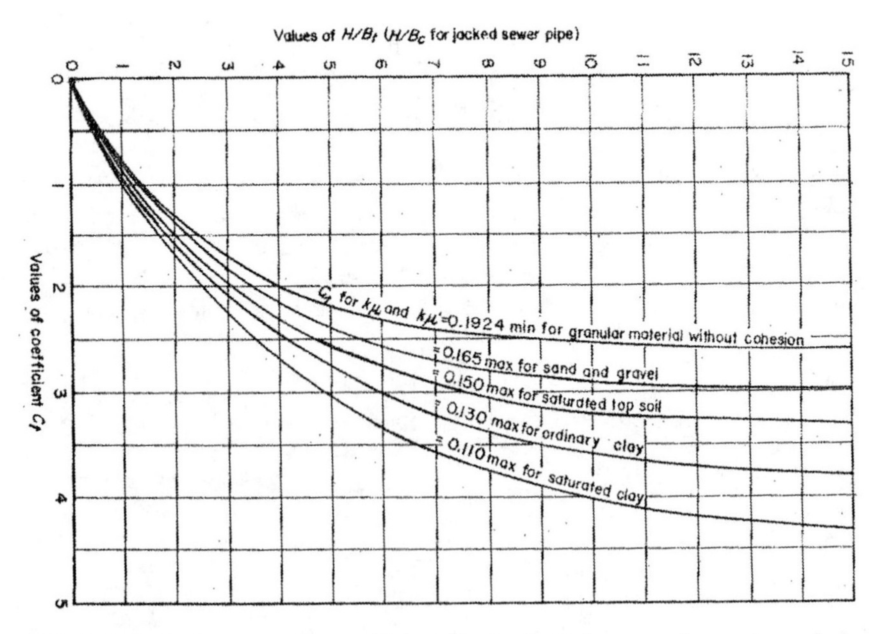

FIGURE 9-13. Diagram for coefficient C_t for jacked sewer pipe or tunnels in undisturbed soil.

To calculate the jacking force required, the forces that arise from the site conditions and the construction methods have to be analyzed. The forces resulting from site conditions include:

- Size, shape, mass and external surface of the pipe.
- The length of the jacking/microtunneling operation.
- Type of soil (or soils) that will be encountered over the length of the drive.
- Position of the water table.
- Stability of the soil.
- Depth and mass of the overburden.
- Vibratory loading.

Factors that can be controlled during the construction process and that impact the jacking force include:

- Amount of overcut of the bore.
- The use of lubricants, such as bentonite.
- Misalignment of the pipeline along the length of the bore.
- Rate of advancement of pipeline.
- Frequency and duration of stoppages.

Once the jacking forces have been calculated, the forces relating to the advancement of the cutting head and shield must be added. The information must be given to the pipe manufacturer to ensure that pipe of proper strength and joint end area is supplied.

9.2.2.7.3. Tunnel Soil Characteristics

The discussion regarding unit weight and coefficient of friction for sanitary sewers in trenches applies equally to the determination of earth loads on jacked sewer pipe or sewer pipe in tunnels through undisturbed soil. The one additional factor that enters into the determination of loads on tunnels is c, the coefficient of cohesion. An examination of Eq. (9-17) shows that the proper selection of c is very important; unfortunately, it can vary widely even for similar types of soils.

It may be possible in some instances to obtain undisturbed samples of the material and to determine the value of c by appropriate laboratory tests. Such testing should be done whenever possible. It is suggested that conservative values of c be used to allow for a saturated condition of the soil or for other unknown factors. Design values should probably be about 33% of the laboratory test value to allow for uncertainties. Recommended safe values of cohesion for various soils (if it is not practicable to determine c from laboratory tests) are shown in Table 9-2.

It is suggested that the value of c be taken as zero in the zone subject to seasonal frost and cracking because of desiccation or loss of strength from saturation. In addition, a minimum value for $(wB_t - 2c)$ should be assumed in cases where $2c$ approaches wB_t. In many cases, the jacking force necessary to install the pipe will govern.

9.2.2.7.4. Effect of Excessive Excavation

Where the tunnel is constructed by a method that results in excessive excavation and where the voids above the sewer pipe or tunnel lining are not backfilled carefully, or packed with grout or other suitable backfill materials, saturation of the soil or vibration eventually may destroy the cohesion of the undisturbed material above the sewer pipe and result in loads in excess of those calculated using Eq. (9-17). If this situation is anticipated, it is suggested that Eq. (9-17) be modified by eliminating the cohesion term. The calculated loads then will be the same as those obtained from Eq. (9-4).

9.2.2.7.5. Loads for Tunnels

When the sanitary sewer is to be constructed in a tunnel through homogeneous soils of low plasticity, design should be based on the theories set forth in the previous section describing jacked sewer pipe. The design of tunnels through other types of materials is discussed in this section. The usual procedure in tunnel construction is to complete the excavation first and then place either a cast-in-place concrete liner or a sewer pipe, and then grout or concrete it in place. Additional strength in such a section can be obtained by means of pressure grouting to strengthen the surrounding material instead of relying totally on the liner or pipe itself. Tunnel loads are therefore usually determined for purposes of selecting supports to be used during excavation, and the sewer pipe or cast-in-place liner is designed primarily to withstand loads from pressure grouting.

A complete discussion of tunnels is not within the scope of this Manual, and the design engineer's attention is called to references listed at the end of this chapter (USACE undated; Proctor and White 1968; "Soil Resistance" 1948; Van Iterson 1948).

Sample Calculation

Example 9-8. Assume the width of excavation, $B_t = 78$ inches $= 6.5$ ft; type of soil is silty sand ($K\mu' = 0.150$, $c = 100$ lb/ft^2, and $w = 110$ lb/ft^3); and the depth of the tunnel, $H = 40$ ft. Then $H/B_t = 40/6.5 = 6.15$ and C_t (from Fig. 9-14) $= 2.83$. Employing Eq. (9-17), $W_t = 2.83 \times 6.5\ (110 \times 6.5 - 2 \times 100)$; or $W_t = 9{,}500$ lb/ft (138,300 N/m).

If the tunnel were very deep, $C_t = 1/(2K\mu') = 3.33$, and $W_t = 11{,}200$ lb/ft (162,700 N/m).

9.2.7.7.6. Load-Producing Forces

When the tunnel is to be constructed through soils which tend to squeeze or swell (such as some types of clay or shale) or through blocky to seamy rock, the vertical load cannot be determined from a consideration of the factors discussed previously, and Eq. (9-17) is not applicable.

The determination of rock pressures exerted against the tunnel lining is largely an estimate based on previous experience of the performance of linings in similar rock formations, although attempts at numerical analysis of stress conditions around a tunnel shaft have been made.

In the case of plastic clay, the full weight of the overburden is likely to come to rest on the tunnel lining some time after construction. The extent of lateral pressures to be expected has not as yet been determined fully, especially the passive resistance which will be maintained permanently by a plastic clay in the case of a flexible, ring-shaped tunnel lining. For normally consolidated clays, suggested lateral pressures are on the order of 2/3 to 7/8 of vertical overburden pressures.

On the other hand, when tunneling through sand, only part of the weight of the overburden will come to rest on the tunnel lining at any time if adequate precautions are taken. The relief will be the result of the transfer of the soil weight immediately above the tunnel to the adjoining soil mass by shearing stresses along the vertical planes. In this case, Marston's formula may be used for estimating the total load which the tunnel lining may have to carry.

Great care must be taken to prevent any escape of sand into the tunnel during its construction. Moist sand will usually arch over small openings and not cause trouble in this respect; however, entirely dry sand (which is sometimes encountered) is liable to trickle into the tunnel through gaps in the temporary lining. Wet sand or sand under the natural water table will flow readily through the smallest gaps. Sand movements of this kind destroy most (if not all) of the arching around the tunnel, with a resulting strong increase of both vertical and horizontal pressures on the supports of the lining. Such cases have been recorded and have caused considerable difficulty. All soil parameters required for design should be obtained from laboratory testing.

9.3. LIVE LOADS AND MINIMUM COVER

In designing sanitary sewers, it is necessary to consider the impact of live loads as well as the earth loads. Live loads become a greater consideration when pipe is installed with shallow cover under unsurfaced roadways, railroads, and/or airport runways and taxiways. The distribution of a live load at the surface on any horizontal plane in the subsoil is shown in Fig. 9-14. The intensity of the load on any plane in the soil mass is greatest

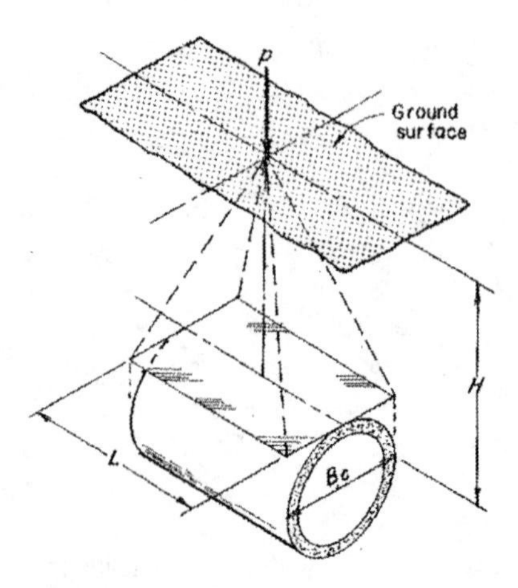

FIGURE 9-14. *Concentrated superimposed load vertically centered over sewer pipe.*

at the vertical axis directly beneath the point of application and decreases in all directions outward from the center of application. As the distance between the plane and the surface increases, the intensity of the load at any point on the plane decreases.

9.3.1. General Pressure Distribution

Concentrated and distributed superimposed loads should be considered in the structural design of sewers, especially where the depth of earth cover is less than 8 ft (2.4 m). Where these loads are anticipated, they are added to the predetermined trench load. Superimposed loads are calculated by use of Holl's and Newmark's modifications to Boussinesq's equation (Spangler 1946).

9.3.1.1. Concentrated Loads

Holl's integration of Boussinesq's solution leads to the following equation for determining loads due to superimposed concentrated load, such as a truck wheel load (Fig. 9-14):

$$W_{sc} = C_s PF/L \qquad (9\text{-}19)$$

where

W_{sc} = the load on the conduit, in lb/ft (kg/m) of length
P = the concentrated load, in lb (kg)
F = the impact factor

C_s = the load coefficient, a function of $B_c/(2H)$ and $L/(2H)$, where H = the height of fill from the top of conduit to ground surface, in ft (m) and B_c = the width of conduit in ft (m)
L = the effective length of conduit, in ft (m).

The effective length of a sewer pipe is defined as the length over which the average load caused by surface traffic wheels produces nearly the same stress in the sewer pipe wall as does the actual load which varies in intensity from point to point. An effective length, L, equal to 3 ft (1 m) for pipe greater than 3 ft (1 m) long, and the actual length for pipe segments shorter than 3 ft (1 m) are recommended.

If the concentrated load is displaced laterally and longitudinally from a vertically centered location over the section of sewer pipe under consideration, the load on the pipe can be computed by algebraically adding the effect of the concentrated load on various rectangles, each with a corner centered under the concentrated load. Values of C_s in Table 9-3 divided by 4 equals the load coefficient for a rectangle whose corner is vertically centered under the concentrated load.

9.3.1.2. Distributed Loads

For the case of a superimposed load distributed over an area of considerable extent (Fig. 9-15), the formula for load on the sewer pipe is:

$$W_{sd} = C_s pFB_c \tag{9-20}$$

where
W_{sd} = the load on the sewer pipe, lb/ft (N/m)
p = the intensity of distributed load, lb/ft^2 (N/m^2)
F = impact factor
B_c = the width of the sewer pipe, ft (m)
C_s = load coefficient, which is a function of $D/(2H)$ and $(M/2H)$ from Table 9-3
H = height from the top of the sewer pipe to the ground surface, ft (m)

and D and M are the width and length, respectively, of the area over which the distributed load acts, ft (m).

For the case of a uniform load offset from the center of the sewer pipe, the loads per unit length of pipe may be determined by a combination of rectangles. The load on the sewer pipe can be computed by algebraically adding the effect of various rectangles of loaded area. It is more convenient to work in terms of load under one corner of a rectangular loaded area rather than at the center. Dividing the tabular values of C_s by 4 will give the effect for this condition. Stresses from various types of surcharge

TABLE 9-3. Values of Load Coefficientss, C_s, for Concentrated and Distributed Superimposed Loads Vertically Centered Over Conduit*

$\frac{D}{2H}$ or $\frac{B_c}{2H}$	$\frac{M}{2H}$ or $\frac{L}{2H}$													
	0.1	0.2	0.3	0.4	0.5	0.5	0.7	0.8	0.9	1.0	1.2	1.5	2.0	5.0
0.1	0.019	0.037	0.053	0.067	0.079	0.089	0.097	0.103	0.108	0.112	0.117	0.121	0.124	0.128
0.2	0.037	0.072	0.103	0.131	0.155	0.174	0.189	0.202	0.211	0.219	0.229	0.238	0.244	0.248
0.3	0.053	0.103	0.149	0.190	0.224	0.252	0.274	0.292	0.306	0.318	0.333	0.345	0.355	0.360
0.4	0.067	0.131	0.190	0.241	0.284	0.320	0.349	0.373	0.391	0.405	0.425	0.440	0.454	0.460
0.5	0.079	0.155	0.224	0.284	0.336	0.379	0.414	0.441	0.463	0.481	0.505	0.525	0.540	0.548
0.6	0.089	0.174	0.252	0.320	0.379	0.428	0.467	0.499	0.524	0.544	0.572	0.596	0.613	0.624
0.7	0.097	0.189	0.274	0.349	0.414	0.467	0.511	0.546	0.584	0.597	0.628	0.650	0.674	0.688
0.8	0.103	0.202	0.292	0.373	0.441	0.499	0.546	0.584	0.615	0.639	0.674	0.703	0.725	0.740
0.9	0.108	0.211	0.306	0.391	0.463	0.524	0.574	0.615	0.647	0.673	0.711	0.742	0.766	0.784
1.0	0.112	0.219	0.318	0.405	0.481	0.544	0.597	0.639	0.673	0.701	0.740	0.774	0.800	0.816
1.2	0.117	0.229	0.333	0.425	0.505	0.572	0.628	0.674	0.711	0.740	0.783	0.820	0.849	0.868
1.5	0.121	0.238	0.345	0.440	0.525	0.596	0.650	0.703	0.742	0.774	0.820	0.861	0.894	0.916
2.0	0.124	0.244	0.355	0.454	0.540	0.613	0.674	0.725	0.766	0.800	0.849	0.894	0.930	0.956

*Influence coefficients for solution of Holl's and Newmark's integration of the Boussinesq equation for vertical stress.

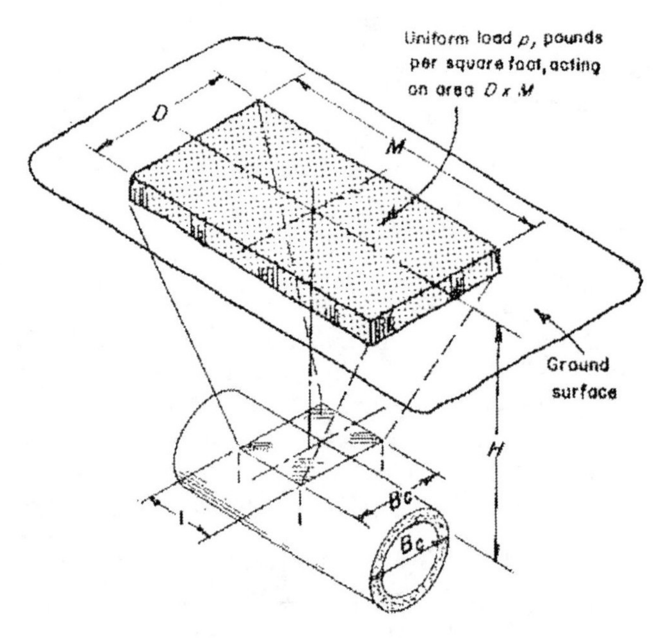

FIGURE 9-15. Distributed superimposed load vertically centered over sewer pipe (lb/ft^2 × 47.9 = N/m^2).

loadings can be computed using computer solutions (Jumikis 1969; 1971). For determination of the stress below a point such as A in Fig. 9-16, as a result of the loading in the rectangle BCDE, the area may be considered to consist of four rectangles: AJDF − AJCG − AHEF + AHBG. Each of these four rectangles has a corner at point A. By computing $D/(2H)$ and $M/(2H)$ for each rectangle, the load coefficient for each rectangle can be taken from Table 9-3. Because point A is at the corner of each rectangle,

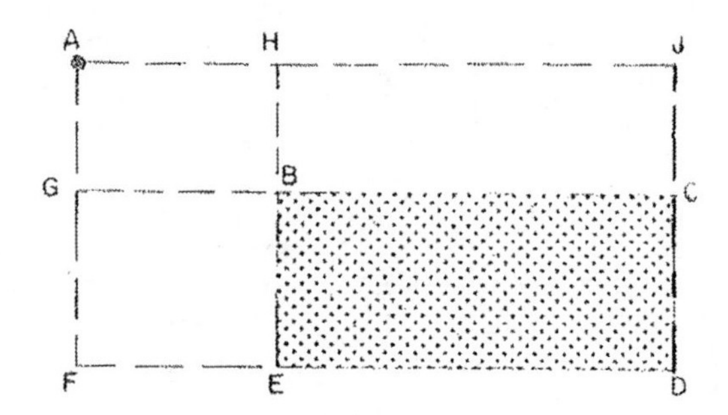

FIGURE 9-16. Diagram for obtaining stress at point A caused by load in shaded area BCDE.

Newmark, N. M. (1935). "Simplified computation of vertical pressures in elastic foundations." Circular 24, Illinois Experiment Station, Urbana, Ill.

the load coefficients from Table 9-3 should be divided by 4. A combination of the stresses from the four rectangles, with signs as indicated above, gives the desired stress.

Values of C_s can be read directly from Table 9-3 if the area of the distributed superimposed load is centered vertically over the center of the sewer pipe under consideration.

9.3.1.3. Highway Truck Loads

Unless other data are available, it is safe to estimate that truck wheel loads are the greatest loads to be supported. HS-20 wheel loadings have been the long-term standard for highway and bridge design and are equally applicable for estimating loads on sewers. With the advent of larger trucks, many states have gone to HS-25 or even HS-32 design vehicles. Figure 9-17 shows the American Association of State Highway and Transportation Officials (AASHTO) loading for both H-20 and HS-20 design vehicles.

If a rigid or flexible pavement designed for heavy-duty traffic is provided, the intensity of a truck wheel load is usually reduced sufficiently so that the live load transmitted to the pipe is negligible. In the case of flexible pavements designed for light-duty traffic but subjected to heavy truck traffic, the flexible pavement should be considered as fill material

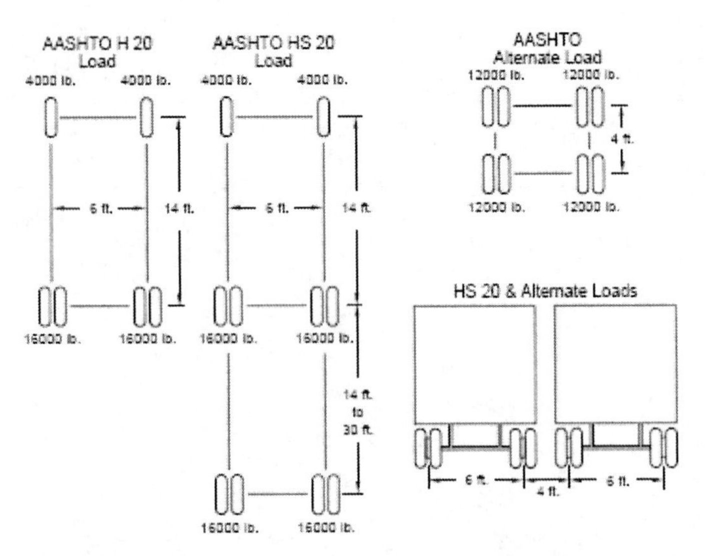

FIGURE 9-17. AASHTO H-20/HS-20 design vehicle loadings.
American Association of State Highway and Transportation Officials (AASHTO). (1997). "Standard specifications for highway bridges," twelfth ed., AASHTO, Washington, D.C.

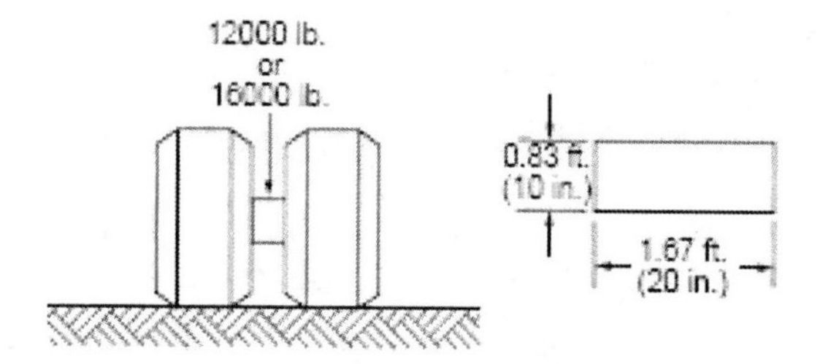

FIGURE 9-18. Wheel load surface contact area.
American Concrete Pipe Association (ACPA). (2000). "Concrete pipe design manual," ACPA, Irving, Tex. Reprinted with permission.

over the top of the pipe. In analyses, the most critical AASHTO loadings shown in Fig. 9-17 are used in either the single mode or passing mode.

Each of these loadings is assumed to be applied through dual-wheel assemblies uniformly distributed over a surface area of 10 inches by 20 inches (254 mm by 508 mm), as shown in Fig. 9-18.

The total wheel load is then assumed to be transmitted and uniformly distributed over a rectangular area on a horizontal plane at the depth, H, as shown in Fig. 9-19 for a single HS-20 dual wheel.

Distributed load areas for the alternate load and the passing mode for either loading are developed in a similar manner. Recommended impact factors, F, to be used in determining live loads imposed on pipe with less than 3 ft (0.9 m) of cover when subjected to dynamic traffic loads are listed in the accompanying Table 9-4.

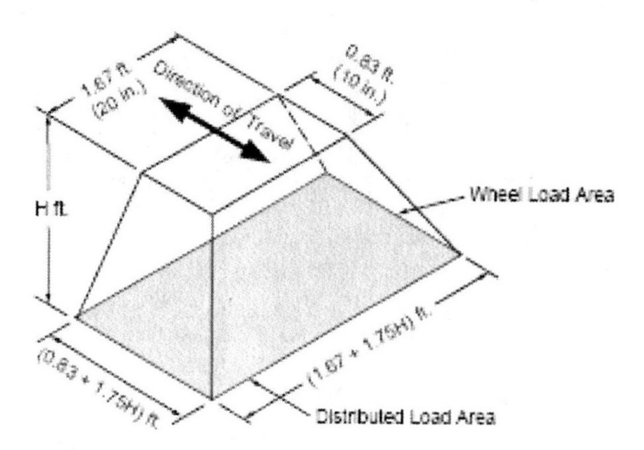

Figure 9-19 Distributed load area, single HS-20 dual wheel.
American Concrete Pipe Association (ACPA). (2000). "Concrete pipe design manual," ACPA, Irving, Tex. Reprinted with permission.

TABLE 9-4. Impact Factors for Highway Truck Loads

H, Height of Cover	F, Impact Factor
0 to 1 ft (0 to 0.30 m)	1.3
1 ft, 1 inch to 2 ft (0.31 to 0.61 m)	1.2
2 ft, 1 inch to 3 ft (0.62 to 0.91 m)	1.1
>3 ft (>0.91 m)	1

American Concrete Pipe Association (ACPA). (2000). "Concrete pipe design manual," ACPA, Irving, Tex. Reprinted with permission.

As the depth, H, increases, the critical loading configuration can be either one HS-20 wheel load, two HS-20 wheel loads in the passing mode, or the alternate load in the passing mode. Since the exact geometric relationship of individual or combinations of surface wheel loads cannot be anticipated, the most critical loading configurations and the outside dimensions of the distributed load areas within the indicated cover depths are summarized in Table 9-5.

9.3.1.4. Railroad Loads

In determining the live load transmitted to a pipe installed under railroad tracks, the weight on the locomotive driver axles plus the weight of the track structure, including ballast, is considered to be uniformly distributed over an area equal to the length occupied by the drivers multiplied by the length of ties. Typically, tie length is assumed to be 8.5 ft (2.6 m). The American Railway Engineering and Maintenance of Way Association (AREMA) recommends a Cooper E80 loading with axle loads and axle spacing, as shown in Fig. 9-20. In addition, 200 lb/ft (2,900 N/m) should be allowed for the weight of the track structure.

Typically, railroads require an impact factor of 1.75 for depth of cover up to 5 ft (1.5 m). Between 5 and 30 ft (1.5 and 9.1 m), the impact factor is reduced by 0.03 per ft (0.1 per m) of depth. Below a depth of 30 ft (9.1 m), the impact factor is 1.

TABLE 9-5. Critical Loading Configurations

H	P	Distributed Load Area
<1.33 ft (<0.40 m)	16,000 lb (71,170 N)	$1.67 + 1.75H$ $(0.83 + 1.75H)$
1.33 to 4.1 ft (0.41 to 1.25 m)	32,000 lb (142,340 N)	$5.67 + 1.75H$ $(0.83 + 1.75H)$
>4.1 ft (>1.25 m)	48,000 lb (213,515 N)	$5.67 + 1.75H$ $(4.83 + 1.75H)$

American Concrete Pipe Association (ACPA). (2000). "Concrete pipe design manual," ACPA, Irving, Tex. Reprinted with permission.

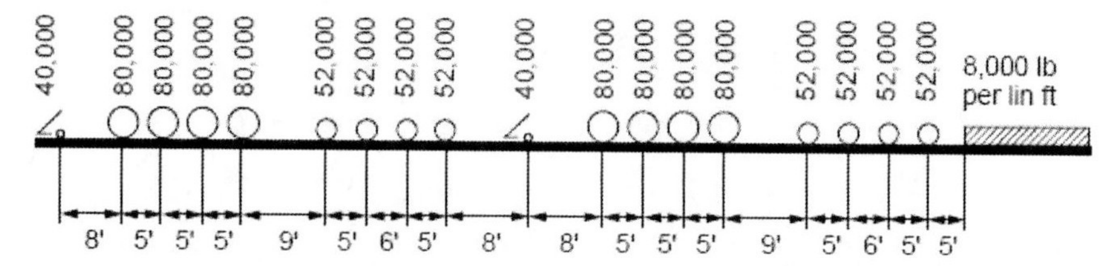

FIGURE 9-20. Cooper E80 load distribution.
American Railway Engineering and Maintenance of Way Association (AREMA). (1981–1982). "Manual for railway engineering," AREMA, Washington, D.C.

9.3.1.5. Aircraft Loads

The distribution of aircraft wheel loads on any horizontal plane in the soil mass is dependent on the magnitude and characteristics of the aircraft loads, the aircraft's landing gear configuration, the type of pavement structure, and the subsoil conditions. Heavier gross aircraft weights, including the advent of New Large Aircraft (NLA) such as the Airbus A380, have resulted in multiple wheel undercarriages consisting of dual-wheel assemblies, dual tandem assemblies, or even "tridem" (three wheels in a row) assemblies. The distribution of wheel loads through rigid and flexible pavements are shown in Figs. 9-21 and 9-22. If a rigid pavement is provided, an aircraft wheel load concentration is distributed

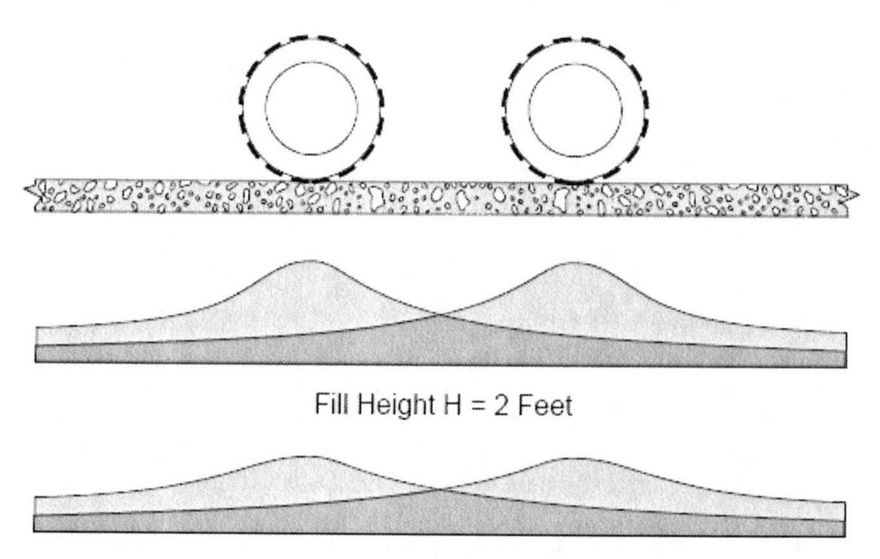

FIGURE 9-21. Aircraft pressure distribution for rigid pavements.
American Concrete Pipe Association (ACPA). (2000). "Concrete pipe design manual," ACPA, Irving, Tex. Reprinted with permission.

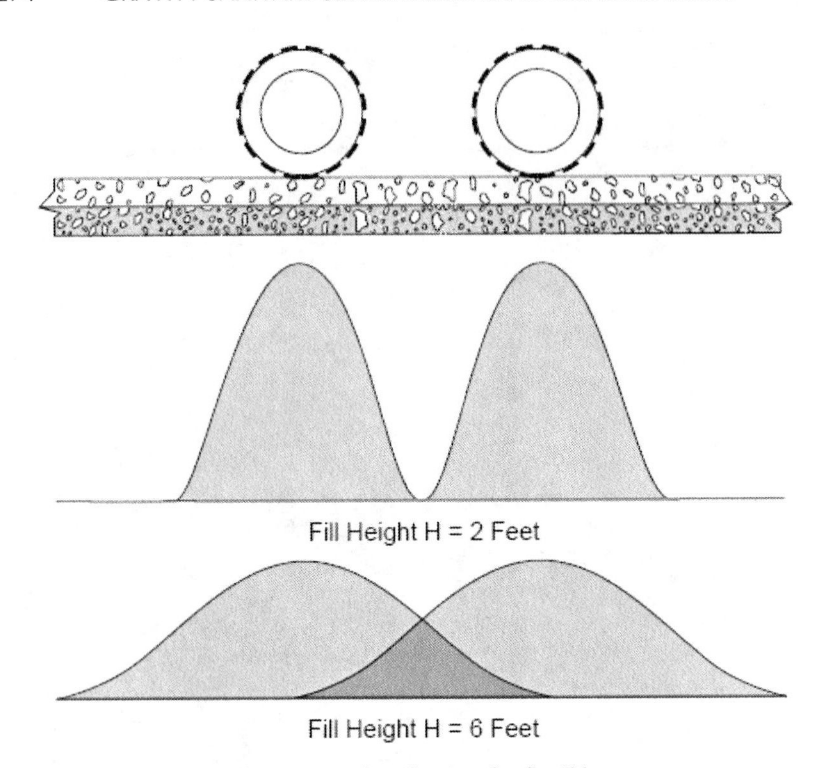

FIGURE 9-22. Aircraft pressure distribution for flexible pavements.
American Concrete Pipe Association (ACPA). (2000). "Concrete pipe design manual," ACPA, Irving, Tex. Reprinted with permission.

over an appreciable area and is substantially reduced in intensity at the subgrade. For multi-wheeled landing gear assemblies, the total pressure intensity is dependent on the interacting pressures produced by each individual wheel. The maximum load transmitted to a pipe varies with the pipe size under consideration, the pipe's relative location with respect to the particular landing gear configuration, and the height of fill between the top of the pipe and the subgrade surface. For a flexible pavement, the area of the load distribution at any plane in the soil mass is considerably less than for a rigid pavement. The interaction of pressure intensities due to individual wheels of a multi-wheeled landing gear assembly is also less pronounced at any given depth of cover. In present airport design practices, the aircraft's maximum takeoff weight is used since the maximum landing weight is usually considered to be about three-quarters of the takeoff weight.

Currently, culvert design for airports is governed by the Federal Aviation Administration (FAA) under Advisory Circular (AC)150/5320-5B, Airport Drainage. AC 150/5320-5B provides information on the required depths and diameters of the culvert system for given overlying pavement surfaces. This AC contains basic parameters for culvert design for aircraft weights up to 1.5 million pounds on traditional-style landing gear configurations, such as the double tandem found on a B747. NLA are predicted to weigh around 1.2 million pounds and feature complex gear designs with triple- and quad-tandem gear posts. If designing for an airport that will be serving NLA, the wheel loading factors associated with these gear configurations will have to be evaluated.

Additional information regarding the calculation of live loads under rigid airport pavement and taxiways can be found in the Portland Cement Association's publication Vertical Pressure on Culverts Under Wheel Loads on Concrete Pavement Slabs (PCA 1951).

9.3.1.6. *Construction Loads*

During grading operations it may be necessary for heavy construction equipment to travel over an installed pipe. Unless adequate protection is provided, the pipe may be subjected to load concentrations in excess of the design loads. Before heavy construction equipment is permitted to cross over a pipe, a temporary earth fill should be constructed to an elevation at least 3 ft (0.9 m) over the top of the pipe. The fill should be of sufficient width to prevent possible lateral displacement of the pipe.

Sample Calculations

Example 9-9. Determine the load on a 24-inch-diameter pipe under 3 ft of cover caused by a 10,000-lb truck wheel applied directly above the center of the pipe.

Assume the pipe section is 2.5 ft long; the wall thickness is 2 inches; and the impact factor = 1. Then $B_c = 24 + 4 = 28$ inches = 2.33 ft; $L = 2.5$ ft; and $H = 3$ ft. Finally, $B_c/2H = 2.33/6 = 0.39$; and $L/2H = 2.5/6 = 0.41$; the load coefficient is 0.240 from Table 9-3. Substituting in Eq. (9-19):

$$W_{sc} = 0.240 \times \frac{10,000 \times 1.0}{2.5} = 960 \text{ lb/ft } (14,000 \text{ N/m})$$

Example 9-10. Determine the load on a 48-inch-diameter concrete pipe under 6 ft of cover (bottom of ties to top of pipe) resulting from the Cooper E80 railroad loading.

Assume the pipe wall thickness is 4 inches The locomotive load consists of four 80,000-lb axles spaced 5 ft center-to-center; the impact factor is 1.4; and the weight of the track structure is 200 lb/ft. Then $B_c = 48 + 9 =$ 56 inches or 4.67 ft; $H = 6$ ft; $D = 8$ ft; and $M = 20$ ft. The unit load plus impact at the base of the ties is:

$$\frac{4 \times 80,000 \times 1.4}{8 \times 20} + \frac{200}{8} = 2,825 \text{ lb/ft}^2; \ \frac{D}{2H} = \frac{8}{12} = 0.$$

$\frac{M}{2H} = \frac{20}{12} = 1.67$. The influence coefficient is 0.641 (Table 9-3).

From Eq. (9-20):

$$W_{sd} = 0.641 \times 2{,}825 \times 4.67 = 8{,}460 \text{ lb/ft } (123{,}420 \text{ N/m}).$$

9.4. DIRECT DESIGN AND INDIRECT DESIGN

The structural design of rigid pipe for installed conditions is defined as Direct Design, in which all stresses induced within the pipe (including moment, thrust, shear, diagonal tension, radial tension, and crack control) are evaluated for a defined external pressure distribution. ASCE Standard 15-98, Standard Practice for Direct Design of Buried Precast Concrete Pipe Using Standard Installations (SIDD) (ASCE 2000), and AASHTO Soil-Reinforced Concrete Structure Interaction Systems, Section 16 (AASHTO 1997) present detailed procedures with applicable structural design equations for the installed direct design condition. A software program titled *PIPECAR*, available through the Federal Highway Administration (FHWA), can also be used for the structural analysis and design of circular reinforced concrete pipe.

Indirect Design procedures for rigid pipe involve the determination of the load-carrying capacity of the pipe under the T.E.B. test load condition and evaluation of the earth and live loads the pipe is subjected to under the installed condition. The load-carrying capacity of the pipe under the test condition of loading is correlated with the installed condition of loading and bedding by means of a bedding factor. The numerical value of the bedding factor is predetermined by the ratio of the maximum tension stress induced within the pipe wall under the test load to the maximum tension stress induced within the pipe wall under the installed load.

A combination of Direct Design pressure distribution for standard installations and correlation with the T.E.B. test load with appropriate bedding factors simplifies the necessary detailed mathematical computation required for the determination of required pipe strength. The Indirect Design method of analysis can also be accomplished utilizing the Marston-Spangler Theory of Loads and Supporting Strengths.

For flexible pipe, the predominant source of structural load-carrying capability is the external pressure on the pipe exerted by the surrounding embedment material as the pipe deforms under load. Therefore, all flexible pipes are evaluated by Direct Design procedures for the installed condition. The structural design of circular flexible pipe is based on compression wall thrust, wall buckling, and limiting deformation/deflection to acceptable limits. For flexible thermoplastic and thermosetting plastic pipe, strain limits are also evaluated.

9.5. RIGID PIPE STRUCTURAL DESIGN

Three methods of design analysis are available for rigid pipe, depending on the pipe material. The methods are:

1. Standard Installation Direct Design (SIDD) utilizing ASCE Standard 15-98, AASHTO Standard Specification for Highway Bridges, Section 16, or the FHWA *PIPECAR* computer program.
2. Standard Installation Indirect Design, as outlined in ACPA's Concrete Pipe Design Manual (ACPA 2000).
3. Marston-Spangler Theory of Loads and Supporting Strength.

Method 1 is applicable only to reinforced concrete pipe. Method 2 is applicable to both concrete pipe and reinforced concrete pipe. Method 3 is applicable to all other rigid pipe materials, including clay pipe, fiber cement pipe, and cast-in-place concrete structures.

9.5.1. Direct Design of Concrete Pipe

ASCE Standard 15-98 provides guidance for the design of standard installations of reinforced precast concrete pipe. This method is automated in the FHWA *PIPECAR* computer program. The SIDD method accounts for the interaction between the pipe and soil envelope in determining loads and distribution of earth pressure on a buried pipe. The loads and pressure distributions are used to calculate the moment, thrust, and shear in the pipe wall, and required pipe reinforcement for ASCE Standard Installations. This standard utilizes two Standard Installations—one in embankments and one in trenches—as well as specific requirements for soils classifications and compaction requirements for four installation types. These are based on the results of pipe–soil interaction, together with an evaluation of current construction practice, equipment, procedures, and experience. The structural design of a concrete pipe is based on a limits state design procedure that accounts for strength and serviceability criteria. To design the concrete pipe, the owner and/or engineer must establish the following design criteria:

- Intended use of the pipeline.
- Pipe inside diameter.
- Pipeline plan and profile drawings with installation cross sections, as required.
- Design earth cover height above the top of the pipe.
- The allowable Standard Installation type as defined in ASCE Standard 15-38. (Type 1 requires the most effort and control in bedding and backfill construction but minimizes the design loads on the pipe. Type 1 is the simplest to construct but imposes greater requirements on the pipe.)

- Soil data sufficient to determine in situ conditions for allowable ASCE and ACPA Standard Installations (including in situ soil classification) and overfill weight per volume.
- Performance requirements for pipe joints.
- Design live and surcharge loadings, if any.
- Design intermittent internal hydrostatic pressures, if required.
- Crack width control criteria.
- Cement type, if different from that in ASTM C1417.

The design of a concrete pipe for a particular Standard Installation type is based on the assumption that the specified design bedding and fill requirements will be achieved during construction of the installation.

The Standard Embankment Installation, and the accompanying terminology, are shown in Fig. 9-23. Table 9-6 shows the varying types of embankment installations and the necessary bedding requirements. The Standard Trench Installation, and accompanying terminology, are shown

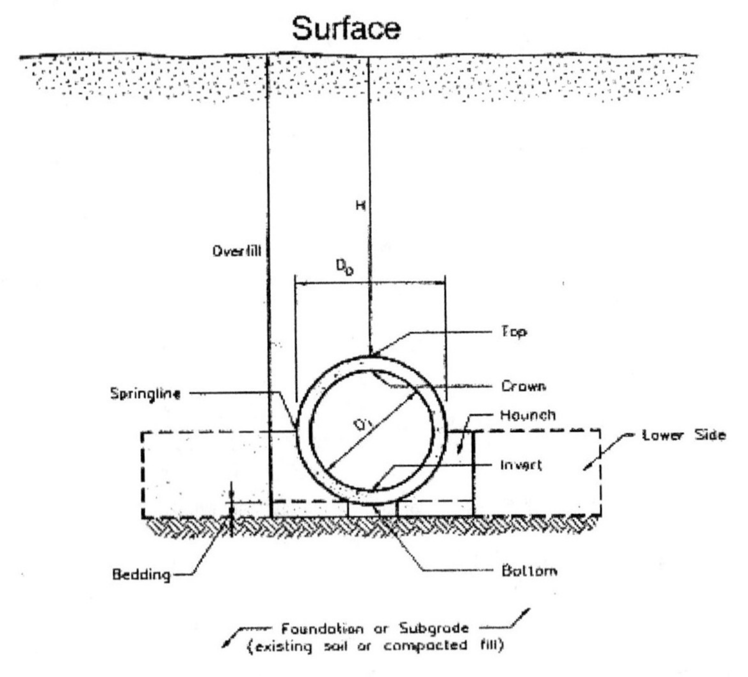

FIGURE 9-23. Standard embankment installation.
From American Society of Civil Engineers (ASCE), Direct Design of Buried Concrete Pipe Standards Committee. (2000). "Standard practice for direct design of buried precast concrete pipe using standard installations (SIDD)," ASCE Standard No. 15-98, ASCE, Reston, Va.

TABLE 9-6. Standard Embankment Installations Soil and Minimum Compaction Requirements

Installation Type	Bedding Thickness	Haunch and Outer Bedding	Lower Side
Type 1	$D_0/24$ minimum, not less than 3 inches (75 mm). If rock foundation, use $D_0/12$ minimum, not less than 6 inches (150 mm)	95% Category I	90% Categogy I, 95% Category II, or 100% Category III
Type 2	$D_0/24$ minimum, not less than 3 inches (75 mm). If rock foundation, use $D_0/12$ minimum, not less than 6 inches (150 mm).	90% Category I or 95% Category II	85% Category I, 90% Category II, or 95% Category III
Type 3	$D_0/24$ minimum, not less than 3 inches (75 mm). If rock foundation, use $D_0/12$ minimum, not less than 6 inches (150 mm).	85% Category I, 90% Category II, or 95% Category III	85% Category I, 90% Category II, or 95% Category III
Type 4	No bedding required, except if rock foundation, use $D_0/12$ minimum, not less than 6 inches (150 mm).	No compaction required, except if Category III, use 85% Category III	No compaction required, except if Category III, use 85% Category III

Notes:

1. Compaction and soil symbols—i.e. "95% Category I" refers to Category 1 soil material with a minimum standard Proctor compaction of 95%. See illustration 4.5 for equivalent modified Proctor values.
2. Soil in the outer bedding, haunch, and lower side zones, except within $D_0/3$ from the pipe springline, shall be compacted to at least the same compaction as the majority of soil in the overfill zone.
3. Substrenches
   3.1 A subtrench is defined as a trench with its top below finished grade by more than 0.1 H or, for roadways, its top is at an elevation lower than 1 inch (0.3 m) below the bottom of the pavement base material.
   3.2 The minimum width of a subtrench shall be 1.33 D_0 or wider if required for adequate space to attain the specified compaction in the haunch and bedding zones.
   3.3 For subtrenches with walls of natural soil, any portion of the lower side zone in the subtrench wall shall be at least as firm as an equivalent soil placed to the compaction requirements specified for the lower side zone and as firm as the majority of soil in the overfill zone, or shall be removed and replaced with soil compacted to the specified level.

American Society of Civil Engineers (ASCE), Direct Design of Buried Concrete Pipe Standards Committee. (2000). "Standard practice for direct design of buried precast concrete pipe using standard installations (SIDD)," ASCE Standard No. 15-98, ASCE, Reston, Va.

in Fig. 9-24, with Table 9-7 indicating the soils and compaction requirements for the trench installation.

In evaluating actual loads on the pipe, a finite element analysis program (*SPIDA*) was programmed with the ASCE and ACPA Standard Installations and many design runs were made (ASCE 2000). An evaluation of the output of the designs by Dr. Frank J. Heger produced a load pressure diagram, Fig. 9-25, significantly different from those proposed by previous theories. This difference is particularly significant under the pipe in the lower haunch area and is due, in part, to the assumption of the existence of partial voids adjacent to the pipe wall in this area.

Table 9-8 relates the generic soil types designated in Tables 9-6 and 9-7 to soil classifications established under the Unified Soil Classification System (USCS) and AASHTO soil classification systems.

Although it is possible to hand-calculate the design of a concrete pipe using this method, the necessity of analyzing multiple states makes the use of a computer program preferable. *PIPECAR* was developed by Simpson Gumpertz & Heger, Inc. for the FHWA and uses the SIDD design method to design concrete pipe to the AASHTO method. Although versions of the software can be downloaded from FHWA web site without

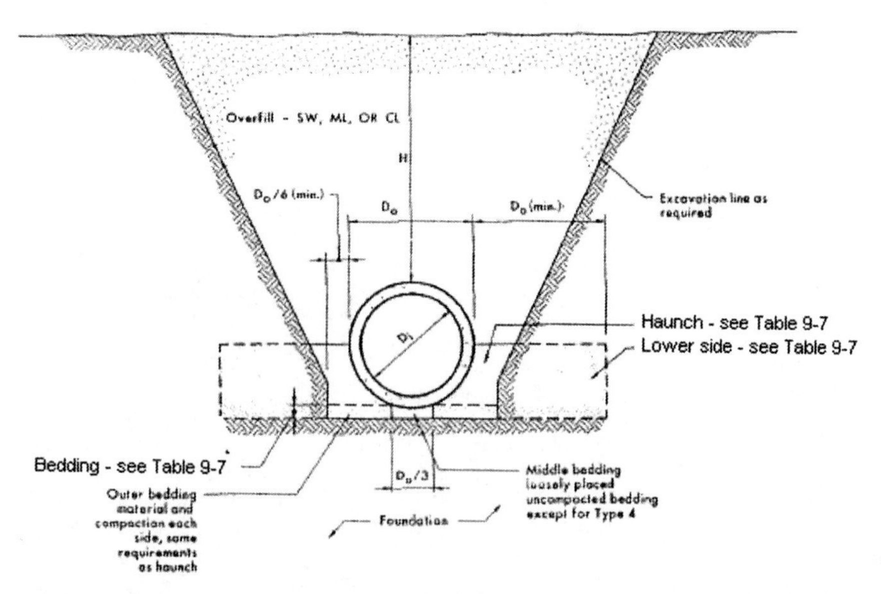

FIGURE 9-24. Standard trench installation.
From American Society of Civil Engineers (ASCE), Direct Design of Buried Concrete Pipe Standards Committee. (2000). "Standard practice for direct design of buried precast concrete pipe using standard installations (SIDD)," ASCE Standard No. 15-98, ASCE, Reston, Va.

TABLE 9-7. Standard Trench Installations Soil and Minimum Compaction Requirements

Installation Type	Bedding Thickness	Haunch and Outer Bedding	Lower Side
Type 1	$D_0/24$ minimum, not less than 3 inches (75 mm). If rock foundation, use $D_0/12$ minimum, not less than 6 inches (150 mm)	95% Category I	90% Categogy I, 95% Category II, or 100% Category III
Type 1	$D_0/24$ minimum, not less than 3 inches (75 mm). If rock foundation, use $D_0/12$ minimum, not less than 6 inches (150 mm).	95% Category I	90% Category I, 95% Category II, or 100% Category III
Type 2	$D_0/24$ minimum, not less than 3 inches (75 mm). If rock foundation, use $D_0/12$ minimum, not less than 6 inches (150 mm).	90% Category I or 95% Category II	85% Category I, 90% Category II, or 95% Category III
Type 3	$D_0/24$ minimum, not less than 3 inches (75 mm). If rock foundation, use D_012 minimum, not less than 6 inches (150 mm).	85% Category I, 90% Category II, or 95% Category III	85% Category I, 90% Category II, or 95% Category III
Type 4	No bedding required, except if rock foundation, use $D_0/12$ minimum, not less than 6 inches (150 mm).	No compaction required, except if Category III, use 85% Category III	No compaction required, except if Category III, use 85% Category III

Notes:

1. Compaction and soil symbols—i.e. 95% Category I"—refers to Category I soil material with minimum standard Proctor compaction of 95%. See illustration 4.8 for equivalent modified Proctor values.
2. The trench top elevation shall be no lower than 0.1 H below finished grade or, for roadways, its top shall be no lower than an elevation of 1 inch (0.3 m) below the bottom of the pavement base material.
3. Soil in bedding and haunch zones shall be compacted to at least the same compaction as specified for the majority of soil in the backfill zone.
4. The trench width shall be wider than shown if required for adequate space to attain the specified compaction in the haunch and bedding zones.
5. For trench walls that are within 10 degrees of vertical, the compaction or firmness of the soil in the trench walls and lower side zone need not be considered.
6. For trench walls with greater than 10 degree slopes that consist of embankment, the lower side shall be compacted to at least the same compaction as specified for the soil in the backfill zone.

American Society of Civil Engineers (ASCE), Direct Design of Buried Concrete Pipe Standards Committee. (2000). "Standard practice for direct design of buried precast concrete pipe using standard installations (SIDD)," ASCE Standard No. 15-98, ASCE, Reston, Va.

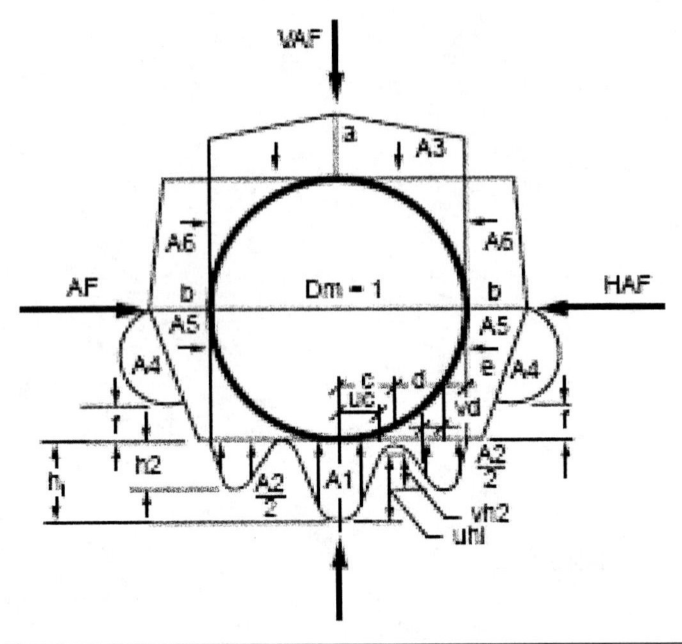

Installation Type	VAF	HAF	A1	A2	A3	A4	A5	A6	a	b	c	e	f	u	v
1	1.35	0.45	0.62	0.73	1.35	0.19	0.08	0.18	1.40	0.40	0.18	0.08	0.05	0.80	0.80
2	1.40	0.40	0.85	0.55	1.40	0.15	0.08	0.17	1.45	0.40	0.19	0.10	0.05	0.82	0.70
3	1.40	0.37	1.05	0.35	1.40	0.10	0.10	0.17	1.45	0.36	0.20	0.12	0.05	0.85	0.80
4	1.45	0.30	1.45	0.00	1.45	0.00	0.11	0.18	1.45	0.30	0.28	0.00	–	0.90	–

Notes:

1. VAF and HAF are vertical and horizontal arching factors. These coefficients represent non-dimensional total vertical and horizontal loads on the pipe, respectively. The actual total vertical and horizontal loads are (VAF) × (PL) and (HAF) × (PL), respectively, where PL is the prism load.
2. PL, the prism load, is the weight of the column of earth cover over the pipe outside diameter and is calculated as:

$$PL = w\left[\left(H + \frac{D_0(4 - x)}{96}\right)\right]\frac{D_0}{12}$$

3. Coefficients A1 through A5 represent the integration of non-dimensional vertical and horizontal components of soil pressure under the indicated portions of the component pressure diagrams (i.e., the area under the component pressure diagrams). The pressures are assumed to vary either parabolically or linearly, as shown, with the non-dimensional magnitudes at governing points represented by h1, h2, uh1, vh2, a and b. Non-dimensional horizontal and vertical dimensions of component pressure regions are defined by c, d, e, vc, vd, and f coefficients.
4. σ is calculated as (0.5−c−e).
   h1 is calculated as (1.5A1)/(c)(1+u).
   h2 is calculated as (1.5A2)/[(d)(1+v)+(2e)]

FIGURE 9-25. Arching coefficients and Heger Earth Pressure Distribution. American Society of Civil Engineers (ASCE), Direct Design of Buried Concrete Pipe Standards Committee. (2000). "Standard practice for direct design of buried precast concrete pipe using standard installations (SIDD)," ASCE Standard No. 15-98, ASCE, Reston, Va.

TABLE 9-8. Equivalent USCS and AASHTO Soil Classifications for SIDD Soil Designations

	Representative Soil Types		Percent Compaction	
SIDD Soil	USCS	AASHTO	Standard Proctor	Modified Proctor
Gravelly Sand (SW)	SW, SP GW, GP	A1, A3	100	95
			95	90
			90	85
			85	80
			80	75
			61	59
Sandy Silt (ML)	GM, SM, ML; also GC, SC with less than 20% passing No. 200 sieve	A2, A4	100	95
			95	90
			90	85
			85	80
			80	75
			49	46
Silty Clay (CL)	CL, MH, GC, SC	A5, A6	100	90
			95	85
			90	80
			85	75
			80	70
			45	40
Silty Clay (CL) but not allowed for haunch or bedding	CH	A7	100	90
			95	85
			90	80
			45	40

American Society of Civil Engineers (ASCE), Direct Design of Buried Concrete Pipe Standards Committee. (2000). "Standard practice for direct design of buried precast concrete pipe using standard installations (SIDD)," ASCE Standard No. 15-98, ASCE, Reston, Va.

charge, updated versions of the software, compatible with graphic user interfaces, can be obtained from the American Concrete Pipe Association (ACPA) for a nominal charge.

9.5.2. Standard Installations Indirect Design

As an alternative to the SIDD method and the use of computer programs, the ACPA developed a design method which combines the Indirect Design methods formerly used with the Marston-Spangler equations with the Standard Installations. This design method is presented both in ACPA Standard Installations and Bedding Factors for the Indirect Design Method, Design Data 40 (ACPA 1993) and in the ACPA Concrete Pipe Design Manual (ACPA 2000).

Although developed for the Direct Design method, the Standard Installations are readily applicable to and simplify the Indirect Design method. The Standard Installations are easier to construct and provide more realistic designs than the historical B, C, and D beddings. Development of bedding factors for the Standard Installations, as presented in the following paragraphs, follows the concepts of reinforced concrete design theories. The basic definition of a bedding factor is that it is the ratio of maximum moment in the T.E.B. test to the maximum moment in the buried condition, when the vertical loads under each condition are equal:

$$B_f = \frac{M_{TEST}}{M_{FIELD}} \quad (9\text{-}21)$$

where

B_f = bedding factor

M_{TEST} = maximum moment in pipe wall under a T.E.B. test load, inch-lb (N-m)

M_{FIELD} = maximum moment in pipe wall under field loads, inch-lb (N-m).

Consequently, to evaluate the proper bedding factor relationship, the vertical load on the pipe for each condition must be equal, which occurs when the springline axial thrusts for both conditions are equal. In accordance with the laws of statics and equilibrium, M_{TEST} and M_{FIELD} are:

$$M_{TEST} = [0.318 \times N_{FS}] \times (D + t) \quad (9\text{-}22)$$

and

$$M_{FIELD} = M_{FI}[0.38tN_{FI}] - [0.125N_{FI} \times c] \quad (9\text{-}23)$$

where

N_{FS} = axial thrust at the springline under a T.E.B. test load, lb/ft (N/m)

D = internal pipe diameter, inch (mm)

t = pipe wall thickness, inch (mm)

M_{FI} = moment at the invert under field loading, inch-lb/ft (N-mm/m)

N_{FI} = axial thrust at the invert under field loads, lb/ft (N/m)

c = thickness of concrete cover over the inner reinforcement, inch (mm).

Substituting equations:

$$B_f = \frac{(0.318N_{FS}) \times (D+t)}{M_{FI} - (0.38tN_{FI}) - (0.125N_{FI} \times c)} \tag{9-24}$$

Using SIDD, bedding factors were determined for a range of pipe diameters and depths of burial. These calculations were based on 1-inch (25.4-mm) cover over the reinforcement, a moment arm of 0.875*d* between the resultant tensile and compressive forces, and a reinforcement diameter of 0.075*t*. Evaluations indicated that for A, B, and C pipe wall thicknesses, there was negligible variation in the bedding factor due to pipe wall thickness or the concrete cover, *c*, over the reinforcement. The resulting bedding factors are presented in Table 9-9.

9.5.2.1. *Determination of Bedding Factor*

For trench installations, as discussed previously, experience indicates that active lateral pressure increases as the trench width increases to the transition width, provided the side fill is compacted. A parameter study of the Standard Installations indicates the bedding factors are constant for all pipe diameters under conditions of zero lateral pressure on the pipe. These bedding factors exist at the interface of the wall of the pipe and the soil and are called the minimum bedding factors, B_{fo}, to differentiate them from the fixed bedding factors developed by Spangler. Table 9-10 presents the minimum bedding factors.

TABLE 9-9. Bedding Factors, Embankment Conditions, B_{fe}

	Standard Installation			
Pipe Diameter	Type 1	Type 2	Type 3	Type 4
12 inches (300 mm)	4.4	3.2	2.5	1.7
24 inches (600 mm)	4.2	3.0	2.4	1.7
36 inches (900 mm)	4.0	2.9	2.3	1.7
72 inches (1,820 mm)	3.8	2.8	2.2	1.7
144 inches (3,650 mm)	3.6	2.7	2.2	1.7

1. For pipe diameters other than listed in Table 9-9, embankment condition factors, B_{fe} can be obtained by interpolation.
2. Bedding factors are based on the soils being placed with the minimum compaction specified in Fig. 9-23 and Table 9-6 for each Standard Installation.

American Concrete Pipe Association (ACPA). (2000). "Concrete pipe design manual," ACPA, Irving, Tex. Reprinted with permission.

TABLE 9-10. Trench Minimum Bedding Factors, B_{fo}

Standard Installation	Minimum Bedding Factor, B_{fo}
Type 1	2.3
Type 2	1.9
Type 3	1.7
Type 4	1.5

1. Bedding factors are based on the soils being placed with the minimum compaction specified in Fig. 9-24 and Table 9-7 for each Standard Installation.
2. For pipe installed in trenches dug in previously constructed embankment, the load and the bedding factor should be determined as an embankment condition unless the backfill placed over the pipe is of lesser compaction than the embankment.

American Concrete Pipe Association (ACPA). (2000). "Concrete pipe design manual," ACPA, Irving, Tex. Reprinted with permission.

The bedding factor is conservatively assumed to vary linearly between the minimum bedding factor and the bedding factor at the transition width, which is the embankment condition bedding factor.

The equation for this variable trench bedding factor is:

$$B_{fv} = \frac{(B_{fe} - B_{fo})(B_d - B_c)}{(B_{dt} - B_c)} + B_{fo} \qquad (9\text{-}25)$$

where

B_c = outside horizontal span of pipe, ft (m)
B_d = trench width at top of pipe, ft (m)
B_{dt} = transition width at top of pipe, ft (m)
B_{fe} = bedding factor, embankment
B_{fo} = minimum bedding factor, trench
B_{fv} = variable bedding factor, trench.

Transition width values, B_{dt} are provided in tables in the ACPA Concrete Pipe Design Manual (ACPA 2000) or can be calculated as identified previously in Fig. 9-5.

For pipe installed with 6.5 ft (2 m) or less of overfill and subjected to truck loads, the controlling maximum moment may be at the crown rather than the invert. Consequently, the use of an earth load bedding factor may produce unconservative designs. Crown and invert moments of pipe for a range of diameters and burial depths subjected to HS-20 truck live loadings were evaluated by the ACPA. The ACPA also evaluated the effect of bedding angle and live load angle (width of loading on the pipe). When HS-20 or other live loadings are encountered to a significant degree, the live load bedding factors, B_{fLL}, presented in Table 9-11 are satisfactory for a Type 4 Standard and become increasingly conservative for Types 3, 2, and 1.

TABLE 9-11. Bedding Factors, B_{fLL}, for HS-20 Live Loadings

Fill Height, ft (m)	Pipe Diameter, inches (mm)										
	12 (300)	24 (600)	36 (900)	48 (1,200)	60 (1,500)	72 (1,800)	84 (2,100)	96 (2,400)	108 (2,700)	120 (3,000)	144 (3,600)
0.5 (0.15)	2.2	1.7	1.4	1.3	1.3	1.1	1.1	1.1	1.1	1.1	1.1
1 (0.3)	2.2	2.2	1.7	1.5	1.4	1.3	1.3	1.3	1.1	1.1	1.1
1.5 (0.45)	2.2	2.2	2.1	1.8	1.5	1.4	1.4	1.3	1.3	1.3	1.1
2.0 (0.6)	2.2	2.2	2.2	2.0	1.8	1.5	1.5	1.4	1.4	1.3	1.3
2.5 (0.75)	2.2	2.2	2.2	2.2	2.0	1.8	1.7	1.5	1.4	1.4	1.3
3.0 (0.9)	2.2	2.2	2.2	2.2	2.2	2.2	1.8	1.7	1.5	1.5	1.4
3.5 (1.05)	2.2	2.2	2.2	2.2	2.2	2.2	1.9	1.8	1.7	1.5	1.4
4 (1.20)	2.2	2.2	2.2	2.2	2.2	2.2	2.1	1.9	1.8	1.7	1.5
4.5 (1.35)	2.2	2.2	2.2	2.2	2.2	2.2	2.2	2.0	1.9	1.8	1.7
5 (1.5)	2.2	2.2	2.2	2.2	2.2	2.2	2.2	2.2	2.0	1.9	1.8
5.5 (1.65)	2.2	2.2	2.2	2.2	2.2	2.2	2.2	2.2	2.2	2.0	1.9
6 (1.8)	2.2	2.2	2.2	2.2	2.2	2.2	2.2	2.2	2.2	2.1	2.0
6.5 (1.95)	2.2	2.2	2.2	2.2	2.2	2.2	2.2	2.2	2.2	2.2	2.2

1. For pipe diameters other than listed in Table 9-11, BfLL values can be obtained by interpolation.

American Concrete Pipe Association (ACPA). (2000). "Concrete pipe design manual," ACPA, Irving, Tex. Reprinted with permission.

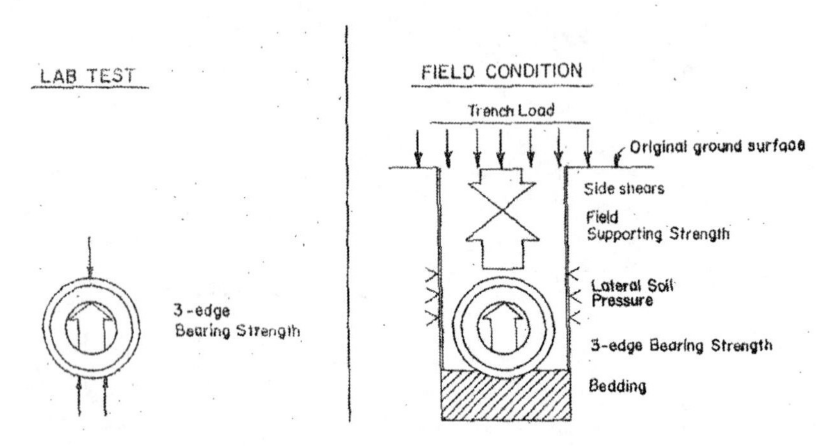

FIGURE 9-26. Both laboratory testing and field conditions should be used for rigid sewer pipe strength determination.

9.5.2.2. Pipe Strengths and Safety Factors

The structural design of rigid sewer pipe systems relates to the product's performance limit, expressed in terms of strength of the installed sewer pipe. Based on anticipated field loadings and concrete and reinforcing steel properties, calculations of shear, thrust, and moment can be made and compared to the ultimate strength of the section.

For precast concrete sewer pipes, the design strength is commonly related to a T.E.B. test strength measured at the manufacturing plant, as illustrated in Fig. 9-26. The plant test is much more severe because it develops shearing forces in the pipe wall that usually are not encountered in the field. Special reinforcement to resist shear in testing may be required in heavily reinforced sewer pipe, which would not be required by the more uniformly distributed field loading. In either case, the design load is equal to the field strength divided by the factor of safety. The design load is the calculated load on the pipe, and the field strength is the ultimate load which the pipe will support when installed under specified conditions of bedding and backfilling.

Since numerous reinforced concrete pipe sizes are available, T.E.B. test strengths are classified by D-loads. The D-load concept provides strength classification of pipe independent of pipe diameter. For reinforced circular pipe, the T.E.B. test load in lb/ft (N/m) equals the D-load times the inside diameter in feet (millimeters). The required T.E.B. strength of nonreinforced concrete pipe is expressed in lb/ft(N/m), not as a D-load, and is computed by the equation:

$$T.E.B. = \left[\frac{W_E}{B_{fe}} + \frac{W_L}{B_{fLL}} \right] \times F.S. \tag{9-26}$$

where
W_E = earth load, lb/ft (N/m)
B_{fe} = bedding factor, earth load
W_L = live load, lb/ft (N/m)
B_{fLL} = bedding factor, live load
$F.S.$ = safety factor.

The required T.E.B. strength of circular reinforced concrete pipe is expressed as D-load and is computed by the equation

$$D\text{-}load = \left[\frac{W_E}{B_{fe}} + \frac{W_L}{B_{fLL}} \right] \times \frac{F.S.}{D} \tag{9-27}$$

where
D = internal pipe diameter, inch (mm) and all other variables are as in Eq. (9-26).

The Indirect Design method employs a safety factor between yield stress and the desired working stress. In the Indirect Method, the factor of safety is defined as the relationship between the ultimate strength D-load and the 0.01-inch- (0.3-mm)-crack D-load. The 0.01-inch- (0.3-mm)-crack D-load (D0.01) is the maximum T.E.B. test load supported by a concrete pipe before a crack occurs having a width of 0.01 inch (0.3 mm) measured at close intervals, throughout a length of at least 1 ft (300 mm). The ultimate D-load (D_u) is the maximum T.E.B. test load supported by a pipe. D-loads are expressed in pounds per linear foot per foot of inside diameter (N/m/mm). The relationship between D_u and the 0.01-inch- (3-mm)-crack D-load is specified in the ASTM Standards C76 and C655 on concrete pipe. The relationship between D_u and the 0.01-inch- (3-mm)-crack D-load is 1.5 for 0.01-inch- (3-mm)-crack D-loads of 2,000 or less; 1.25 for 0.01-inch- (3-mm)-crack D-loads of 3,000 or more; and a linear reduction from 1.5 to 1.25 for 0.01-inch- (3-mm)-crack D-loads greater than 2,000 and less than 3,000. Therefore, a factor of safety of 1 should be applied if the 0.01-inch- (3-mm)-crack strength is used as the design criterion rather than the ultimate strength. The 0.01-inch- (3-mm)-crack width is an arbitrarily chosen test criterion and not a criterion for field performance or service limit.

When an HS-20 truck live loading is applied to the pipe, the live load bedding factor, B_{fLL}, is utilized as indicated in Eq. (9-27), unless the earth load bedding factor, B_{fe}, is of lesser value. If the earth load bedding factor is lower, use the lower B_{fe} value in place of B_{fLL}. For example, with a Type 4 Standard Installation of a 48-inch- (1,200-mm)-diameter pipe under 1 ft (0.3 m) of fill, the factors used would be $B_{fe} = 1.7$ and $B_{fLL} = 1.5$; but under 2.5 ft (0.75 m) or greater fill, the factors used would be $B_{fe} = 1.7$ and $B_{fLL} = 1.7$, rather than 2.2. For trench installations with trench widths less than transition width, B_{fLL} would be compared to the variable trench bedding factor, B_{fv}.

9.5.3. Marston-Spangler Supporting Strength Calculations

9.5.3.1. *General Relationships*

As noted previously, the inherent strength of a rigid sewer pipe usually is given by its strength in the T.E.B. test. Although this test is both convenient and severe, it does not reproduce the actual field load conditions. Thus, to select the most economical combination of bedding and sewer pipe strength, a relationship must be established between calculated load, laboratory strength, and field strength for various installation conditions (Fig. 9-26).

Field strength, moreover, depends on the distribution of the reaction against the bottom of the sewer pipe and on the magnitude and distribution of the lateral pressure acting on the sides of the pipe. These factors therefore make it necessary to qualify the term "field strength" with a description of conditions of installation in a particular case, as they affect the distribution of the reaction and the magnitude and distribution of lateral pressure. Just as for sewer pipe load computations, it is convenient when determining field strength to classify installation conditions as either trench or embankment.

9.5.3.2. *Laboratory Strength*

Rigid sewer pipe is tested for strength in the laboratory by the T.E.B. test (Fig. 9-26). Methods of testing are described in detail in ASTM C301 for vitrified clay sewer pipe and in C497 for concrete and reinforced concrete sewer pipe. The minimum strengths required for the T.E.B. tests of the various types of rigid sewer pipe are stated in the ASTM specifications for the pipe.

As noted above, in the case of reinforced concrete sewer pipe, laboratory strengths are divided into two categories: the load that will produce a 0.01-inch (0.3-mm) crack and the ultimate load the sewer pipe will withstand. Nonreinforced concrete and clay sewer pipe class designations indicate the ultimate strength in T.E.B. directly in pounds per foot (newtons per meter) for any diameter.

9.5.4. Design Relationships

The field strength is equal to the T.E.B. test strength of the pipe times the bedding factor. The ratio of the strength of a sewer pipe under any stated condition of loading and bedding to its strength measured by the T.E.B. test is called the bedding factor.

Bedding factors for trenches and embankments were determined experimentally between 1925 and 1935 at Iowa State College. They can also be calculated for most classes of bedding (Spangler 1956). The experimentally determined and calculated relationships may be combined as follows to determine the required bedding factor for any given pipe class:

$$\text{Bedding Factor} = \frac{\text{Design Load} \times \text{Factor of Safety}}{\text{T.E.B. Strength}} \quad (9\text{-}28)$$

By rearranging this expression, the required T.E.B. strength for a given class of bedding can be calculated as follows:

$$\text{Required T.E.B. Strength} = \frac{\text{Design Load} \times \text{Factor of Safety}}{\text{Bedding Factor}} \quad (9\text{-}29)$$

The strength of reinforced concrete sewer pipe in pounds per foot (newtons per meter) at either the 0.01-inch- (0.3-mm)-crack or ultimate load, divided by the nominal internal diameter of the sewer pipe in inches (millimeters), is defined as the D-load strength. Different classes of reinforced concrete pipe per ASTM C76 and other reinforced concrete pipe specifications are defined by D-load strengths. For example, ASTM C76 Class IV reinforced concrete sewer pipe is manufactured to a D-load of 2,000 lb/ft/ft of diameter (100 N/m/mm) to produce the 0.01-inch (0.3-mm) crack, and 3,000 lb/ft/ft (150 N/m/mm) to produce the ultimate load. Consequently, a 48-inch- (1,200-mm)-diameter Class IV reinforced concrete sewer pipe per ASTM C76 would have a minimum laboratory strength of 8,000 lb/ft (120,000 N/m) to produce the 0.01-inch (0.3-mm) crack and 12,000 lb/ft (180,000 N/m) at ultimate strength.

$$D\text{-}load = \frac{\text{Design Load} \times \text{Safety Factor}}{\text{Bedding Factor} \times \text{Diameter}} \quad (9\text{-}30)$$

In considering the design requirements, the engineer should evaluate the options of different types of rigid sewer pipe and different bedding classes, keeping in mind the differences in the relationship between test loadings and field loadings. It is customary when designing for 0.01-inch- (0.3-mm)-crack loads to use a factor of safety of 1, and when designs are based on ultimate strengths to use a factor of safety varying between 1.25 and 1.5.

9.5.5. Rigid Sewer Pipe Installation—Classes of Bedding and Bedding Factors for Trench Conditions

Four classes of beddings, shown in Fig. 9-27, are used most often for sewer pipes in trenches. They are described in the following subsections.

9.5.5.1. *Class A Bedding—Concrete Cradle*

The sewer pipe is bedded in a cast-in-place cradle of plain or reinforced concrete having a thickness equal to one-quarter of the inside pipe diameter, with a minimum of 4 inches (100 mm) and a maximum of 15 inches (380 mm) under the pipe barrel and extending up the sides for a height equal to

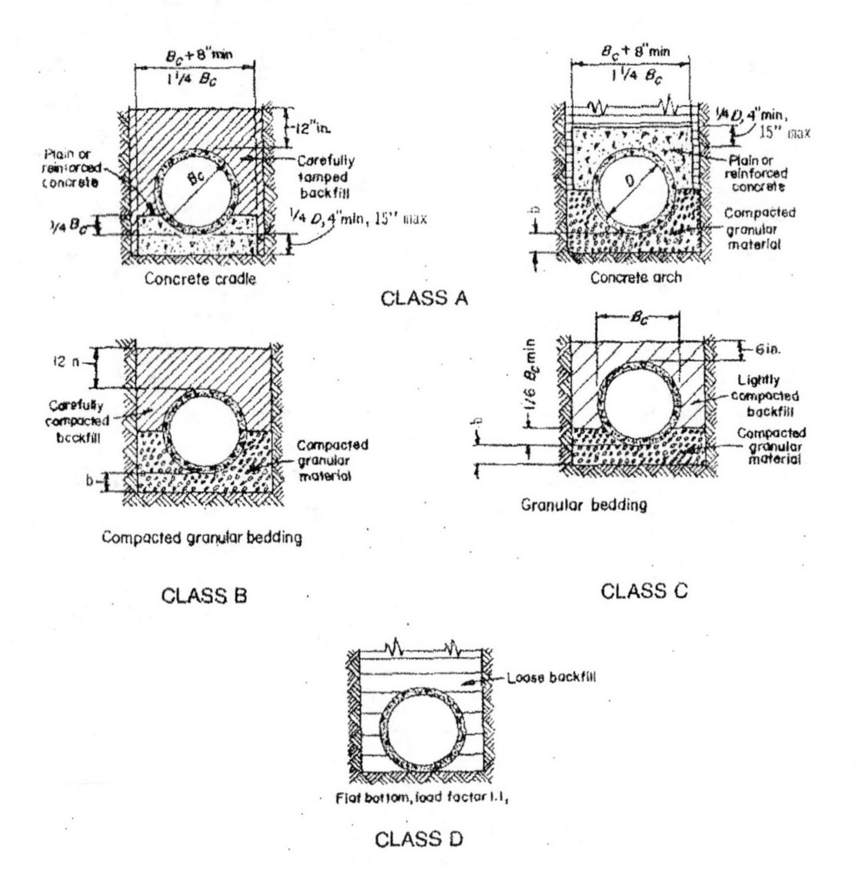

FIGURE 9-27. Classes of bedding for rigid sewer pipes in trench.

In rock trench, excavate at least 6 inches (152.4 mm) below bell of pipe except where a concrete cradle is used (inch × 25.4 = mm). Bedding thickness under pipe barrel, *b*, shall be ⅛ B_c, 100 mm (4") min, 150 mm (6") max. American Concrete Pipe Association (ACPA). (2000). "Concrete pipe design manual," ACPA, Irving, Tex. Reprinted with permission.

one-quarter of the outside diameter. The cradle shall have a width at least equal to the outside diameter of the sewer pipe barrel plus 8 inches (200 mm). Construction procedures must be executed carefully to prevent the sewer pipe from floating off line and grade during placement of the cradle concrete.

If the cradle is made of reinforced concrete, the reinforcement is placed transverse to the pipe and 3 inches (76 mm) clear from the bottom of the cradle. The percentage of reinforcement, p, is the ratio of the area of transverse reinforcement to the area of concrete cradle at the pipe invert above the centerline of the reinforcement.

Consideration must be given to the points at which the cradle (or arch, described in the next section) begins and terminates with respect to the pipe joints. In general, the concrete cap, cradle, or envelope starts and terminates at the face of a pipe bell or collar to avoid shear cracks.

Haunching and initial backfill above the cradle to 12 inches (300 mm) above the crown of the sewer pipe should be placed and compacted as described in later sections of this chapter. The cradle must be cured sufficiently to develop full bedding support prior to final backfilling.

The bedding factor for Class A concrete cradle bedding is 2.2 for plain concrete with lightly tamped backfill; 2.8 for plain concrete with carefully tamped backfill; up to 3.4 for reinforced concrete with the percentage of reinforcement, p, equal to 0.4%; and up to 4.8 with p equal to 1%.

9.5.5.2. *Class A Bedding—Concrete Arch*

The sewer pipe is bedded in carefully compacted granular material having a minimum thickness of one-eighth of the outside sewer pipe diameter but not less than 4 inches (100 mm) or more than 6 inches (150 mm) between the sewer pipe barrel and bottom of the trench excavation. Granular material is then placed to the springline of the sewer pipe and across the full breadth of the trench. The haunching material beneath the sides of the arch must be compacted so as to be unyielding. Crushed stone in the 0.25- to 0.75-inch (5- to 20-mm) size range is the preferred material. The top half of the sewer pipe is covered with a cast-in-place plain or reinforced concrete arch having a minimum thickness of 4 inches (100 mm) or one-quarter of the inside pipe diameter but not to exceed 15 inches (380 mm), and having a minimum width equal to the outside sewer pipe diameter plus 8 inches (200 mm).

If the arch is made of reinforced concrete, the reinforcement is placed transverse to the pipe and 2 inches (50 mm) clear from the top of the arch. The percentage of reinforcement is the ratio of the transverse reinforcement to the area of concrete arch above the top of the pipe and below the centerline of the reinforcement.

The bedding factor for Class A concrete arch-type bedding is 2.8 for plain concrete; up to 3.4 for reinforced concrete with p equal to 0.4%; and up to 4.8 for p equal to 1%.

9.5.5.3. *Class B Bedding*

The sewer pipe is bedded in carefully compacted granular material having a minimum thickness of one-eighth of the outside sewer pipe diameter but not less than 4 inches (100 mm) or more than 6 inches (150 mm) between the pipe barrel and the bottom of the trench excavation. Granular material is then placed to the springline of the pipe and across the full width of the trench. The haunching material beneath the arches of the pipe must be compacted so as to be unyielding. Crushed stone in the 0.25- to 0.75-inch (5- to 20-mm) size range is the preferred material. Both haunching and initial backfill to a minimum depth of 12 inches (300 mm) over the top of the sewer pipe should be placed and compacted.

The bedding factor for Class B bedding is 1.9.

9.5.5.4. *Class C Bedding*

The sewer pipe is bedded in compacted granular material. The bedding has a minimum thickness of one-eighth of the outside sewer pipe diameter but not less than 4 inches (100 mm) or more than 6 inches (150 mm), and shall extend up the sides of the sewer pipe to one-sixth of the pipe outside diameter. The remainder of the side fills, to a minimum depth of 6 inches (150 mm) over the top of the pipe, consists of lightly compacted backfill.

The bedding factor for Class C bedding is 1.5.

9.5.5.5. *Class D Bedding*

For this class of bedding the bottom of the trench is left flat, as cut by excavating equipment. Successful Class D installations of rigid conduits can be achieved in locations with appropriate soil conditions, trench width control, and light superimposed loads. Care must be taken to prevent point loading of sewer pipe bells. The excavation of bell holes will prevent such loading. Existing soil should be shovel-sliced or otherwise compacted under the haunching of the sewer pipe to provide some uniform support. In poor soils, granular bedding material is generally a more practical cost-effective installation.

Since the field conditions with Class D bedding can approach the load conditions for the T.E.B. test, the bedding factor for Class D bedding is 1.1.

9.5.6. Encased Pipe

Total encasement of rigid sewer pipe in concrete may be necessary where the required field strength cannot be obtained by other bedding. A typical concrete encasement detail is shown in Fig. 9-28. The bedding factor for concrete encasement varies with the thickness of concrete and the use of reinforcing, and may be greater than that for a concrete cradle or arch. Concrete thickness and reinforcing should be determined by the application of conventional structural theory and analysis. The bedding factor for the encasement shown in Fig. 9-28 is 4.5.

Concrete encasement also may be required for sanitary sewers built in deep trenches to ensure uniform support or for sewer pipe built on comparatively steep grades where there is the possibility that earth beddings may be eroded by currents of water under and around the pipe. Flotation of the pipe during concrete placement should be prevented by suitable means.

Sample Calculations

Example 9-11. Refer to Example 9-1, which describes the backfill load on a 24-inch- (610-mm)-diameter pipe with 14 ft (4.3 m) of cover; the load was found to be 4,720 lb/ft. If vitrified clay pipe is to be specified and a factor of safety of 1.5 is selected, the design load will be 4,720 × 1.5 or 7,080 lb/ft (103,300 N/m).

The crushing strength requirement of 24-inch- (610-mm)-diameter extra-strength clay sewer pipe (ASTM C700) based on the T.E.B. method is 4,400 lb/ft. Dividing this into 7,080 lb/ft (the design load), the minimum required bedding factor of 1.61 is obtained. Figure 9-27 and the accompanying information indicates that a Class B bedding is required for this installation.

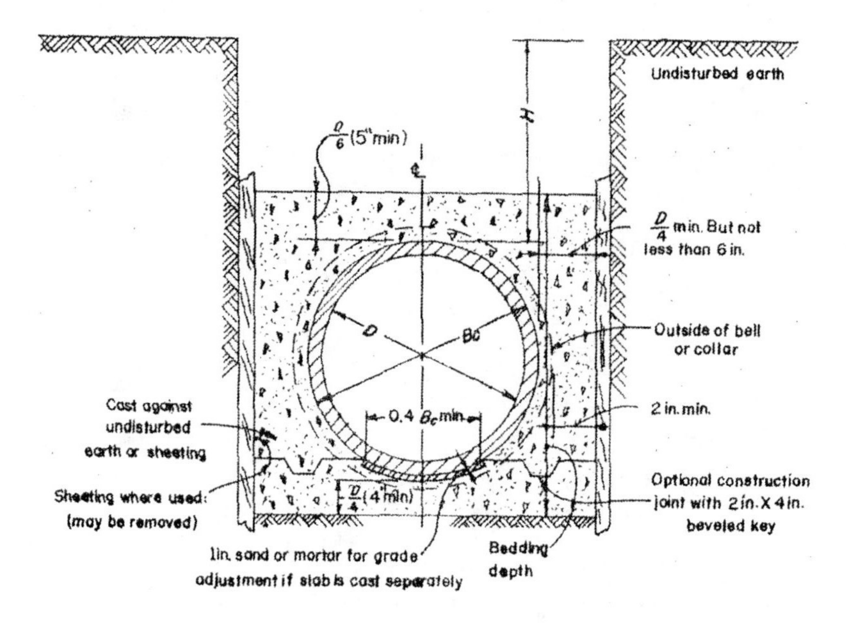

FIGURE 9-28. Typical concrete encasement details (inch × 25.4 = mm).

Example 9-12. Refer to Example 9-3. If a 24-inch- (610-mm)-diameter reinforced concrete sewer pipe (one line of reinforcement near center of wall) and a factor of safety of 1.5 based on the minimum ultimate test strength are selected, the design load will be 1.50 × 5,750 or 8,620 lb/ft (125,900 N/m).

If the minimum ultimate test strength of the pipe is 6,000 lb/ft (ASTM C76, Class IV-3,000 D), the required bedding factor will be 8,620/6,000 or 1.44. According to Fig. 9-27, this installation will require a Class C bedding.

9.5.7. Field Strength in Embankments

It is possible for the active soil pressure against the sides of rigid sewer pipe placed in an embankment to be a significant factor in the resistance of the structure to vertical load. This factor is important enough to justify a separate examination of the field strength of embankment sewer pipe.

The following discussion of field strength in embankments is based on the theory developed by Marston and Spangler. Given the potential for variations from the ideal conditions assumed in this theory, the design engineer should approach designs based on this theory with some caution. With time, lateral pressures for trench installation usually will approach "at rest" conditions, which correspond to vertical overburden times a factor (usually between 0.5 and 1). However, for a negative-projecting conduit, a positive-projecting conduit, and induced trench embankment conditions, the lateral pressure magnitude and distribution may be much different, and these may control the structural design of the sewer pipe.

9.5.8. Positive-Projecting Sewer Pipe

The bedding factor for rigid sewer pipes installed as projecting sewer pipe under embankments or in wide trenches depends on the class of bedding in which the sewer pipe is laid, the magnitude of the active lateral soil pressure against the sides of the sewer pipe, and the area of the sewer pipe over which the active lateral pressure is effective.

For projecting sewer pipe, the bedding factor L_f is:

$$L_f = \frac{A}{N - xq} \tag{9-31}$$

in which A is a sewer pipe shape factor; N is a parameter which is a function of the bedding class; x is a parameter dependent on the area over which lateral pressure effectively acts; and q is the ratio of total lateral pressure to total vertical load on the sewer pipe.

Classes of bedding for projecting sewer pipe are shown in Fig. 9-29. The values of A for circular, elliptical, and arch sewer pipe are shown

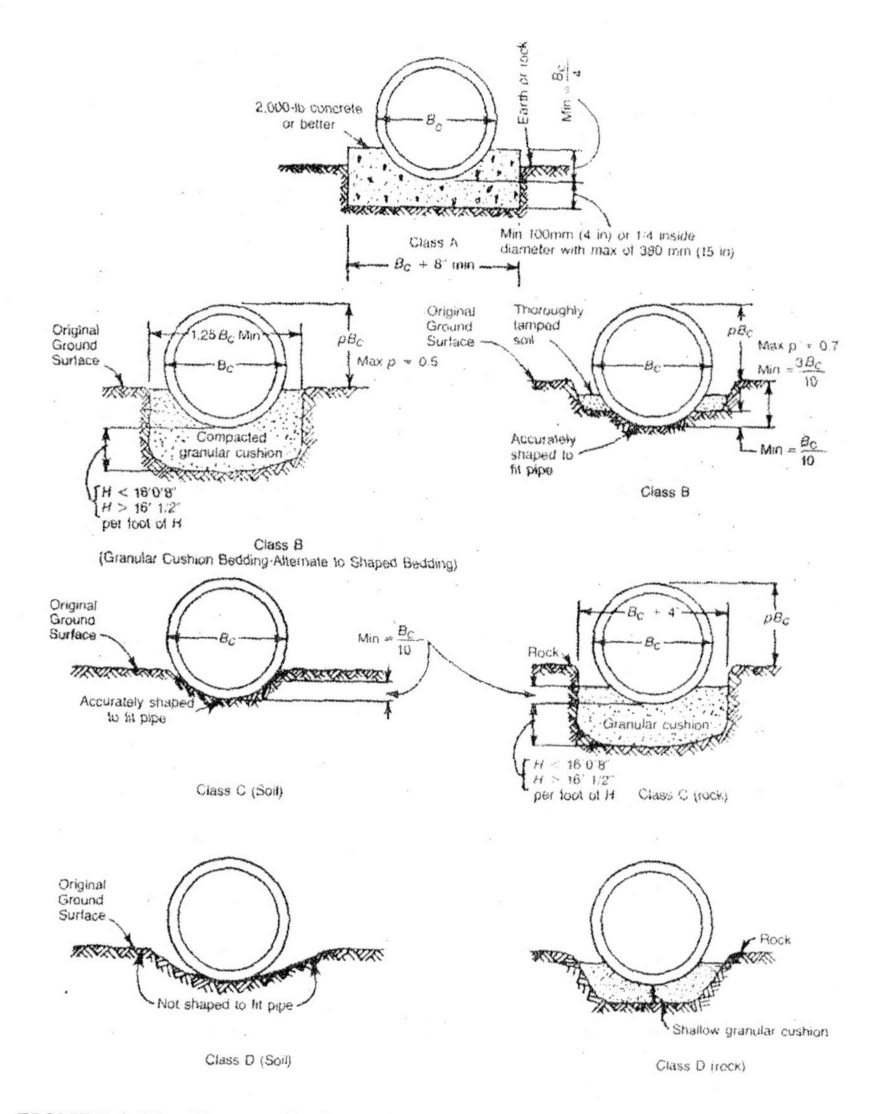

FIGURE 9-29. Classes of bedding for projecting sewer pipes (ft × 0.3 = m; inch × 25.4 = mm).

in Table 9-12. Values of N for various classes of bedding are given in Table 9-13. Values of x for circular, elliptical, and arch sewer pipe are listed in Table 9-14.

The projection ratio, m, in Eq. (9-32) refers to the fraction of the sewer pipe diameter over which lateral pressure is effective. For example, if lateral pressure acts on the top half of the sewer pipe above the horizontal diameter, m equals 0.5. The ratio of total lateral pressure to total

TABLE 9-12. Values of A for Circular, Elliptical, and Arch Sewer Pipe

Sewer Pipe Shape	A
Circular	1.431
Elliptical:	
Horizontal elliptical and arch	1.337
Vertical elliptical	1.021

TABLE 9-13. Values of N

	Value of N Sewer Pipe Shape		
Class of Bedding	Circular	Horizontal Elliptical	Vertical Elliptical
A (reinforced cradle)	0.421–0.505	—	—
A (unreinforced cradle)	0.505–0.636	—	—
B	0.707	0.630	0.516
C	0.840	0.763	0.615
D	1.310	—	—

TABLE 9-14. Values of x

	x			
	Class A Bedding	Other than Class A Bedding		
Fraction of Sewer Pipe Subjected to Lateral Pressure	Circular Sewer Pipe	Circular Sewer Pipe	Horizontal Elliptical Sewer Pipe	Vertical Elliptical Sewer Pipe
0	0.150	0	0	0
0.3	0.743	0.217	0.146	0.238
0.5	0.856	0.423	0.268	0.457
0.7	0.811	0.594	0.369	0.639
0.9	0.678	0.655	0.421	0.718
1.0	0.638	0.638	—	—

vertical load, q, for positive-projecting sewer pipe may be estimated by the formula:

$$q = \frac{mK}{C_c}\left(\frac{H}{B_c} + \frac{m}{2}\right)K \qquad (9\text{-}32)$$

where

K = the ratio of unit lateral pressure to unit vertical pressure (Rankine's ratio)
H = height of cover
B_c = outside diameter of pipe.

A value of K equal to 0.33 usually will be sufficiently accurate for use in Eq. (9-32). Values of C_c are found in Fig. 9-8.

9.5.9. Negative-Projecting Sewer Pipe

The bedding factor for negative-projecting sewer pipe may be the same as that for trench conditions corresponding to the various classes of bedding given in Section 9.5.5. The bedding factors for Class B, C, and D bedding do not take into account lateral pressures against the sides of the sewer pipe because unfavorable construction conditions often prevail at the bottom of a trench. However, in the case of negative-projecting sewer pipe, conditions may be more favorable and it may be possible to compact the side fill soils to the extent that some lateral pressure against the sewer pipe can be relied on. If such favorable conditions are anticipated, it is suggested that the bedding factor be computed by means of Eqs. (9-31) and (9-32), using a value of K equal to 0.15 for estimating the lateral pressure on the sewer pipe.

9.5.10. Induced Trench Conditions

Induced trench sewer pipes usually are installed as positive-projecting sewer pipes before the overlying soil is compacted and the induced trench is excavated. Therefore, lateral pressures are effective against the sides of the sewer pipe and the bedding factor should be calculated using Eqs. (9-31) and (9-32).

9.6. FLEXIBLE PIPE STRUCTURAL DESIGN

Several types of flexible pipe are available for use as sanitary sewer pipe material. Among the most common are ductile iron pipe (DIP), acrylonitrile-butadiene-styrene (ABS) composite pipe, and ABS solid wall pipe, polyvinyl chloride (PVC) pipe, polyethylene (PE) pipe, fiberglass-reinforced plastic (FRP), and reinforced plastic mortar (RPM) pipe. Coated corrugated metal pipe (CMP) has been used for sanitary sewer pipe, but its use is not common.

9.6.1. General Method

Flexible sewer pipes under earth fills and in trenches derive their ability to support load from their inherent strength plus the passive resistance of the soil as the pipe deflects and the sides of the sewer pipe move outward against the soil side fills. Proper compaction of the soil side fills is important to the long-term structural performance of flexible sewer pipe. This type of pipe fails by excessive deflection and by collapse, buckling, cracking, or delamination rather than by rupture of the sewer pipe walls, as in the case of rigid pipes. The extent to which flexible pipe deflects is most commonly used to judge performance and as a basis for design. The amount of deflection considered permissible is dependent on physical properties of the pipe material used and on project limitations.

The limiting buckling stress for flexible pipes takes into account the restraining effect of the soil structure around the pipe and the properties of the pipe wall. Equations for the critical stress in the pipe wall can be found in manufacturers' handbooks for the various types of pipe.

Empirical data on long-term deflection in different burial conditions generally can be obtained from the manufacturer. When such design data are not available, the approximate long-term deflection of flexible sewer pipe can be calculated using the Modified Iowa Formula developed by Watkins and Spangler (1958):

$$\Delta X = \frac{D_L K_b W_c r^5}{EI + 0.061E' r^3} \qquad (9\text{-}33)$$

where

ΔX = horizontal deflection, in inches (mm)
K_b = bedding factor
D_L = deflection lag factor
W_c = load, in lb/ft (N/m)
r = mean radius of pipe, in inches (mm)
E = modulus of tensile elasticity, in lb/inch^2 (N/m^2)
I = moment of inertia per length, in inches (mm) to fourth power per inch (mm)
E' = modulus of soil reaction, in lb/inch^2 (N/m^2).

Equation (9-33) was developed primarily for flexible corrugated metal sewer pipe under embankments, but has been shown to be applicable to most flexible pipe materials.

For small deflections, the vertical deflection ΔY may be assumed to approximately equal the horizontal deflection ΔX in Eq. (9-33). Much research is being conducted on the application of Eq. (9-33) to pipe having a low ratio of pipe stiffness to soil stiffness where the vertical deflection may not be assumed to approximately equal the horizontal deflection. The result of this research is beyond the scope of this chapter but should be reviewed by the designer as a part of the design process.

Flexible sewer pipes that are to support a fill should not be placed directly on a cradle or on pile bents. If such supports are necessary, they should have a flat top and be covered with a compressible earth cushion. In those instances where flexible pipe is to be encased in concrete, the pipe manufacturer should be consulted. Flexible pipes should not be encased in concrete unless the encasement is designed for the full vertical load as a rigid pipe.

The passive resistance of the soil at the sides of the pipe greatly influences pipe deflection. This passive resistance is expressed as the modulus of soil reaction, E'. It is generally recognized as being related to the degree of compaction of the soil and to the type of soil. Extensive laboratory and field tests by the U.S. Bureau of Reclamation have resulted in the establishment of an empirical relationship between modulus of soil reaction, degree of compaction of bedding, and type of bedding material. Recommended values for E' in Eq. (9-33) have been developed and are given in Table 9-15 (Howard 1977). (See Table 9-16 for soil symbol definitions.)

The deflection lag factor, empirically determined, compensates for the time consolidation characteristics of soil, which may permit flexible sewer pipes to continue to deform for some period after installation. Long-term deflection will be greater with light or moderate degrees of compaction of side fills when compared to values for heavy compaction. The better the compaction, the lower the initial deflection and the greater the magnitude of the long-term lag factor. Lag factors greater than 2.5 have been recorded in dry soil. Recommended values of this factor range from 1.25 to 1.5. Bedding requirements for flexible sewer pipe installation are discussed below.

Values of the bedding constant, K_b, depending on the width of the sewer pipe bedding, are shown in Table 9-17.

In the deflection formula [Eq. (9-33)], the first term in the denominator, EI (stiffness factor), reflects the influence of the inherent stiffness of the sewer pipe on deflection. The second term, $0.061\ E'r^3$, reflects the influence of the passive pressure on the sides of the pipe. With flexible pipes, the second term is normally predominant.

It should be noted that the E' values in Table 9-15 are average values. Using them results in a 50% chance that the actual deflections will be higher than calculated. A conservative approach would be to use 75% of the E' values given to calculate maximum deflections.

TABLE 9-15. Average Values of Modulus of Soil Reaction E', for Initial Flexible Pipe Deflection

Soil Type Pipe Bedding Material (United Classification System[a])	E′ for Degree of Compaction of Bedding, in psi (Pa)			
	Dumped	Slight, <85% Proctor, 40% Relative Density	Moderate, 85%–95% Proctor, 40%–70% Relative Density	High, >95% Proctor, >70% Relative Density
Fine-grained soils (LL > 50)[b] Soils with medium to high plasticity CH, MH, CH-MH	No data available; consult a competent soils engineer; otherwise use $E' = 0$			
Fine-grained soils (LL < 50) Soils with medium to no plasticity CL, ML, ML-CL, with less than 25% coarse-grained particles	50 (340)	200 (1,380)	400 (2,760)	1,000 (6,890)
Fine-grained soils (LL < 50) Soils with medium to no plasticity CL, ML, ML-CL, with more than 25% coarse-grained particles				
Coarse-grained soils with fines GM, GC, SM, SC, with more than 12% fines	100 (690)	400 (2,760)	1,000 (6,890)	2,000 (13,790)
Coarse-grained soils with little or no fines GW, GP, SW, SP[c] (contains less than 12% fines)	200 (1,380)	1,000 (6,890)	1,000 (6,890)	2,000 (13,790)
Crushed rock	1,000 (6,890)	3,000 (20,680)	3,000 (20,680)	3,000 (20,680)

1. Standard Proctors in accordance with ASTM D698 are used with this table.
2. Values applicable only for fills less than 50 ft (15 m). Table does not include any safety factor. For use in predicting initial deflections only, the appropriate deflection lag factor must be applied for long-term deflections. If bedding falls on the borderline between two compaction categories, select lower E′ value or average the two values.
3. It should be noted that the E' values in Table 9-15 are average values. Using them results in a 50% chance that the actual deflections will be higher than calculated. A conservative approach would be to use 75% of the E' values given to calculate maximum deflections.

[a] ASTM Designation D2487, USBR Designation E-3.

[b] LL, liquid limit.

[c] Or any borderline soil beginning with one of these symbols (e.g., GM-GC, GC-SC).

TABLE 9-16. Soil Classifications[a]

Soil Class	Group Symbol	Typical Names	Comments
I		Crushed rock	Angular, 6–40 mm
II	GW	Well-graded gravels	40 mm maximum
	GP	Poorly graded gravels	
	SW	Well-graded sands	
	SP	Poorly graded sands	
III	GM	Silty gravels	
	GC	Clayey gravels	
	SM	Silty sands	
	SC	Clayey sands	
IV	MH, ML	Inorganic silts	Not recommended for
	CH, CL	Inorganic clays	bedding, haunching,
V	OL, OH	Organic silts and clays	or initial backfill
	PT	Peat	

[a]Adapted, with permission, from ASTM D2321-89, "Standard Practice for Underground Installation of Thermoplastic Pipe for Sewers and Other Gravity-Flow Applications." Copyright © 2005 ASTM International, West Conshohocken, Penn.

The performance of the flexible sewer pipe in retaining its shape and integrity is largely dependent on the selection, placement, and compaction of the envelope of soil surrounding the structure. For this reason, as much care should be taken in the design of the bedding and initial backfill as is used in the design of the sewer pipe. The backfill material selected should preferably be of a granular nature to provide good shear characteristics. Cohesive soils are generally less suitable because of the importance of proper moisture content and the difficulty of obtaining proper compaction in a limited work space. If the pipe–soil envelope is subject to variations in the groundwater level, the engineer will need to take steps to prevent migration of fines from the surrounding material into the granular backfill, resulting in a loss of sidewall support.

TABLE 9-17. Values of Bedding Constant, K_b

Bedding Angle, in degrees	K_b
0	0.110
30	0.108
45	0.105
60	0.102
90	0.096
120	0.090
180	0.083

If, under embankment conditions, the material placed around the sewer pipe is different from that used in the embankment, or if for construction reasons fill is placed around the sewer pipe before the embankment is built, the compacted backfill should cover the pipe by at least 1 ft (0.3 m) and extend at least one diameter to either side.

9.6.2. Loads on Flexible Plastic Pipe

The load carried by a buried flexible pipe in a narrow trench may be calculated using the Marston formula [Eq. (9-3)]. A conservative design approach may be used by assuming the dead load carried by a flexible pipe–soil system in any installation to be the prism load. For normal installations, the prism load is the maximum load that can he developed.

The load on a projecting flexible sewer pipe is calculated using Eq. (9-8). As before, the load coefficient (C_c) depends on the projection ratio (p), settlement ratio (r_{sd}), and the ratio of fill height to pipe width (H/B_c), and can be determined from Fig. 9-8. For flexible projecting pipe, the product $r_{sd}p$ is negative or zero. As shown in Fig. 9-8, when the product is zero the load coefficient, C_c, equals H/B_c. Equation (9-8) then becomes equivalent to Eq. (9-1), which gives the prism load:

$$W_d = HwB_c$$

9.6.3. Thermoplastic Pipe

A wide range of physical properties is available from various plastic sewer pipe materials. Many have ASTM or ANSI specifications. Thermoplastic pipe materials (ABS, PE, and PVC) are all affected by temperature. Specified test requirements are applicable at 73.4 °F (23 °C). At higher temperatures, the plastic becomes less rigid but impact strength is increased. Lower temperatures result in increased brittleness but greater pipe stiffness. Pipe being installed at low temperatures requires careful handling. Certain chemicals will increase the possibility of environmental stress cracking (ESC) for certain thermoplastics. Typically, these compounds include concentrated oxidizing agents, organic chemicals, and oils, fats, and waxes. If any of these compounds are expected in high concentrations, the engineer should review the proposed application with the pipe manufacturer. It may be possible to compensate by utilizing a different plastic formulation or by utilizing another material.

9.6.3.1. Laboratory Load Test

The standard test to determine pipe stiffness or load deflection characteristics of plastic pipe is the parallel-plate loading test. This test is conducted in accordance with ASTM D2412, Standard Test Method for

External Loading Properties of Plastic Pipe by Parallel-Plate Loading. In the test, a short length of pipe is loaded between two rigid, parallel flat plates as they are moved together at a controlled rate. Load and deflection are noted. The parallel-plate loading test determines the pipe stiffness (PS) at a prescribed deflection, which for convenience in testing is arbitrarily set at 5%. This is not to be considered a limitation on field deflection. The pipe stiffness is defined as the value obtained by dividing the force (F) per unit length by the resulting deflection in the same units at the prescribed percentage deflection, and is expressed in pounds per square inch (newtons per square millimeter):

$$PS = \frac{F}{\Delta Y} = \frac{EI}{0.149r^3} \tag{9-34}$$

where

F = force per unit length, in lb/inch (N/mm)
ΔY = deflection, in inches (mm)
E = modulus of elasticity, in lb/inch^2 (N/m^2)
r = mean radius of pipe, in inches (mm)
$I = t^3/12$
t = wall thickness, in inches (mm).

Minimum required pipe stiffness values are stated in plastic sewer pipe specifications. Table 9-18 lists the ASTM specifications for the various types of plastic pipe and the corresponding pipe stiffness values.

The stiffness factor, SF, is the pipe stiffness multiplied by the quantity $0.149r^3$.

$$SF = EI = \frac{F}{\Delta Y} 0.149r^3 \tag{9-35}$$

$$SF = 0.149r^3(PS) \tag{9-36}$$

The stiffness factor, EI, is used in the Modified Iowa Formula [Eq. (9-33)] to determine approximate field deflections under earth loads. It is the engineer's responsibility to establish the acceptable field deflection limit and to design the installation accordingly. The manufacturer should be consulted for recommended field installation deflection limits.

Specifications also may require that some types of plastic sewer pipe withstand extreme deflections, such as 40% of the original diameter in a parallel-plate loading test, without evidence of splitting, cracking, or breaking. There is no corresponding load requirement. These extreme deflection tests are instantaneous and are intended for production quality control during pipe manufacturing. Although these extreme deflections (observed

TABLE 9-18. Stiffness Requirements for Plastic Sewer Pipe Parallel-Plate Loading

Material	ASTM Specification	Nominal diameter, D, inches (mm)	Required stiffness at 5% deflection, lb/inch² (kPa)
ABS Composite	D2680	8–15 (200–380)	200 (1,380)
RPM	D3262	8–18 (200–450)	17–99 (Varies)
		20–108 (500–2,750)	10 (69)
PVC	D2729	2 (50)	59 (407)
	(PVC-12454)	3 (75)	19 (131)
		4 (100)	11 (76)
		5 (125)	9 (62)
		6 (150)	8 (55)
	D2729	2 (50)	74 (510)
	(PVC-13364)	3 (75)	24 (165)
		4 (100)	13 90)
		5 (125)	12 (83)
		6 150)	10 (69)
	D3034	4–15 (100–375)	
	SDR 41		28 (193)
	SDR 35		46 (317)
	SDR 26		115 (792)
	SDR 23.5		135 (930)
	F679	18–48 (450–1,200)	46 (317)
			115 (792)
	F794	4–48 (100–1,200)	46 (317)
	F949	4–36 (100–900)	317 (46)
		8 and 10 (200 and 250)	792 (115)
	F1803	18–60 (450–1,500)	46 (317)
PVC Composite	D2680	8–15 (200–375)	200 (1,380)

in ASTM D2412 testing) are limiting in terms of duration and extent of pipe distortion, they can be important to the engineer in selecting sewer pipe materials of construction.

9.6.3.2. Methods of Analysis

Plastic pipe analysis requires the engineer to check values that include pipe stiffness, pipe deflection, ring buckling strength, hydrostatic wall buckling, wall crushing strength, and wall strain cracking.

Pipe Stiffness. When plastic pipe is installed in granular backfills, the stiffness of the plastic pipe selected will affect the end performance. Stiffness for plastic pipes is most widely discussed in terms of pipe stiffness ($F/\Delta Y$), which must be measured by the ASTM D2412 test. Most plastic pipe standards have specific minimum required pipe stiffness levels. Although pipe stiffness is used to estimate deflections due to service loads, stiffness is also the primary factor in controlling installation deflections.

AASHTO controls installation deflection with a flexibility factor (FF) limit indicated in Eqs. (9-37) and (9-38).

$$FF = \frac{D^2}{EI} \times 1000 \le C_{FF} \tag{9-37}$$

$$PS = \frac{EI}{0.149r^3} \ge \frac{C_{PS}}{D} \tag{9-38}$$

where

D = mean pipe diameter, inch (m)
E = the initial modulus (Young's modulus) of the pipe wall material, lb/inch^2 (N/m^2)
I = pipe wall moment of inertia, inch^4/inch (m^4/m)
C_{FF} = constant: 95 English (0.542 metric)
PS = pipe stiffness, lb/inch^2 (N/m^2)
r = mean pipe radius, inch (m)
C_{PS} = constant: 565 English (98,946 metric).

Deflection. Excessive pipe deflections should not occur if the proper pipe is selected and it is properly installed and backfilled with granular materials. However, when pipes are installed in cohesive soils, the deflection can be excessive. Deflections occur from installation loadings (the placement and compaction of backfill) and service loads due to soil cover and live loads.

In installations in cohesive soils, where heavy compaction equipment is often used or when difficult to compact backfill materials (GP, SP, CL, ML, etc.; refer to Table 9-8) are used, specifying a minimum pipe stiffness of 46 psi (317 N/m^2) or twice that required by Eq. (9-38), whichever is less, is desirable to facilitate backfill compaction and control installation deflections.

Deflections under service loads depend mostly on the quality and compaction level of the backfill material in the pipe envelope. Service load deflections are generally evaluated by using Spangler's Iowa Formula:

$$\frac{\Delta Y}{D} = \frac{KWr^3}{EI + (0.061E'r^3)}$$

where

$\Delta Y/D$ = pipe deflection, percent
K = bedding constant
W = service load on crown of pipe, lb/inch (N/mm)
r = mean pipe radius, inches (mm)
E and I have meanings previously given
E' = modulus of soil reaction, lb/inch^2 (N/m^2).

However, it significantly overpredicts deflections for stiffer pipes [pipe stiffnesses greater than 100 lb/inch^2 (4,790 N/m^2)] and underpredicts deflections for less stiff pipes [pipe stiffnesses less than 20 lb/inch^2 (960 N/m^2)]. In both cases, the error is roughly a factor of 2. The form of the Iowa Formula (Eq. 9-28) easiest to use is shown in Eq. (9-39):

$$\frac{\Delta Y}{D} = \left(\frac{D_L KP}{0.149(PS) + 0.061(E')}\right)100 \tag{9-39}$$

where

$\Delta Y/D$ = pipe deflection, percent
D_L = deflection lag factor
= 1 minimum value for use only with granular backfill and if the full soil prism load is assumed to act on the pipe
= 1.5 minimum value for use with granular backfill and assumed trench loadings
= 2.5 minimum value for use with CL and ML backfills, for conditions where the backfill can become saturated, etc.
K = bedding constant (typically 0.11)
P = service load on the crown of the pipe, lb/inch^2 (N/m^2)
PS = pipe stiffness, lb/inch^{22} (N/m^2)
E' = modulus of soil reaction, lb/inch^2 (N/m^2).

Table 9-15 provides generally accepted values that may apply to specific site conditions and backfill materials if they do not become saturated or inundated.

Wall Stress (Crushing). Wall stress is evaluated on the basis of conventional ring compression formulas. Because of the time-dependent strength levels of plastic materials, long-term loads such as soil and other dead loads must be evaluated against the material's long-term (50-year) strength. Very short-term loads, such as rolling vehicle loads, may be evaluated using initial properties. The following equations are used to evaluate wall stress:

$$T_{ST} = \frac{DP_{ST}}{2} \tag{9-40}$$

$$T_{LT} = \frac{DP_{LT}}{2} \tag{9-41}$$

$$A \geq 2\left(\frac{T_{ST}}{f_i} + \frac{T_{LT}}{f_{50}}\right) \times 10^6 \tag{9-42}$$

where

T_{ST} = thrust due to short-term loads
D = pipe diameter or span, ft (m)
P_{ST} = short-term load at the top of the pipe, psf (N/m^2)
T_{LT} = thrust due to long-term loads
P_{LT} = long-term load at the top of the pipe, psf (N/m^2)
A = required wall area using a minimum factor of safety of 2 ($A/10^6$ $inch^2/ft$)
f_i = initial tensile strength level, psi (N/m^2) (Table 9-19)
f_{50} = 50-year tensile strength level, psi (N/m^2) (Table 9-19).

Ring Buckling. The backfilled pipe may buckle whether the groundwater table is above the bottom of the pipe or not. The critical buckling stress may be evaluated by the AASHTO formula shown in Eq. (9-43):

$$f_{cr} = 0.77\frac{r}{A}\sqrt{\frac{BM_sEI}{0.149r^3}} \tag{9-43}$$

where

f_{cr} = maximum, critical stress in the pipe wall, psi (N/m^2), using a factor of safety of 2
r = mean pipe radius, inch (m)
A = pipe wall area, $inch^2/inch$ (mm^2/m)
B = water buoyancy factor
$= 1 - 0.33\, h_w/h$
h_w = height of water surface above the top of pipe, ft (m)
h = height of cover above the top of pipe, ft (m)
M_s = soil modulus of the backfill material, psi (N/m^2)
E = 50-year modulus of elasticity of the pipe wall material, psi (N/m^2)
I = pipe wall moment of inertia, $inch^4/inch$ (mm^4/m)

Hydrostatic Buckling. When pipes are submerged but not adequately backfilled, the critical hydrostatic pressure to cause buckling can be evaluated by the Timoshenko buckling formula provided in Eq. (9-44). The variable C is used to account for the decrease in buckling stress due to pipe out-of-roundness, P_{cr}.

TABLE 9-19. Mechanical Properties for Plastic Pipe Design

Type of Pipe	Initial Minimum Tensile Strength MPa (psi)	Initial Minimum Modulus of Elasticity MPa (psi)	Standard Cell Class	50-Year Minimum Tensile Strength MPa (psi)	50-Year Minimum Modulus of Elasticity MPa (psi)	Strain Limit Percent (%)	Pipe Stiffness kPa (psi)
Smooth Wall, PE	20.7 (3,000)	758 (110,000)	ASTM D 3350, 335434C ASTM F 714	9.93 (1,440)	152 (22,000)	5	Varies
Corrugated PE	20.7 (3,000)	758 (110,000)	ASTM D 3350, 335412C AASHTO M 294	6.21 (900)	152 (22,000)	5	Varies
Ribbed, PE	20.7 (3,000)	758 (110,000)	ASTM D 3350, 335434C AASHTO M 278 ASTM F 679	9.93 (1,440)	152 (22,000)	5	320 (46)
Smooth Wall, PVC	48.3 (7,000)	2,758 (400,000)	ASTM D 1754, 12454C AASHTO M 278 ASTM F 679	25.51 (3,700)	965 (140,400)	5	320 (46)
Smooth Wall, PVC	41.4 (6,000)	3,034 (440,000)	ASTM D 1784, 12364C ASTM F 679	17.93 (2,600)	1,092 (158,400)	3.5	320 (46)
Ribbed, PVC	41.4 (6,000)	3,034 (440,000)	ASTM D 1784, 12454C ASTM F 794	17.93 (2,600)	1,092 (158,400)	3.5	70 (10) 320 (46)
Ribbed, PVC	48.3 (7,000)	2,758 (400,000)	ASTM D 1784, 12454C ASTM F 794 & ASTM F 949	25.51 (3,700)	965 (140,000)	5	348 (50)
PVC Composite	48.3 (7,000)	2,758 (400,000)	ASTM D 1784, 12454C ASTM D 2680	25.51 (3,700)	965 (140,000)	5	1,380 (200)

U.S. Army Corp of Engineers (USACE). (1998). "Engineering and design: Conduits, culverts, and pipes, " Engineer Manual 1110-2-2902, USACE, Washington, D.C.

$$P_{cr} = C\left(\frac{K(EI)}{(1-\nu^2)r^3}\right) \tag{9-44}$$

where

P_{cr} = critical buckling pressure, psf (N/m^2)
C = ovality factor
0% deflection $C = 1$
1% deflection $C = 0.91$
2% deflection $C = 0.84$
3% deflection $C = 0.76$
4% deflection $C = 0.70$
5% deflection $C = 0.64$
K = constant 216 English [$1.5(10)^{-12}$ metric],
r = mean pipe radius, inch (m)
ν = Poisson's ratio for the pipe wall material (typically 0.33 to 0.45).

A factor of safety of 2 is typically applied for round pipe. However, note that 5% pipe deflection reduces P_{cr} to 64% of its calculated value. Equation (9-44) can be conservatively applied to hydrostatic uplift forces acting on the invert of round pipes.

Wall Strain Cracking. Wall strain cracking is a common mode of failure in plastic pipe. AASHTO provides information on the allowable long-term strain limits for many plastics. Excessive wall strain will lead to an accelerated premature failure of the pipe. The typical long-term strain value for PE and PVC is 5% at a modulus of 400,000 psi (2,760 MPa), or 3.5% for PVC with a modulus of 440,000 psi (3,030 MPa)

$$\varepsilon_b = \frac{t_{max}}{D}\left(\frac{0.03\frac{\Delta Y}{D}}{1-0.02\frac{\Delta Y}{D}}\right) \le \frac{\varepsilon_{limit}}{F.S.} \tag{9-45}$$

where

ε_b = bending strain due to deflection, percent
t_{max} = pipe wall thickness, inch (m)
D = mean pipe diameter, inch (m)
$\Delta Y/D$ = pipe deflection, percent
ε_{limit} = maximum long-term strain limit of pipe wall, percent
$F.S.$ = safety factor, minimum 2 recommended.

9.6.4. Thermosetting Resin (Fiberglass) Pipe

Reinforced thermosetting resin pipe (RTRP), commonly referred to as fiberglass pipe, is designed and manufactured in accordance with standards presented by the American National Standards Institute (ANSI), the American Water Works Association (AWWA), and ASTM International (ASTM). The standards used for designing and manufacturing RTRP for sanitary sewage and other liquids under pressure are ANSI/AWWA C950. Gravity sewer pipe is designed under ASTM D3262.

The design procedures within these standards are conservative. The basis of the design standard is that fiberglass behaves as a flexible conduit when subjected to internal pressure and external loading conditions. The pipe is designed separately to withstand external loads and internal pressures.

9.6.4.1. Design Conditions

The following conditions should be established before performing structural design calculations:

- *Nominal Pipe Size.* Nominal pipe sizes are given in Tables 1 through 6 of ANSI/AWWA Standard C950.
- *Working Pressure.* The pressure class of the pipe should be equal to or greater than the working pressure in the system. The working pressure will be minimal, or zero, in gravity lines.
- *Surge Pressure.* Surge pressure should be calculated using recognized hydraulic equations. The surge pressure is equal to zero in gravity lines.
- *Soil Conditions.* The stresses in the soil at a depth below the ground surface, which includes those due to the weight above the pipe and any buoyant forces that the water in the soil exerts.
- *Pipe-Laying Conditions.* Soil conditions for the pipe zone embedment and native material at pipe depth.
- *Depth of Cover.* The minimum and maximum cover measured from the burial depth to the top of the pipe.
- *Vehicular Traffic Load.* Vehicular traffic loads are calculated as outlined earlier in this chapter.

9.6.4.2. Other Design Considerations

Fiberglass pipe systems are resistant to corrosion, both inside and out, in a wide range of fluid-handling applications. As a result, additional linings and exterior coatings are not required. Fiberglass composite piping systems have excellent strength-to-weight properties. The ratio of strength per unit of weight of fiberglass composites is greater than that of iron, carbon, and stainless steels. Fiberglass composites are lightweight, ranging from one-sixth the weight of similar steel products to one-tenth the weight of similar concrete products.

9.6.4.3. *Design Method*

The design method for fiberglass pipe is based on ANSI/AWWA C950 or AWWA M45. The most important property of the pipe is to determine the wall thickness that will withstand external trench loadings and internal pressures. From the determined wall thickness, the designer can then select the nominal thickness and standard pressure class from those specified for fiberglass pipe.

9.6.4.3.1. Determining Internal Pressure

The hydrostatic design basis (HDB) of the fiberglass pipe is defined in terms of reinforced wall hoop stress or hoop strain on the inside surface of the pipe as per ANSI/AWWA C950. The HDB of fiberglass pipe is based on a 50-year product performance and will vary for different manufacturers depending on the materials and composition used in the reinforced wall.

For stress basis HDB:

$$P_c < \left(\frac{HDB}{F.S.}\right)\left(\frac{2t}{D}\right) \quad \text{(9-46, English)}$$

$$P_c < \left(\frac{HDB}{F.S.}\right)\left(\frac{2t}{D}\right) \times 10^3 \quad \text{(9-46, SI)}$$

For strain basis HDB:

$$P_c < \left(\frac{HDB}{F.S.}\right)\left(\frac{2tE_H}{D}\right) \quad \text{(9-47, English)}$$

$$P_c < \left(\frac{HDB}{F.S.}\right)\left(\frac{2tE_H}{D}\right) \times 10^6 \quad \text{(9-47, SI)}$$

where

P_c = pressure class, lb/in² (kPa)
HDB = hydrostatic design basis, lb/in² (N/m²)
$F.S.$ = factor of safety, 1.8
t = pipe reinforced wall thickness, inch (mm)
E_H = hoop tensile modulus of elasticity for pipe, lb/in² (GPa) (See Table 9-20)
D = mean pipe diameter, inch (mm)
= $(OD - t)$
OD = outside diameter, inch (mm).

There are two design factors required in ANSI/AWWA Standard C950 for internal pressure design. The first design factor is the ratio of short-term ultimate hoop tensile strength, S_i, to hoop tensile stress, S_r, at pressure class P_c. The hoop tensile strength values for typical sewer pipe are given in Table 9-20. Hoop tensile strength for sizes not shown in Table 9-20 can be found in Table 10 of ANSI/AWWA Standard C950. All of these requirements are based on a minimum design factor of 4 on initial hydrostatic strength. The second design factor is the ratio of HDB to hoop stress or strain, S_r, at pressure class P_c. This minimum design factor is 1.8 for fiberglass pipe.

9.6.4.3.2. Determining Working Pressure

The pressure class of the pipe should be equal to or greater than the working pressure in the system.

$$P_c \geq P_w$$

where P_w = working pressure, psi (Pa).

9.6.4.3.3. Determining Bending Stress and Strain

Ring bending strain (or stress) should not occur when the pipe reaches the allowable long-term pipe deflection. The following equations verify that this condition is not met.

TABLE 9-20. Hoop Tensile Strength for Fiberglass Pipe

Nominal Pipe Size, inch (mm)	Pressure Class, psi (kPa)			
	100 (689)	150 (1,034)	200 (1,379)	250 (1,724)
	Minimum Hoop Tensile Strength lbf/inch of width (kN/m of width)			
8 (200)	1,600 (280)	2,400 (420)	3,200 (560)	4,000 (700)
12 (300)	2,400 (420)	3,600 (630)	4,800 (840)	6,000 (1,050)
15 (375)	3,000 (525)	4,500 (788)	6,000 (1,050)	7,500 (1,313)
18 (450)	3,600 (630)	5,400 (945)	7,200 (1,260)	9,000 (1,575)
24 (600)	4,800 (840)	7,200 (1,260)	9,600 (1,680)	12,000 (2,100)
30 (750)	6,000 (1,050)	9,000 (1,575)	12,000 (2,100)	15,000 (2,625)
36 (900)	7,200 (1,260)	10,800 (1,890)	14,400 (2,520)	18,000 (3,150)
42 (1,100)	8,400 (1,470)	12,600 (2,205)	16,800 (2,940)	21,000 (3,675)
48 (1,200)	9,600 (1,680)	14,400 (2,520)	19,200 (3,360)	24,000 (4,200)
60 (1,500)	12,000 (2,100)	18,000 (3,150)	24,000 (4,200)	30,000 (5,250)
72 (1,800)	14,400 (2,520)	21,600 (3,780)	28,800 (5,040)	36,000 (6,300)
90 (2,300)	18,000 (3,150)	27,000 (4,725)	36,000 (6,300)	45,000 (7,875)

Reprinted from AWWA Standard C950-01: *Fiberglass pressure pipe*, by permission. Copyright 2001 American Water Works Association.

For stress basis:

$$\sigma_b = D_f E\left(\frac{\Delta y_A}{D}\right)\left(\frac{t}{D}\right) \le \frac{S_b E}{F.S.} \qquad \text{(9-48, English)}$$

$$\sigma_b = 10^3 D_f E\left(\frac{\Delta y_A}{D}\right)\left(\frac{t}{D}\right) \le 10^3 \frac{S_b E}{F.S.} \qquad \text{(9-48, SI)}$$

For strain basis (most common check):

$$\varepsilon_b = D_f\left(\frac{\Delta y_A}{D}\right)\left(\frac{t}{D}\right) \le \frac{S_b}{F.S.} \quad \text{or} \quad \varepsilon_b = D_f\left(\frac{t}{D}\right)\left(\frac{\delta d}{D}\right) \qquad \text{(9-49)}$$

where

σ_b = maximum ring bending stress due to deflection, psi (MPa)
D_f = shape factor, dimensionless (See Table 9-21)
E = ring flexural modulus of elasticity for the pipe, psi (GPa)
Δy_A = maximum allowable long-term vertical pipe deflection, inch (mm)
δd = maximum permitted long-term installed deflection, inch (mm)—typically chosen to be less than Δy_A
D = mean pipe diameter, inch (mm)
t = total wall thickness, inch (mm)
S_b = long-term ring bending strain for the pipe, inch/inch (mm/mm)
$F.S.$ = factor of safety, 1.5
ε_b = maximum ring-bending strain due to deflection, inch/inch (mm/mm).

The shape factor is dependent on both the pipe stiffness and the installation (e.g., backfill material, backfill density, compaction method, haunching, trench configuration, native soil characteristics, and vertical loading). A conservative chart assuming that tamped compaction limits the long-term deflections to 5% will give the shape factors in Table 9-21.

TABLE 9-21. Shape Factors for Fiberglass Pipe[a]

Pipe Stiffness, psi (kPa)	D_f
9 (62)	8.0
18 (124)	6.5
36 (248)	5.5
72 (496)	4.5

[a]Products may have use limits of other than 5% long-term deflection. In such cases, the requirements should be proportionally adjusted. For example, a 4% long-term limiting deflection would result in a 50-year requirement of 80% of the values given in Table 9-21, whereas a 6% limiting deflection would yield a requirement of 120% of the values given in Table 9-21.

9.6.4.3.4. Determining Deflection

Sewer pipe should be installed to ensure that external loading would not cause a long-term decrease in the vertical diameter of the pipe. The pipe must be capable of withstanding a 5% long-term deflection; therefore, the maximum allowable deflection may be stated as follows:

$$\frac{\Delta y}{D} \leq \frac{\delta \cdot d}{D} \leq \frac{\Delta y_a}{D} \tag{9-50}$$

where

$\frac{\Delta y}{D}$ = predicted vertical pipe deflection, fraction of mean diameter; maximum deflection for fiberglass is typically 5%

$\frac{\delta \cdot d}{D}$ = permitted vertical pipe deflection, fraction of mean diameter

$\frac{\Delta \cdot d}{D}$ = maximum allowable vertical pipe deflection, fraction of mean diameter.

The amount of deflection is a function of the soil load, live load, native soil characteristics at pipe elevation, pipe embedment material and density, trench width, haunching, and pipe stiffness.

$$\Delta y = \frac{(D_L W_e + W_L)K_x}{0.149PS + 0.061M_s} \qquad \text{(9-51, English)}$$

$$\Delta y = \frac{(D_L W_e + W_L)K_x}{149PS + 61{,}000M_S} \qquad \text{(9-51, SI)}$$

where

D_L = deflection lag factor, dimensionless; for long-term deflection use a D_L greater than 1

W_e = vertical soil load on pipe, psi (N/m^2)

W_L = live load on pipe, psi (N/m^2)

K_x = bedding coefficient, dimensionless, 0.1 for typical direct bury

PS = pipe stiffness, psi (kPa)

M_s = composite soil constrained modulus, psi (MPa).

The lag factor converts the immediate deflection of the pipe to the deflection of the pipe after many years. For long-term deflection prediction, a value of 1 or greater is appropriate. The bedding coefficient reflects the degree of support provided by the soil at the bottom of the pipe and over which the bottom reaction is distributed. Assuming a typical direct bury condition, a K_x value of 0.1 is appropriate.

The long-term vertical soil load for pipe is given by Eq. (9-1).

With the common acceptance that fiberglass pipe must be capable of withstanding 5% long-term deflection, the maximum installed bending strain may be expressed as:

$$\varepsilon_b \max = (0.05)D_f\left(\frac{t}{D}\right) \tag{9-52}$$

9.6.4.3.5. Pipe Stiffness

The pipe stiffness can be determined by conducting parallel-plate loading tests in accordance with ASTM D2412 using the following equations:

$$PS = \frac{EI}{0.149(r_m + \Delta y_t/2)^3} \quad \text{or} \quad PS = F/\Delta y_t \tag{9-53}$$

where

PS = pipe stiffness, lb/inch^2 (N/m^2)
E = ring flexural modulus, lb/inch^2 (N/m^2)
I = moment of inertia of unit length, inch^4/inch (mm^4/mm) = $(t)^3/12$, where t = average thickness of pipe wall
r_m = mean pipe radius, inch (mm) $(OD - t)/2$, where OD = outside diameter, inch (mm) and t = average wall thickness, inches (mm).
Δy_t = vertical pipe deflection, inch (mm), when tested by ASTM D2412 with a vertical diameter reduction of 5%
F = load per unit length, lb/ft (N/m).

9.6.4.3.6. Constrained Soil Modulus

The vertical loads on a flexible pipe cause a decrease in the vertical diameter and an increase in the horizontal diameter. The horizontal movement develops a passive soil resistance that helps support the pipe. The passive soil resistance varies depending on the soil type and the degree of compaction of the pipe backfill material, soil characteristics, cover depth, and trench width.

The following equation is used to determine the soil modulus:

$$M_s = S_c M_{sb} \tag{9-54}$$

where

M_s = composite constrained soil modulus, psi (kPa)
S_c = soil support combining factor from Table 9-23, dimensionless

TABLE 9-22. M_{sb} based on Soil Type[a–h]

Vertical Stress Level,[a] psi (kPa)	Depth for γ_s = 120 pcf, ft (m)	SPD100, psi (kPa)	SPD95, psi (kPa)	SPD90, psi (kPa)	SPD85, psi (kPa)
		Stiffness Categories 1 and 2 (SC1, SC2)			
1 (6.9)	1.2 (0.36)	2,350 (16,200)	2,000 (13,800)	1,275 (8,800)	470 (3,200)
5 (34.5)	6 (1.8)	3,450 (23,800)	2,600 (17,900)	1,500 (10,300)	520 (3,600)
10 (68.9)	12 (3.6)	4,200 (29,000)	3,000 (20,700)	1,625 (11,200)	570 (3,900)
20 (138)	24 (7.2)	5,500 (37,900)	3,450 (23,800)	1,800 (12,400)	650 (4,500)
40 (276)	48 (14.4)	7,500 (51,700)	4,250 (29,300)	2,100 (14,500)	825 (5,700)
60 (414)	72 (21.9)	9,300 (64,100)	5,000 (34,500)	2,500 (17,200)	1,000 (6,900)
			Stiffness Categories 3 (SC3)		
1 (6.9)	1.2 (0.36)		1,415 (9,800)	670 (4,600)	360 (2,500)
5 (34.5)	6 (1.8)		1,670 (11,500)	740 (5,100)	390 (2,700)
10 (68.9)	12 (3.6)		1,770 (12,200)	750 (5,200)	400 (2,800)
20 (138)	24 (7.2)		1,880 (13,000)	790 (5,400)	430 (3,000)
40 (276)	48 (14.4)		2,090 (14,400)	900 (6,200)	510 (3,500)
60 (414)	72 (21.9)		2,300 (15,900)	1,025 (7,100)	600 (4,100)
			Stiffness Categories 4 (SC4)		
1 (6.9)	1.2 (0.36)		530 (3,700)	255 (1,800)	130 (900)
5 (34.5)	6 (1.8)		625 (4,300)	320 (2,200)	175 (1,200)
10 (68.9)	12 (3.6)		690 (4,800)	355 (2,400)	200 (1,400)
20 (138)	24 (7.2)		740 (5,100)	395 (2,700)	230 (1,600)
0 (276)	48 (14.4)		815 (5,600)	460 (3,200)	285 (2,000)
60 (414)	72 (21.9)		895 (6,200)	525 (3,600)	345 (2,400)

[a]Vertical stress level is the vertical effective soil stress at the springline elevation of the pipe. It is normally computed as the design soil unit weight times the depth of fill. Buoyant unit weight should be used below the groundwater level.

[b]SC1 soils have the highest stiffness and require the least amount of compactive energy to achieve a given density. SC5 soils, which are not recommended for use as backfill, have the lowest stiffness and require substantial effort to achieve a given density.

[c]SC1 soils have higher stiffness than SC2 soils, but data on specific soil stiffness values are not available at the current time. Therefore, the soil stiffness of dumped SC1 soils can be taken as equivalent to SC2 soils compacted to 90% of maximum Standard Proctor Density (SPD90), and the soil stiffness of compacted SC1 soils can be taken as equivalent to SC2 soils compacted to 100% of maximum Standard Proctor Density (SPD100).

[d]The soil types SC1 to SC5 are defined in Table 9-25.

[e]The numerical suffix to the SPD indicates the compaction level of the soil as a percentage of maximum dry density determined in accordance with ASTM D698 or AASHTO T-99.

[f]Engineers may interpolate intermediate values of M_{sb} for vertical stress levels not shown in the table.

[g]For pipe installed below the water table, the modulus should be corrected for reduced vertical stress due to buoyancy and by an additional factor of 1 for SC1 and SC2 soils with SPD of ≥95, 0.85 for SC2 soils with SPD of 90, 0.70 for SC2 soils with SPD of 85, 0.50 for SC3 soils, and 0.30 for SC4 soils.

[h]It is recommended to embed pipe with stiffness of 9 psi (62 kPa) or less only in SC1 or SC2 soils.

M_{sb} = constrained soil modulus of the pipe zone embedment, psi (kPa) (See Table 9-22).

To find the Soil Support Combining Factor, Table 9-23 must be used with the following values:

M_{sn} = constrained soil modulus of native soil at pipe elevation, psi (kPa) (see Table 9-24)
B_d = trench width at pipe springline, inch (mm).

9.6.4.3.7. Hydrostatic Design Basis for Stress and Strain

The maximum stress or strain resulting from the combined effects of internal pressure and deflections should meet the following requirements.

For stress basis HDB:

$$\frac{\sigma_{pr}}{HDB} \leq \frac{1-\left(\frac{\varepsilon_b r_c}{S_b E}\right)}{F.S._{pr}} \tag{9-55}$$

$$\frac{\sigma_b r_c}{S_b E} \leq \frac{1-\left(\frac{\sigma_{pr}}{S_b E}\right)}{F.S._b} \tag{9-56}$$

TABLE 9-23. Values for S_c^a

M_{sn}/M_{sb}	B_d/D = 1.25	B_d/D = 1.5	B_d/D = 1.75	B_d/D = 2	B_d/D = 2.5	B_d/D = 3	B_d/D = 4	B_d/D = 5
0.005	0.02	0.05	0.08	0.12	0.23	0.43	0.72	1.00
0.01	0.03	0.07	0.11	0.15	0.27	0.47	0.74	1.00
0.02	0.05	0.10	0.15	0.20	0.32	0.52	0.77	1.00
0.05	0.10	0.15	0.20	0.27	0.38	0.58	0.80	1.00
0.1	0.15	0.20	0.27	0.35	0.46	0.65	0.84	1.00
0.2	0.25	0.30	0.38	0.47	0.58	0.75	0.88	1.00
0.4	0.45	0.50	0.56	0.64	0.75	0.85	0.93	1.00
0.6	0.65	0.70	0.75	0.81	0.87	0.94	0.98	1.00
0.8	0.84	0.87	0.90	0.93	0.96	0.98	1.00	1.00
1	1.00	1.00	1.00	1.00	1.00	1.00	1.00	1.00
1.5	1.40	1.30	1.20	1.12	1.06	1.03	1.00	1.00
2	1.70	1.50	1.40	1.30	1.20	1.10	1.05	1.00
3	2.20	1.80	1.65	1.50	1.35	1.20	1.10	1.00
≥5	3.00	2.20	1.90	1.70	1.50	1.30	1.15	1.00

[a]In-between values of S_c may be determined by straight-line interpolation from adjacent values.

Reprinted from AWWA Manual of Practice M45: *Fiberglass pipe design*, by permission. Copyright 2005 American Water Works Association.

TABLE 9-24. Constrained Soil Modulus of Native Soil

Native In Situ Soils[a]						
		Cohesive				
Granular		q_u[b]			M_{sn}[c]	
Blows/ft[d] (0.3 m)	Description	Tons/ft²	kPa	Description	psi	MPa
>0–1	Very, very loose	>0–0.125	0–13	Very, very soft	50	0.34
1–2	Very loose	0.125–0.25	13–25	Very soft	200	1.4
2–4		0.25–0.50	25–50	Soft	700	4.8
4–8	Loose	0.50–1.00	50–100	Medium	1,500	10.3
8–15	Slightly compact	1.0–2.0	100–200	Stiff	3,000	20.7
15–30	Compact	2.0–4.0	200–400	Very stiff	5,000	34.5
30–50	Dense	4.0–6.0	400–600	Hard	10,000	69.0
>50	Very dense	>6.0	>600	Very hard	20,000	138.0

[a]The constrained modulus M_{sn} for rock is ≥50,000 psi (345 MPa).
[b]q_u = unconstrained compressive strength, tons/ft² (kPa).
[c]For embankment installation, $M_{sb} = M_{sn} = M_s$.
[d]Standard penetration test per ASTM D1586.

Reprinted from AWWA Manual of Practice M45: *Fiberglass pipe design*, by permission. Copyright 2005 American Water Works Association.

For strain basis HDB:

$$\frac{\varepsilon_{pr}}{HDB} \le \frac{1-\left(\frac{\varepsilon_b r_c}{S_b}\right)}{F.S._{pr}} \tag{9-57}$$

$$\frac{\varepsilon_b r_c}{S_b} \le \frac{1-\left(\frac{\varepsilon_{pr}}{HDB}\right)}{F.S._b} \tag{9-58}$$

where

σ_{pr} = working stress due to internal pressure, lb/inch² (kPa)
HDB = hydrostatic design basis, psi (kPa)
ε_b = bending strain due to maximum permitted deflection, inch/inch (mm/mm)—See Eq. (9-49)
ε_{pr} = working strain due to internal pressure, inch/inch (mm/mm)
$= \frac{P_w D}{2tH_H}$
r_c = re-rounding coefficient, dimensionless
$P_w/435$ (where $P_w \le 435$ psi) (English)
$= 1 - P_w/3{,}000$ (where $P_w \le 3{,}000$ kPa) (SI)
P_w = working pressure in pipe

TABLE 9-25. Soil Stiffness Categories

Soil Stiffness Category	Unified Soil Classification System Soil Groups[a]	AASHTO Soil Groups[b]
SC1	Crushed rock: ≤15% sand, maximum 25% passing the ⅜ inch sieve and maximum 5% passing No. 200 sieve[c]	
SC2	Clean, coarse-grained soils: SW, SP, GW, GP, or any soil beginning with one of these symbols with 12% or less passing a No. 200 sieve[d]	A1, A3
SC3	Coarse-grained soils with fines: GM, GC, SM, SC, or any soil beginning withone of these symbols with more than 12% fines Sandy or gravelly fine-grained soils: CL, ML (or CL-ML,CL/ML, ML/CL) with more than 30% retained on a No. 200 sieve	A-2-4, A-2-5, A-2-6, or A-4 or A-6 soils with more than 30% retained on a No.200 sieve
SC4	Fine-grained soils: CL, ML (or CL-ML, CL/ML, ML/CL) with30% retained on a No. 200 sieve	A-2-7, or A-4 or A-6 soils with 30% or less retained on a No. 200 sieve
SC5	Highly plastic and organic soils: MH, CH, OL, OH, PT	A5, A7

[a]ASTM D2487, Standard Classification of Soils for Engineering Purposes.
[b]AASHTO M145, Classification of Soils and Soil Aggregate Mixtures.
[c]SC1 soils have higher stiffness than SC2 soils, but data on specific soil stiffness values are not available at the current time. Until such data are available, the soils stiffness of dumped SC1 soils can be taken to be equivalent to SC2 soils compacted to 90% of maximum standard Proctor density, and the stiffness of compacted SC1 soils can be taken to be equivalent to SC2 soils compacted to 100% of maximum standard Proctor density. Even if dumped, SC1 materials should be worked into the haunch zone.
[d]Uniform fine sands (SP) with more than 50% passing a No. 100 sieve are very sensitive to moisture and should not be used as backfill for fiberglass pipe, unless the engineer has given this specific consideration. If use of these materials is permitted, compaction and handling procedures should follow the guidelines for SC3 materials.

S_b = long-term ring bending strain for the pipe, inch/inch (mm/mm)
E = ring flexural modulus, psi (kPa)
$F.S._{pr}$ = pressure safety factor, 1.8
σ_b = bending stress due to the maximum permitted deflection, psi (kPa)
$F.S._b$ = bending safety factor, 1.5.

9.6.4.3.8. Buckling

Due to the restraining influence of the soil, external radial pressure can buckle pipe at a high pressure. The external loads should be equal to or less than the allowable buckling pressure. The allowable buckling pressure is determined by the following equation:

$$q_a = \frac{(1.2C_n)(EI)^{0.33}(\varphi_s M_s k_\nu)^{0.67} R_h}{(F.S.)r} \qquad (9\text{-}59)$$

where

q_a = allowable buckling pressure, lb/inch² (kPa)
M_s = composite constrained soil modulus, psi (kPa)
$F.S.$ = safety factor, 2.5
C_n = scalar calibration factor to account for nonlinear effects = 0.55
φ_s = factor to account for variability in stiffness of compacted soil; suggested value is 0.9
k_ν = modulus correction factor for Poisson's ratio, commonly assumed as = 0.74
r = mean pipe radius, inches (mm)
R_h = correction factor for depth of fill = $11.4/(11 + D/h)$ (English) or = $11.4/(11 + D/1{,}000h)$ (SI); where h = depth of pipe, inches (m) and D = mean diameter of the pipe, inches (mm).

$$\gamma_w h_w + R_w(W_c) + P_v \le q_a \qquad (9\text{-}60)$$

If live loads are considered, as in gravity sewers, the buckling requirement is checked by the following equation:

$$\gamma_w h_w + R_w(W_c) + W_L \le q_a \qquad (9\text{-}61,\ \text{English})$$

$$[\gamma_w h_w + R_w(W_c) + W_L] \times 10^{-3} \le q_a \qquad (9\text{-}61,\ \text{SI})$$

where

γ_w = specific weight of water = 0.0361 lb/inch³ (9800 N/m³)
P_ν = internal vacuum pressure, psi (kPa)
R_w = water buoyancy factor = $1 - 0.33(h_w)/h$ $[0 \le h_w \le h]$
W_L = live load (as calculated previously)
W_c = soil load (as calculated previously)
h_w = height of water surface above the pipe top, inch (m)
h = depth of soil, inch (m).

The total load for gravity sewers combines the live loads, water load and the soil load as follows:

$$\text{Total Load} = W_c(R_w) + W_w + W_L \qquad (9\text{-}62)$$

where

$$W_w = \frac{\gamma_w h_w}{144} \quad \text{(9-63, English)}$$

$$W_w = \gamma_w h_w \quad \text{(9-63, SI)}$$

9.6.4.3.9. Axial Loads

Axial stresses in sanitary sewer pipe can occur in pipes that have hoop expansion due to internal pressure, restrained thermal expansion, and uneven bedding. The minimum requirements for axial strengths are specified in Sections 5.1.2.4 and 5.1.2.5 and Tables 11, 12, and 13 of ANSI/AWWA C950.

9.6.4.3.10. Special Design Considerations

Special consideration should be made for shallow bedding, where $H <$ 2 ft (0.6 m), uneven bedding, differential settlement of unstable native soils, restrained tension joints, unusually high surface or constructions loads, or any extremely difficult construction conditions.

Sample Calculations

Example 9-13. Given the following parameters, verify whether the pipe thickness will be adequate for a gravity sewer main to be installed under a railroad.

Pipe properties:

- Nominal pipe diameter = 60 inches
- Outside diameter of pipe = 62.9 inches
- Outside diameter of joint = 62.9 inches
- Minimum pipe stiffness = 238 psi
- Minimum total wall thickness = 2.16 inches
- Minimum liner thickness = 0.04 inch
- Ultimate compressive strength = 10,500 psi
- Hoop flexural modulus of pipe = 1.40E+06 psi

Installation conditions:

- Soil weight = 120 lb/ft^3
- Water weight = 62.4 lb/ft^3
- Soil cover depth = 16 ft
- Live load (will be under railroad) = 3.05 psi

- Water table depth = 16 ft
- Modulus of soil reaction (native material) = 1,500 psi

Pipe geometry:

- Outside diameter = 62.9 inches
- Mean diameter = 60.78 inches
- Total wall thickness = 2.16 inches
- Liner thickness = 0.04 inch
- Structural wall thickness = 2.12 inches
- Nominal interior diameter = 58.2 inches

Factors and coefficients:

- Bedding coefficient = 0.083
- $D_f = 3$, obtained from Table 9-12
- Deflection lag factor = 1

Step 1. Check pipe stiffness:

$$PS = \frac{EI}{0.149(r_m)^3}$$

$$I = (t_t)^3/12 = (2.12)^3/12 = 0.794 \text{ inch}^4/\text{inch}$$

$$r_m = (OD - t)/2 = (62.9 - 2.12)/2 = 30.39 \text{ inches}$$

$$r_m(5\%) = \text{from ASTM D2412} = r_m \times 1.025$$

$$PS = \frac{(1.4E+06)\times(0.794)}{(0.149)(1.025\times 30.39)^3}$$

$$PS = \quad 246.81$$

Step 2. Determine total load:

$$W_c = \frac{Y_s H}{144} = (120)(16)/144 = 13.33 \text{ psi}$$

$$W_L = 3.05 \text{ psi}$$

$$W_W = (62.4)(16)/144 = 6.93 \text{ psi}$$

$$R_w = 1 - 0.33(16/16) = 0.67$$

$$\text{Total Load} = W_c\,(R_w) + W_w + W_L = 18.92 \text{ psi}$$

Step 3. Check deflection:

$$\Delta_y = \frac{(D_L W_c + W_L)K_x}{0.149PS + 0.061M_s} = 1.07\% < 3\%$$

Step 4. Check bending strain:

$$\varepsilon_b \max = (0.05)Df\left(\frac{t}{D}\right) = 0.32\% < 0.62\%.$$

Step 5. Check buckling:

$$q_a = \frac{(1.2C_n)(EI)^{0.33}(\varphi_s M_s k_v)^{0.67} R_h}{(FS)r} = 243 \text{ psi}$$

$W_c\,(R_w) + W_w$
$+ W_L \leq q_a$
$243 \text{ psi} \geq 18.92 \text{ psi}$

Since all checks are correct, the design is acceptable.

9.6.5. Design of Ductile Iron Sewer Pipes

The design standard for ductile iron pipe (DIP) is based on its behavior as a flexible conduit, capable of re-rounding under pressure. Therefore, the pipe is designed separately to withstand external loads and internal pressure. The current design approach, as outlined by the Ductile Iron Pipe Research Association (DIPRA), includes the following:

- Design for internal pressures (static pressure plus surge pressure allowance).
- Design for bending stress due to external loads (earth load plus truck loads).
- Select the larger resulting net wall thickness.
- Add a 0.08-inch (2-mm) service allowance.
- Check deflection.
- Add a standard casting tolerance.

This procedure results in the total calculated design thickness, from which the appropriate pressure class is chosen. It should be noted that DIPRA has developed a computer program (dipra.exe) which will automatically perform these calculations, and that can be downloaded from the DIPRA web site (www.dipra.org).

9.6.5.1. Standard Laying Conditions

Several of the factors necessary to calculate the bending stress and deflection are dependent on the type of laying condition and the width of the bedding at the pipe bottom. To expedite the design calculations, the five Standard Laying Conditions have been identified by DIPRA and ASTM for installation of DIP. These Standard Laying Conditions are shown in Fig. 9-30.

Laying Condition	Description	E′	Bedding Angle	K_b	K_x
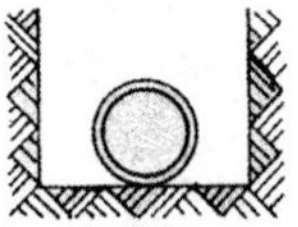 Type 1*	Flat-bottom trench.† Backfill not tamped.	150	30°	0.235	0.108
 Type 2	Flat-bottom trench.† Backfill lightly tamped** to centerline of sewer pipe.	300	45°	0.210	0.105
 Type 3	Sewer pipe bedded in 4-in. minimum loose soil.†† Backfill lightly tamped** to top of sewer pipe.	400	60°	0.189	0.103
 Type 4	Sewer pipe bedded in sand, gravel or crushed stone to depth of 1/8 pipe diameter, 4-in minimum. Backfill lightly compacted** to top of sewer pipe.	500	90°	0.157	0.096
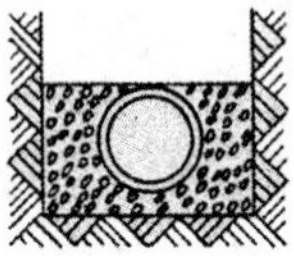 Type 5	Sewer pipe bedded in compacted granular material to centerline of pipe. Carefully compacted** granular or select †† material to top of sewer pipe.	700	150°	0.128	0.085

*Laying condition Type 1 is limited to 16-in. and smaller pipe.

†"Flat-bottom" is defined as undisturbed earth.

††"Loose soil" or "select material" is defined as native soil excavated from the trench, free of rocks, foreign materials and frozen earth.

**These laying conditions can be expected to develop the following backfill densities (Standard Proctor): Types 2 and 3, approximately 70%; Type 4, approximately 75%; Type 5, approximately 85%

FIGURE 9-30. Standard laying conditions for ductile iron pipe. Reprinted with permission, © ASTM International, 100 Barr Harbor Drive, West Conshohocken, PA 19428.

9.6.5.2. Stress Design

As noted previously, the stress design for DIP is based on the larger of the values obtained from internal pressure and external ring bending stress. By definition, the internal pressure of a gravity sanitary sewer would be essentially 0 psi (Pa). Therefore, the designer can proceed to evaluation of the ring bending stress.

When a trench load of sufficient magnitude is applied, DIP behaves similarly to other flexible conduits, deflecting to develop passive resistance from the side fill soil, thereby transmitting part of the trench load to the side fill soil. Thus, the load-carrying capacity of DIP is a function of soil and ring stiffness. In addition, an upward reaction to the vertical trench load exerted on the pipe develops in the trench embedment below the pipe. Design for flexure is based on the Spangler equation for bending stress (Spangler 1941). This equation utilizes both the bending stress coefficient, K_b, and the deflection coefficient, K_x. Both coefficients are functions of the width of the bedding under the pipe.

The design ring-bending stress that is recommended by DIPRA (2006) is 48,000 psi (330 MPa). This provides safety factors under trench loading of at least 1.5 based on ring yield strength and at least 2 based on ultimate ring stress. The following equation is used to calculate the trench load required to develop a bending stress of 48,000 psi (330 MPa) at the pipe invert.

$$P_V = \frac{f}{3\left(\frac{D}{t}\right)\left(\frac{D}{t}-1\right)\left[K_b - \frac{K_x}{\frac{8E}{E'\left(\frac{D}{t}-1\right)^3}+0.732}\right]} \tag{9-64}$$

where

P_V = trench load = $P_e + P_t$, psi (MPa)
P_e = earth load, psi (MPa)
P_t = live load, psi (MPa)
f = design maximum bending stress, 48,000 psi (330 MPa)
D = outside diameter, inch (m)
t = net thickness, inch (m)
K_b = bending moment coefficient, see Figure 9-30
K_x = deflection coefficient, see Figure 9-30
E = modulus of elasticity, 24×10^6 psi (165×10^3 MPa)
E' = modulus of soil reaction, see Figure 9-30.

Based on this equation, a net thickness can be calculated for the pipe. A service allowance of 0.08 inches (2 mm) is then added to the net thickness. The resulting thickness is the minimum thickness, t_1.

9.6.5.3. *Deflection Analysis*

Deflection analysis of the DIP is based on the Modified Iowa Formula [Eq. (9-33)]. The maximum recommended ring deflection for cement-mortar-lined DIP is 3% of the outside diameter. This provides a minimum safety factor of at least 2 with regard to the failure of the cement-mortar lining. The following equation is used to calculate the trench load required to develop a ring deflection of 3% of the outside diameter.

$$P_V = \frac{\Delta x/D}{12K_X}\left[\frac{8E}{\left(\frac{D}{t_1}-1\right)^3} + 0.732E'\right] \tag{9-65}$$

where

t_1 = minimum thickness, inch (m)
Δx = design deflection, inch (m) ($\Delta x/D = 0.03$).

The other variables have the same meaning as in Eq. (9-64). The t_1 required for deflection is compared to the t_1 resulting from the bending stress design. The greater t_1 is used and is called the minimum manufacturing thickness.

9.6.5.4. *Allowance for Casting Tolerance*

Once the minimum manufacturing thickness is determined, an allowance for casting tolerance is added to provide the latitude required by the manufacturing process and to ensure that any negative deviations from design thickness do not adversely impact the pipe. The casting tolerances are dependent on pipe size, as shown in Table 9-26.

9.6.5.5. *Design Aids*

Manual use of the equations above for determining pipe thickness requires significant calculations. If the design conditions can be limited to

TABLE 9-26. Allowances for Casting Tolerance

Size, inches (mm)	Casting Tolerance, inches (mm)
3–8 (75–200)	0.05 (1.3)
10–12 (250–300)	0.06 (1.5)
14–42 (350–1,050)	0.07 (1.8)
48 (1,200)	0.08 (2.0)
54–64 (1,350)	0.09 (2.3)

Reprinted from AWWA Standard C150/A21. 50-02: *Thickness design of ductile-iron pipe*, by permission.

those included in the AWWA standards, then simplified design tables provided by DIPRA can be utilized. The design criteria included in ANSI/AWWA C150/A21.50 include:

- Yield strength in tension, 42,000 psi (289.59 MPa).
- Ring bending stress, 48,000 psi (330.96 MPa).
- Ring deflection, 3%.
- AASHTO H-20 truck loading at all depths, with 1.5 impact factor.
- Minimum depth of cover, 2.5 ft (0.7 m).
- Prism earth load for all pipe sizes.
- The DIPRA-defined five types of laying conditions.

In this case, the simplified design tables from DIPRA can be utilized to calculate a maximum depth of cover for each size of pipe.

9.7. INSTALLATION

9.7.1. Recommendations for Field Procedures

The factor of safety against ultimate collapse of sewer pipe is about the same as that used in the design of most engineered structures of monolithic concrete. However, the design of sewer pipe is based on calculated loads, bedding factors, and experimental factors, which are less well-defined than the dead and live loads used in building design. It is therefore important that the loads imposed on the sewer pipe be not greater than the design loads.

To obtain the objective of imposed loads being less than design loads, the following procedures are recommended:

- *Specifications.* Construction specifications should set forth limits for the width of trench below the top of sewer pipe. The width limits should take into account the minimum width required to lay and join sewer pipe, and the maximum allowable width for each class of sewer pipe and bedding to be used. Where the depth is such that a positive-projecting condition will be obtained, maximum width should be specified as unlimited unless the width must be controlled for some reason other than to meet structural requirements of the sewer pipe. Appropriate corrective measures should be specified in the event the maximum allowable width is exceeded. These measures may include provision for a higher class of bedding or concrete encasement. Maximum allowable construction live loads should be specified for various depths of cover if appropriate for the project.
- *Construction Inspection.* Construction should be observed by an experienced engineer or field representative who reports to a competent field engineer.

- *Testing.* Sewer pipe testing should be under the supervision of a reliable testing laboratory, and close liaison should be maintained between the laboratory and the field engineer.
- *Field Conditions.* The field engineer should be furnished with sufficient design data to enable the intelligent evaluation of unforeseen conditions. The field engineer should be instructed to confer with the design engineer if changes in design appear advisable.

9.7.2. Effect of Trench Sheeting

Because of the various alternative methods employed in sheeting trenches, generalizations on the proper construction procedure to follow (to ensure that the design load is not exceeded) are risky and dangerous. Each method of sheeting and bracing should be studied separately. The effect of a particular system on the sewer pipe load, as well as the consequences of removing the sheeting or the bracing, must be estimated.

When trench sheeting is necessary, it should be driven at least to the bottom of the pipe bedding or foundation material, if used. In general, in a constantly wet or dry area, sheeting and bracing should be left in place to prevent reduction in lateral support at the sides of the pipe because of voids formed by removal of the sheeting. Sheeting left in place should be cut off as far below the surface as practicable, but in no case less than 3 ft (0.9 m) below final ground elevation.

It is difficult to obtain satisfactory filling and compaction of the void left when wood sheeting is pulled. If granular materials are used for backfill, it is possible to fill and compact the voids left by the wood sheeting if the material is placed in lifts and jetted as the sheeting is pulled. If cohesive materials are used for backfill, a void will be left when the wood sheeting is pulled and the full weight of the prism of earth contained between the sheeting will come to bear on the sewer pipe. Therefore, wood sheeting driven alongside the sewer pipe should generally not be pulled, but should be cut off and left in place to an elevation of 1.5 ft (450 mm) above the top of the sewer pipe.

Steel sheeting to be removed should be pulled in increments as the trench is backfilled, and the soil should he compacted to prevent formation of voids. The portion of wood sheeting to be removed should be handled similarly.

Skeleton sheeting or bracing should be cut off and left in place to an elevation of 1.5 ft (450 mm) above the top of the sewer pipe if removal of the trench support might cause a collapse of the trench wall and a widening of the trench at the top of the conduit. Entire skeleton sheeting systems should be left in place if removal would cause collapse of the trench before backfill can be placed.

Where steel soldier beams with horizontal lagging between the beam flanges are used for sheeting trenches, efforts to reclaim the steel beams before the trench is backfilled may damage pipe joints. It is recommended that use of this type of sheeting be allowed only with stipulation that the beams be pulled after backfilling and that the lagging be left in place.

Steel sheeting may be used and reused many times, so the relative economy of this type of sheeting compared with timber or timber and soldier beams should be explored. Because of the thinness of the sheeting, it is often feasible to achieve reasonable compaction of backfill so that the steel sheeting may be withdrawn with about the same factor of safety against settlement of the surfaces adjacent to the trench as that for other types of sheeting left in place.

9.7.3. Trench Boxes

Several backfilling techniques are possible when a trench box is used. Granular material can be placed between the box and the trench wall immediately after placing the box, and the box is advanced by lifting it slightly before moving it forward. Boxes can be made with a step at the rear which makes the trench wall accessible for compacting the embedment against the walls. Also, the box can be used where the sewer pipe is laid in a small subtrench, and the box is utilized only in the trench above the top of the sewer pipe. Advancement of the box must be done carefully to avoid pulling pipe joints apart.

9.7.4. OSHA Requirements

In addition to the structural design requirements for sewer installation, the federal government and many state governments have passed specific safety requirements for excavation and trenching. A review of state requirements is beyond the scope of this Manual. In addition, although the following information is current as of the date of the writing of this Manual, engineers and contractors should review the appropriate state and federal regulations to ensure that there have not been intervening changes.

The standard covering excavation safety is Title 29 Code of Federal Regulations, Part 1926.650-652 (Subpart P), OSHA's Rules and Regulations for Construction Employment (Standard 29CFR1926.650). The standard covers all excavations made in the Earth's surface, including trenches, and the requirements for protective systems to be used.

OSHA defines an excavation as any man-made cut, cavity, trench, or depression in the Earth's surface as formed by earth removal. This can include anything from excavations for home foundations to a new highway. A trench refers to a narrow excavation made below the surface of the ground in which the depth is greater than the width, and the width does not exceed 15 ft (4.5 m). Trenching is common in utility work, where underground piping or cables are being installed or repaired.

If an excavation is more than 5 ft (1.5 m) in depth, there must be a protective system in place while workers are in the excavation. Excavations more than 4 ft (1.2 m) in depth must have a way to get in and out (usually a ladder) for every 25 ft (7.5 m) of horizontal travel.

OSHA standards require that no matter how deep the excavation is, a "competent person" must inspect conditions at the site on a daily basis and as frequently as necessary during the progress of work to make sure that the hazards associated with excavations are eliminated before workers are allowed to enter. A competent person has the following qualifications:

- Has a thorough knowledge of Title 29 Code of Federal Regulations, Part 1926.650-652 (Subpart P), OSHA's Rules and Regulations for Construction Employment.
- Understands how to classify soil types.
- Knows the different types and proper use of excavation safety equipment (e.g., protective systems).
- Has the ability to recognize unsafe conditions, the authority to stop the work when unsafe conditions exist, the knowledge of how to correct the unsafe conditions, *and does it!*

If someone else must be called in order to stop the work, or the designated competent person does not stop unsafe acts and conditions, the person is not acting "competently" within the meaning of the standard.

It is the responsibility of the competent person to conduct daily inspections prior to the start of any work and as needed throughout the shift. Part of this inspection process includes determining the soil classification. OSHA has included Standard 29CFR1926, Subpart P in Appendix A of its excavation rule standard methods to make it easier for a competent person to classify soils. The ability to determine soil type correctly is critical because soil type is one of the determining factors in specifying protective systems.

When a protective system is required, the three most commonly used kinds of protective systems are shoring, shielding, and sloping. Each of these protective systems is acceptable to OSHA; it is up to the competent person to determine which method will be most effective for the job. The competent person must inspect these systems regularly to ensure they are functioning properly.

9.7.5. Pipe Bedding and Backfilling

9.7.5.1. General Concepts

The ability of a sewer pipe to safely resist the calculated soil load depends not only on its inherent strength but also on the distribution of the bedding reaction and on the lateral pressure acting against the sides of the sewer pipe.

Construction of the sewer pipe–soil system focuses attention on the pipe zone, which is made up of five specific areas: foundation, bedding, haunching, initial backfill, and final backfill (see Fig. 9-31 for definitions and limits of these five areas). However, all of these areas are not necessarily referred to in all pipe design standards. The discussion in this section is in general terms and is intended to describe the effect of the various areas on the pipe–soil system. More detailed requirements are given in the previous sections on pipe design and in Chapter 10.

For all sewer pipe materials, the calculated vertical load is assumed to be uniformly distributed over the width of the pipe. This assumption was originated by Marston's work and is part of the bedding factors developed by Spangler and presented in this chapter. Many years of field experience indicate this assumption results in conservative designs.

The load capacity of sewer pipes of all materials is influenced by the sewer pipe–soil system, although the importance of the specific areas may

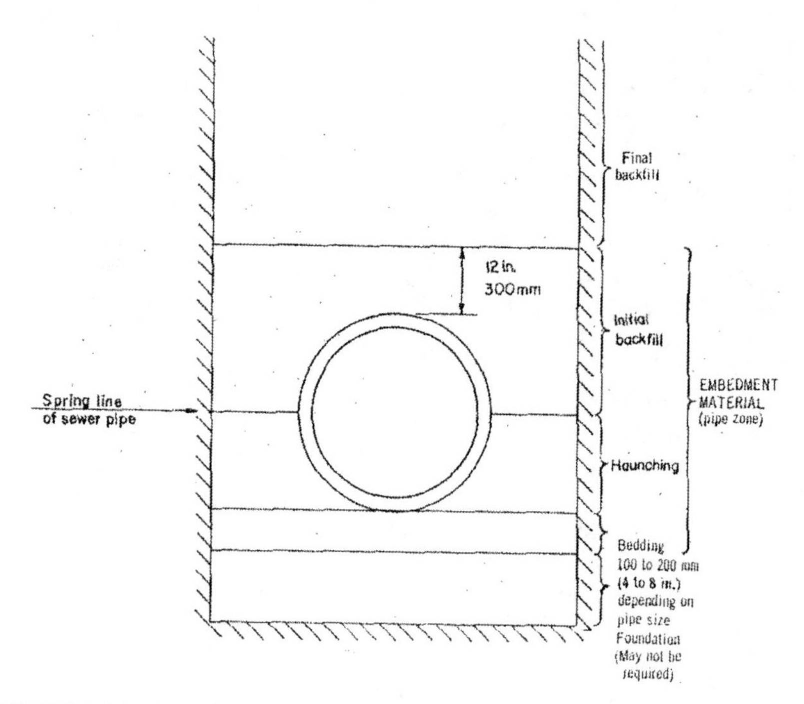

FIGURE 9-31. Trench cross section illustrating terminology.
Reprinted with permission, copyright ASTM International, 100 Barr Harbor Drive, West Conshohocken, PA 19428.

vary with different pipe materials. The information in this chapter includes descriptions of pipe beddings for the following:

- Rigid sewer pipe in trench.
- Rigid sewer pipe in embankment.
- Flexible nonmetallic pipe.
- Ductile iron pipe.

Detailed information on pipe bedding classes is contained in the various ANSI and ASTM specifications and industry literature for each material. The engineer should consult the applicable specifications or literature for information to be used in design.

9.7.5.2. Foundation

The foundation provides the base for the sewer pipe–soil system. In trench conditions, the total weight of the pipe and soil backfill will normally be no more than the weight of the excavated soil. In this case, foundation pressures are not increased from the initial condition, and the designer should be concerned primarily with the presence of unsuitable soils, such as peat or other highly organic or compressible soils, and with maintaining a stable trench bottom.

If the full benefit of the bedding is to be achieved, the bottom of the trench or embankment must be stable. Methods for achieving this condition are discussed in Chapter 10.

To ensure that the sewer pipe is properly bedded or embedded, it is suggested that compaction tests be made at selected or critical locations, or that the method of material placement be observed and correlated to known results. Where compaction measurement or control is desired or required, the recommended references are:

- Standard Method of Test for Relative Density of Cohesionless Soils, ASTM D2049
- Standard Method of Test for Moisture Density Relations of Soils Using 5.5-lb (2.5-kg) Hammer and 12-in (204.8-mm) Drop, ASTM D698
- Standard Method of Test for Density of Soil in Place by the Rubber-Balloon Method, ASTM D2167
- Standard Method of Test for Density of Soil in Place by the Sand-Cone Method, ASTM D1556
- Standard Method of Test of Density of Soil and Soil-Aggregate in Place by Nuclear Methods (Shallow Depth), ASTM D2922

It is recommended that the in-place density of Class I and Class II embedment materials be measured by ASTM D2049 by percentage of relative density, and Class III and Class IV measured by ASTM D2167, D1556, or D2922, by percentage of Standard Proctor Density according to ASTM D698, or AASHTO T-99.

9.7.5.3. *Bedding*

The contact between a pipe and the foundation upon which it rests is the sewer pipe bedding. The bedding has an important influence on the distribution of the reaction against the bottom of the sewer pipe, and therefore influences the supporting strength of the pipe as installed.

Some research (Griffith and Keeney 1967; Sikora 1980) has indicated that well-graded crushed stone is a more suitable material for sewer pipe bedding than well-graded gravel. Both materials, however, are better suited than uniformly graded pea gravel.

Even though larger particle sizes give greater stability, the maximum size and shape of granular embedment should also be related to the pipe material and the recommendations of the manufacturer. For example, sharp, angular embankment material larger than 0.5 to 0.75 inch (12 to 20 mm) should not be used against corrosion protection coatings. For small sewer pipes, the maximum size should be limited to about 10% of the pipe diameter.

Soil classifications under the United Soil Classification System, including manufactured materials, are grouped into five broad categories according to their ability to develop an interacting sewer pipe–soil system (Table 9-16). These soil classes are described in ASTM D2321. In general, crushed stone or gravel meeting the requirements of ASTM Designation C33, Gradation 67 [0.75 inch to No. 4 (19 to 4.8 mm)] will provide the most satisfactory sewer pipe bedding.

In some locations the natural soils at the level of the bottom of the sewer pipe may be sands of suitable grain size and density to serve as both foundation and bedding for the pipe. In such situations, as determined by the design engineer, it may not be necessary to remove and replace these soils with the special bedding materials described above. If the natural soil is left in place, it should be properly shaped for the class of bedding required.

9.7.5.4. *Haunching*

The soil placed at the sides of a pipe from the bedding up to the springline is the haunching. The care with which this material is placed has a significant influence on the performance of the sewer pipe, particularly in the space just above the bedding. Poorly compacted material in this space will result in a concentration of reaction at the bottom of the pipe.

For flexible pipe, compaction of the haunching material is essential. For rigid pipe, compaction can ensure better distribution of the forces on the pipe. Material used for sewer pipe haunching should be shovel-sliced or otherwise placed to provide uniform support for the pipe barrel and to completely fill all voids under the pipe. Because of space limitations, haunching material is often compacted manually. Results should be checked to verify that the class of bedding or installation criteria are achieved.

Material used in haunching may be crushed stone or sand, or a well-graded granular material of intermediate size. If crushed stone is used, it should be subject to the same size limitations and cautions regarding use against corrosion protection coatings. Sand should not be used if the pipe zone area is subject to a fluctuating groundwater table or where there is a possibility of the sand migrating into the pipe bedding or trench walls.

9.7.5.5. Initial Backfill

Initial backfill is the material that covers the sewer pipe and extends from the haunching to some specific point [6 to 12 inches (15 to 30 cm)] above the top of the pipe, depending on the class of bedding. Its function is to anchor the sewer pipe, protect the pipe from damage by subsequent backfill, and ensure the uniform distribution of load over the top of the pipe.

The initial backfill is usually not mechanically tamped or compacted since such work may damage the sewer pipe, particularly if done over the crown of the pipe. Therefore, it should be a material that will develop a uniform and relatively high density with little compactive effort. Initial backfill should consist of suitable granular material but not necessarily as select a material as that used for bedding and haunching. Clayey materials requiring mechanical compaction should not be used for initial backfill.

The fact that little compaction effort is used on the initial backfill should not lead to carelessness in choice or placement of material. Particularly for large sewer pipes, care should be taken in placing both the initial backfill and final backfill over the crown to avoid damage to the sewer pipe.

9.7.5.6. Final Backfill

The choice of material and placement methods for final backfill are related to the site of the sewer line. Generally, they are not related to the design of the sewer pipe. Under special embankment conditions or induced trench conditions, final backfill may play an important part in the sewer pipe design. However, for most trench installations, final backfill does not affect the pipe design.

The final backfill of trenches in traffic areas, such as under improved existing surfaces, is usually composed of material that is easily densified to minimize future settlement. In undeveloped areas, the final backfill often will consist of the excavated material placed with little compaction and left mounded over the trench to allow for future settlement. Studies have indicated that with some soils, this settlement may continue for more than 10 years.

Trench backfilling should be done in such a way as to prevent dropping material directly on top of a sewer pipe through any great vertical distance. When placing material with a bucket, the bucket should be lowered so that the shock of falling earth will not cause damage.

An economical indicator of proper bedding and backfill execution on flexible sewer pipes may be by testing the inside pipe deflection by a go/no-go mandrel. Mandrel dimensions can be determined by applying the initial deflection to the base inside diameter as determined from the appropriate ASTM pipe standard. The initial deflection does not include the deflection lag factor, as is applied to ultimate long-term design deflection.

REFERENCES

American Association of State Highway and Transportation Officials (AASHTO). (1997). "Standard specifications for highway bridges," twelfth ed., AASHTO, Washington, D.C.

American Concrete Pipe Association (ACPA). (1960). "Jacking reinforced concrete pipe lines," ACPA, Arlington, Va.

ACPA. (1993). "Standard Installations and Bedding Factors for the Indirect Design Method, Design Data 40," ACPA, Irving, Tex. Available online at www.concrete-pipe.org/pdfs1/DD_40.pdf, accessed March 18, 2007.

ACPA. (2000). "Concrete pipe design manual," ACPA, Irving, Tex.

American Railway Engineering and Maintenance of Way Association (AREMA). (1981–1982). "Manual for railway engineering," AREMA, Washington, D.C., 1-4-26 and 8-10-10.

American Society of Civil Engineers (ASCE), Direct Design of Buried Concrete Pipe Standards Committee. (2000). "Standard practice for direct design of buried precast concrete pipe using standard installations (SIDD)," *ASCE Standard No. 15-98, ASCE*, Reston, Va.

Ductile Iron Pipe Research Association (DIPRA), (2006). "Design of ductile iron pipe." *DIPRA*, Birmingham, AL

American Water Works Association (AWWA). (2005). "Manual M45, Fiberglass pipe design manual," second ed., AWWA, Denver, Colo.

Griffith, J. S., and Keeney, C. (1967). "Load bearing characteristics of bedding materials for sewer pipe." *J. Water Poll. Control Fed.*, 39, 561.

Howard, A. K. (1977). "Modulus of soil reaction (E′) values for buried flexible pipe." *J. Geotech. Engrg. Div., ASCE, Vol. 103, No. GT, Proc. Paper* 12700, ASCE. Reston, Va.

"Jacked-in-place pipe drainage." (1960). *Contractors and Engr. Monthly*, 45.

Jumikis, A. R. (1969). "Stress distribution tables for soil under concentrated loads." *Engineering Res. Pub. No. 48*, Rutgers University, New Brunswick, N.J., 233.

Jumikis, A. R. (1971). "Vertical stress tables for uniformly distributed loads on soil." *Engineering Res. Pub. No. 52*, Rutgers University, New Brunswick, N.J., 495.

Marston, A. (1930). "The theory of external loads on closed conduits in the light of the latest experiments," *Bull. No. 96,* Iowa Engineering Experiment Station, Ames, Iowa.

Marston, A., and Anderson, A. O. (1913). "The theory of loads on pipes in ditches and tests of cement and clay drain tile and sewer pipe," *Bull. No. 31,* Iowa Engineering Experiment Station, Ames, Iowa.

Portland Cement Association (PCA). (1951). "Vertical pressure on culverts under wheel loads on concrete pavement slabs," Pub. No. ST-65, PCA, Skokie, Ill.

Proctor, R. V., and White, T. L. (1968). "Rock tunneling with steel supports," Commercial Shearing and Stamping Co., Youngstown, Ohio.

Schlick, W. J. (1932). "Loads on pipe in wide ditches," *Bull. No. 108,* Iowa Engineering Experiment Station, Ames, Iowa.

Schrock, B. J. (1978). "Installation of fiberglass pipe." *J. Transp. Div., ASCE, Vol. 104, No. TE6,* Proc. Paper 14175, ASCE, Reston, Va.

Sikora, E. J. (1980). "Load factors and non-destructive testing of clay pipe." *J. Water Poll. Control Fed., 53,* 2964.

"Soil resistance to moving pipes and shafts." (1948). *Proc., 2nd Intl. Conf., Soil Mech. and Found. Eng., 7,* 149.

Spangler, M. G. (1956). "Stresses in pressure pipelines and protective casing pipes." *J. Structural Div., ASCE, Vol. 82, No. ST5, Proc. Paper 1054,* ASCE, Reston, Va.

Spangler, M. G. (1941). "The structural design of flexible pipe culverts," *Bull. No. 153,* Iowa Engineering Experiment Station, Ames, Iowa.

Spangler, M. G., and Hennessy, R. L. (1946). "A method of computing live loads transmitted to underground conduits." *Proc., 26th Ann. Mtg., Highway Research Board, Transportation Research Board,* Washington, D.C., 179.

Taylor, R. K. (1971). "Final report on induced trench method of culvert installation," Project 1HR-77, State of Illinois, Dept. of Public Works and Buildings, Division of Highways, Springfield, Ill.

U.S. Army Corps of Engineers (USACE). (Undated). "Report of test tunnel," Part 1, Vol. 1 and 2, Garrison Dam and Reservoir, USACE Publication Depot, Hyattsville, Md.

Van Iterson, F., K. Th. (1948). "Earth pressure in mining." *Proc., 2nd Intl. Conf. Soil Mech. and Found. Eng., 3,* 314.

BIBLIOGRAPHY

Harell, R. F., and Keeney, C. (1977). "Loads on buried conduit—A ten-year study." *J. Water Poll. Control Fed., 48,* 1988.

Moser, A. P., Watkins, R. K., and Shupe, O. K. (1977). "Design and performance of PVC pipes subjected to external soil pressure," Buried Structure Laboratory, Utah State University, Logan, Utah.

Seaman, D. J. (1979). "Trench backfill compaction control bumpy streets." *Water and Sewage Works, 53*, 67.

Watkins, R. K., and Spangler, M. G. (1958). "Some characteristics of the modulus of passive resistance of soil: A study in similitude." *Proc., Highway Research Board, Vol. 37*, 576–583, Transportation Research Board, Washington, D.C.

Wenzel, T. H., and Parmelee, R. A. (1977). "Computer-aided structural analysis and design of concrete pipe, concrete pipe and the soil-structural system," ASTM STP 630, ASTM International, West Conshohocken, Penn., 105–118.

CHAPTER 10

CONSTRUCTION CONTRACT DOCUMENTS

10.1. INTRODUCTION

The purpose of the contract documents is to portray clearly, by words and drawings, the nature and extent of the work to be performed, the conditions known or anticipated under which the work is to be executed, and the basis for payment. Most sewer construction projects are accomplished by contracts entered into between an owner and a construction contractor. The contract documents, consisting of the contract forms, project forms, conditions of the contract, specifications, drawings, addenda, and change orders, if any, constitute the construction contract. In addition, other exhibits may be included in the construction contract, such as the bidding requirements and bid forms.

The contract drawings and the specifications together define the work to be done by the contractor under the terms of the construction agreement. These documents are complementary. What is called for by one is to be executed as if called for by both.

The contract documents establish the legal relationship between the owner and the contractor, as well as the duties and responsibilities of the engineer. It follows that the engineer's potential liability is influenced by the contract documents, although the engineer is not a party to the construction contract.

Experience in the courts over many years has revealed a number of sensitive areas of potential liability which may not be covered in the documents developed internally by the owner and/or engineer. Among these are:

- Means and methods of construction.
- Right to stop the work.

- Safety.
- Insurance.
- Supervision.
- Indemnification.

In general terms, the contractor is responsible for means and methods of construction, and safety and supervision of workmen. Only the owner has the right to stop the work, and its insurance adviser makes final decisions on insurance provisions. Subject to limitations of state law, the engineer should be included with the owner as a party indemnified.

These provisions and many others are covered in an adequate manner in the standard documents of the Engineers' Joint Contract Documents Committee (EJCDC 2002a through 2002e). These standard documents are recommended as the basis for developing the contract documents. They are further discussed in Section 10.3 of this chapter. Of course, good legal advice is prudent in all contractual matters.

Prior to bidding, the project usually requires approval by the regulatory agencies. When approved, a permit to construct the project may be issued. After the project has been approved and the contract documents are in final form, bids are solicited.

For ease in bidding and administration, frequently the work is divided into various items, with either unit or lump sum prices received for each item of work. The contract documents must clearly describe and limit these items to obviate all possible confusion in the mind of the bidder with regard to methods of measurement and payment. The subdivision of the work is often based on local customs, the customs and conventions of the engineer, or the specific requirement of the owner.

Unit price bids have been used most generally where quantities of work are likely to be variable and adjustment is found necessary during construction. Linear feet (meters) of sewer, numbers of manholes, and cubic yards (cubic meters) of rock excavation or concrete cradle are examples of such unit price items.

Lump sum bids have been applied most generally to special structures that are completely defined and not subject to alteration or quantity changes during construction. Lump sum bids may also be taken for an entire sewer construction contract where the contract documents define the work with sufficient completeness to permit the bidder to make an accurate determination of the quantities of work. Such contracts may contain unit adjustment prices for items of work, such as rock excavation, piles, additional excavation, selected fill material, and sheeting requirements which cannot be determined precisely beforehand. An appropriate quantity of the unit price work may be included for comparison of bids. Both unit costs and unit adjustment prices, as applicable, are typically reflected in the Schedule of Values. The purpose of the Schedule of Values is to provide a basis for monthly progress payments. The administration of the project, provided extensive changes are not made during construction, is simplified in the lump sum type of contract.

A final caveat: Administration of the contract is controlled by the contract documents that must include applicable administrative contract provisions required by federal, state, and local governments. The more the contract documents put the engineer in control of the contractor's activities, rather than of the *results* of the contractor's work, the greater the duties and legal obligation imposed on the engineer.

10.2. CONTRACT DRAWINGS

10.2.1. Purpose

The purpose of the contract drawings is to convey graphically the work to be done, first to the owner, then to project reviewers, to the bidders, and later to the construction observers/inspectors/engineer and the contractor. All information that can best be conveyed graphically, including configurations, dimensions, and notes, should be shown on the contract drawings. Lengthy word descriptions are best included only in the specifications and need not be repeated on the contract drawings.

As a general rule, contract drawings should be prepared carefully in a neat, legible fashion. Hastily produced, sloppy drawings can lead to mistakes during construction which might be blamed on the condition of the drawings. The drawings are an extension of the contract and should be carefully considered. The following summarizes key points in obtaining design information and preparing the contract drawings.

10.2.2. Field Data

A survey and investigation of the route of the sewer are required to obtain information as to the existing topography, underground utilities, and property boundaries to be shown on the contract drawings. The route may be mapped from data obtained by conventional ground surveys and/or by aerial photogrammetry. Field verification of dated photographs should be performed in order to properly represent the proposed construction area. Survey work is discussed in some detail in Chapter 2.

When the location of the sewer has been well-defined by preliminary studies, it may be possible to run the ground survey baseline directly on the centerline of the proposed alignment. This procedure will facilitate office plotting of field data and will later simplify stakeout of the sewer for construction. If the actual alignment is not established by field surveys, baselines or reference marks must be established in the field.

10.2.3. Preparation

Contract drawings generally are prepared using graphical computer-aided design software to facilitate design updates and reduce drafting time. There are typically several revisions of the drawings as the project moves from the design phase through construction. The following represent typical revisions experienced during a sewer design project from concept through construction:

- *60% Detail Drawings.* Provided to communicate general concept of sewer layout and obtain confirmation prior to more detailed design.
- *90% Detail Drawings.* Detailed design produced for comment by all reviewers prior to submitting to contractors for a bid price on construction.
- *Issue for Bid.* Provided to contractors interested in performing the proposed construction.
- *Issue for Construction.* Final adjusted drawings before construction begins.
- *Record Drawings.* Final drawings of actual construction layout.

To track updates, a log of revisions should be maintained and is typically shown in the title block tagged with a revision number.

Typically, through the 60% design phase the sewer layout is provided on plan sheets only. As the proposed layout is established, profile details are drawn to coordinate with the plan view. The plan view is usually drawn on the top half of the sheet and the profile is plotted directly beneath it, on the bottom half at the same horizontal scale, facilitating coordination of the two views.

The plan view should communicate detailed information on the proposed sewer pipe location and relevant site conditions that would affect the construction. Typical features include the following:

- Site topography.
- Property information and location of acquired rights-of-way or easements.
- Existing aboveground and underground utilities.
- Existing pavement and edge of pavement locations.
- Contract limit lines.
- Additional site-specific details which may affect construction work.

When the plan view utilizes background aerial photographs, the topography may be plotted directly on the aerial photographs. The site topography and record data obtained from utility companies as to underground utilities are also plotted on the plan. The plan showing existing conditions is completed by plotting property and easement boundaries from tax map information, deed descriptions, property maps of individual parcels, and physical evidence obtained by field survey. In many cases, Geographic Information System (GIS) data on property information can be downloaded directly from the county tax appraiser's office. In this case, boundary information can be imported directly onto the sewer plan. When applicable, proposed temporary and permanent easement boundaries are determined and drawn on the sewer drawings. Additional relevant information may include basement elevations and inverts of existing utilities or other structures that may affect the work and should also be shown. Contract limit lines are added to complete the preparation of the drawings.

A profile of the ground surface along the proposed sewer alignment is drawn and proposed sewer invert elevations are determined and plotted on the profile.

The scale of the drawings should be large enough to show all of the necessary surface and subsurface information without excessive crowding. A horizontal scale of 40 feet to the inch to 100 feet to the inch (1:400 to 1:1000, or 5 to 10 meters to the centimeter) is suitable for many drawings. However, in urban areas or smaller area details, 20 feet to the inch (1:250, or 2.5 meters to the centimeter) should be considered. In such areas, large-scale drawings of street intersections are quite useful. Generally, a common vertical scale for the profile is 10 feet to the inch (1:100, or meter to the centimeter) or 2 to 5 feet to the inch (1:20 to 1:50, or 0.2 to 0.5 meter to the centimeter) based on the height variation of the terrain. Larger scales are used for sections and details.

Contract drawings are sometimes reduced to approximately one-half scale and issued to bidders in this size for convenience. This practice requires careful preparation of the full-size drawings to produce clear and readable reductions. Reduced-size drawings should contain a note stating the magnitude of size reduction (if it is an exact reduction, such as half-size) and should always have a graphical scale.

Drawings are typically prepared using the U.S. Convention units. Stations will typically be represented in 100-foot increments followed by subincrements of one foot. Benchmark elevations and design details, including proposed sewer inverts, are typically presented in feet with two decimal places of accuracy. Nominal sewer sizes should always be shown in inches.

When plan and profiles are drawn using metric dimensions, stations each will be 100-meter; horizontal dimensions should only be shown to 0.01 meter. Benchmark elevations will be given to 1 millimeter (i.e., three places of decimals). However, it will generally be appropriate to show sewer inverts, etc., to only two decimal places. Nominal sewer sizes should always be shown in millimeters.

Lettering on contract drawings falls into three general categories:

- Labeling and dimensioning
- Notes
- Titles

The text font should be one that is easily read (such as the standard Simplex font) with the recommended height of the text sized for the final plotted height on the drawing document. Labeling and dimensioning of existing conditions should have a text height of 0.06 inches. If the drawings are later to be reduced in size, a minimum letter size of 0.08 inches should be considered. Proposed facilities are labeled and dimensioned with the same size or larger lettering, preferably 0.08 to 0.10 inches, and should stand out from the lettering for existing facilities either by size or by line weight. Additional notes should be lettered in the same size used for labels of proposed facilities. Titles should be larger in size and be consistent throughout the drawing set, preferably 0.12 to 0.14 inches. Finally, assuming computer-aided design software is used, the drawing file text should be placed in the proper corresponding layers according to their function within the drawing, thus alleviating confusion during subsequent revisions of the drawing file.

Sections and details of the proposed sewer and appurtenances are typically inserted following the plan and profile sheets. Typically, each detail has a detail number and reference page associated with it. Reference numbers are used throughout the plan and profile sheets to refer to a specific detail number and page.

10.2.4. Contents

10.2.4.1. Arrangement

The most logical arrangement for a set of contract drawings develops the project from general views to more specific views, and finally to more minute details. The subsections below are arranged to follow this generally accepted order of drawing presentation.

10.2.4.2. Title Sheet

The title sheet should identify the project by presenting the following information:

- Project name
- Contract number
- Federal or state agency project number (if applicable)
- Owner's name

- Owner's officials, key people, or dignitaries
- Engineer's name
- Engineer's project number
- Engineer's address
- Engineer's engineering business number
- Drawing set number (for distribution records)
- Professional engineer's seal and signature

10.2.4.3. Title Blocks

Each sheet except the title sheet should have a title block containing the following:

- Sheet title
- Project name
- Federal or state agency project number (if applicable)
- Owner's name
- Engineer's name
- Engineer's seal and signature
- Sheet number
- Engineer's project number
- Scale
- Date
- Designer, drafter, and checker identification
- Revisions identification block
- Sign-off by owner's chief engineer or district superintendent (as applicable)

10.2.4.4. Index/Legend

Contract drawings should contain an index that lists all of the drawings in the set by title and drawing number in order of presentation. It also is useful to provide a general plan map or key map sheet to identify the sheets that show the details for each length of proposed sewer proposed. These indices should be located on the drawing following the title sheet. A legend showing a set of symbols for elements of topographic abbreviations and the various items of the sewer works indicated in the sewer drawings should be included on the index sheet. An example of a legend for sewer drawings is shown in Figure 10-1.

10.2.4.5. Location Map

There should be a general location map showing the location of all work in the contract in relation to the community, either on the title sheet or on the index/legend sheet. This location map also may be used as an index map as outlined in the preceding paragraph.

LEGEND:

BM	BENCH MARK
BOB	BOTTOM OF BANK
BOC	BACK OF CURB
CM	CONCRETE MONUMENT
CMP	CORRUGATED METAL PIPE
CONC.	CONCRETE
D.O.T.	DEPARTMENT OF TRANSPORTATION
E./ELEC.	ELECTRIC
–E–	ELECTRICAL LINE
EL./ELEV.	ELEVATION
EOP	EDGE OF PAVEMENT
GRD	GROUND
HDPE	HIGH DENSITY POLYETHYLENE PIPE
INV	INVERT
LB	LICENSED BUSINESS
LS	LICENSED SURVEYOR
M.B.	MAP BOOK
MH	MANHOLE
N&D	NAIL AND DISK
N&F	NAIL AND FLASHER
–OHW–	OVERHEAD WIRE
PLS	PROFESSIONAL LAND SURVEYOR
PSM	PROFESSIONAL SURVEYOR & MAPPER
PVC	POLYVINYL CHLORIDE PIPE
R	RADIUS
RCP	REINFORCED CONCRETE PIPE
RNG.	RANGE
R/W	RIGHT–OF–WAY
RRS	RAILROAD SPIKE
SEC.	SECTION
(SP)	STATE PLANE COORDINATE SYSTEM
T./TELE.	TELEPHONE
TOB	TOP OF BANK
(TYP)	TYPICAL
TWP.	TOWNSHIP
U.G.S.	UNDERGROUND SERVICE
VCP	VITREOUS CLAY PIPE

Symbol	Meaning
	UTILITY POLE
	CONCRETE UTILITY POLE
	METAL UTILITY POLE
	GUY ANCHOR
	BENCH MARK
	LIGHT POLE
	EXISTING MANHOLE (GENERAL)
D	DRAINAGE MANHOLE
S	SEWER MANHOLE
T	TELELPHONE MANHOLE
GV	GAS VALVE
	GATE VALVE
WM	WATER METER
TJB	TELEPHONE JUNCTION BOX
	FIRE HYDRANT
––––C––––	EXISTING CABLE TV LINE
–––FOC–––	EXISTING FIBER OPTICS CABLE
––––G––––	EXISTING GAS LINE
––––T––––	EXISTING TELEPHONE LINE
––––W––––	EXISTING WATER MAIN
–– FM ––	EXISTING FORCE MAIN
–– SS ––	EXISTING STORM SEWER
–––SAN–––	EXISTING SANITARY SEWER
	PROPOSED GRAVITY SEWER
	PROPOSED FORCE MAIN
	PROPOSED MANHOLE
	PROPOSED LIFT STATION
	FLOW ARROW
	EXISTING GRADE (PROFILE)
	EXISTING R/W
	EXISTING CONTOUR
	BUILDING TO BE SERVED UNDER THIS CONTRACT
	BUILDING NOT SERVED UNDER THIS CONTRACT
	ASPHALT
	CONCRETE

FIGURE 10-1. Typical legend for sewer drawings.

10.2.4.6. General Notes

Notes that pertain to more than one drawing should be presented on the index/legend sheet. An example is a note warning that the location and sizes of existing underground utilities shown on the contract drawings are only approximate and that it is the responsibility of the contractor to confirm or locate all underground utilities in the area of his work.

10.2.4.7. Subsoil Information

Whether or not the locations of soil borings made during the design phase of a project and the boring logs should be included in the contract documents is a decision that should be made only after proper legal advice and consideration. In any event, whatever subsurface information has been obtained should be made available to bidders. Drawings and specifications should indicate where special construction is required because of known unfavorable subsoil conditions. Neither the owner nor the engineer should guarantee subsoil conditions as a known element of the contract agreement. If soil boring information is provided, location of the soil borings should be shown on the plan sheet.

10.2.4.8. *Survey Control and Data*

Survey control information may be shown on the sewer plan/profile sheets or on a general plan on a separate sheet. Baseline bearings and distances should be included with reference ties to permanent physical features. Vertical control points, or benchmarks, should be indicated, and the datum plane used for determining these elevations must be defined. A note indicating the dates of the ground survey and aerial photography should be included.

10.2.4.9. *Sewer Drawings*

A continuous strip map, drawn directly above the profile, to indicate the plan locations of all work in relation to surface topography and existing facilities is an integral part of each sewer plan and profile drawing. The width of the strip map should be such that only topography which directly affects construction or access to the work is indicated.

Drawings for sewers to be constructed within easements on private property should show survey baseline and sewer alignment data. Widths of proposed temporary and permanent easements should be dimensioned.

Sewer drawings should generally be oriented so that the flow in the sewer is from right to left on the sheet. Each sewer plan should include a North arrow consistent with this arrangement. Stationing should typically be upgrade from left to right, generally along the sewer centerline. Survey baseline stationing also may be provided on the drawings but it should not be substituted for stationing along the centerline of the sewer.

Stationing indicated on construction drawings for location of manholes and wye-branches or house connections is to be considered approximate only and should be so noted. Record drawings after construction must, however, give accurate locations of all features of a completed sewer system.

Locations of junction structures must be held firm as given on construction drawings. Match lines should be used and should be easily identifiable. Special construction requirements, such as sheeting to be left in place, should be shown on the drawings.

An example of a plan and profile for a sewer to be constructed is provided in Figure 10-2.

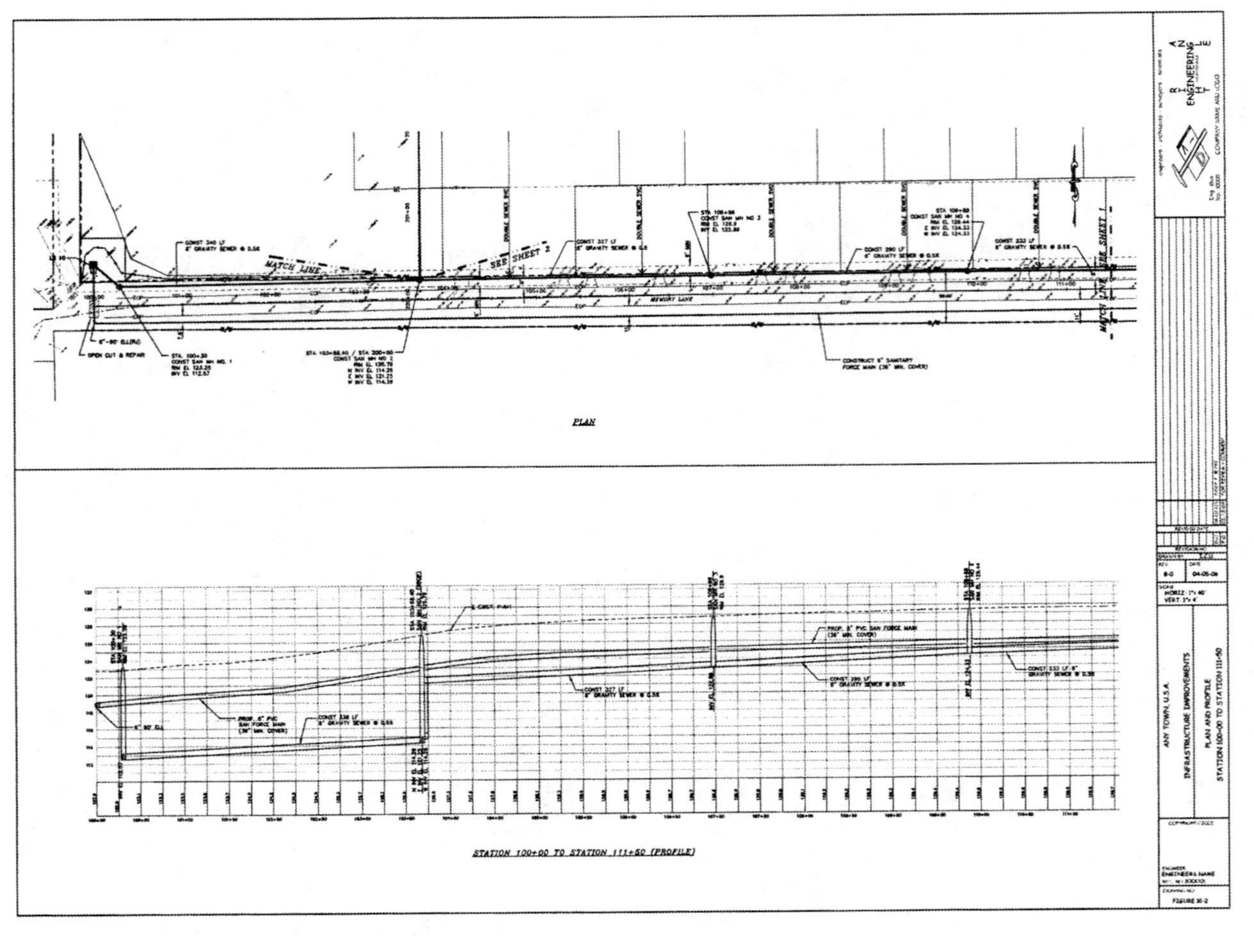

FIGURE 10-2. Typical plan and profile for sewer drawings.

10.2.4.10. Sewer Profile

Contract drawings should include a continuous profile of all sewer runs indicating centerline ground surface and sewer elevations and grades. Stationing shown on the plan should be repeated on the profile.

The profile is also a convenient place to show the size, slope, and type of pipe; the limits of each size, pipe strength, or type; the locations of special structures and appurtenances; and crossing utilities and drainage pipes. Where interference with other structures is known to exist, explanatory cross sections and notes should be included. Such cross sections, often enlarged in scale, should be identified as to specific location and, if practicable, should be placed on the plan/profile drawing near where the section is cut. Examples of scenarios where cross sections are suitable include major roadway, waterway, railroad, and/or utility crossings. By best judgment, if the additional detail causes the plan/profile sheet to be cluttered, the detail may be placed on attached sheets with an appropriate reference. If similar utility crossings occur at multiple locations, a typical detail drawing depicting the separation distance may also be included as a detail attachment.

10.2.4.11. Sewer Sections

When sewers consist of pipes of commonly known or specified dimensions, materials, or shapes, no sewer sections need to be shown. For cast-in-place concrete sections, complete dimensions with all reinforcement steel shown should be included in the drawings.

10.2.4.12. Sewer Details

In unique cases, sewer details may be provided on the specific plan/profile sheet showing the respective location of the detail. However, separate sheets of sewer details sheets that follow the plan/profile sheets are most common. The details may be stand-alone, referring to common structures detailed in multiple locations, or referenced details from the plan/profile sheets.

The following details, when applicable, should be included:

- *Trenching and Backfilling.* Payment limits including those for rock excavation and types of backfill materials.
- *Pipe Bedding and Cushion.* Dimensions, material types, and payment limits.
- *House Lateral Connections.* Type and arrangement of fittings and minimum pipe grade.
- *Special Connections.* Type and configuration of fittings and dimensions.
- *Manholes.* Foundation, base, barrel, top slab, frame and cover, and invert details.

- *Sewer/Water Main Crossings.* Separation requirements.
- *Waterways, Highway, or Railroad Crossings.* Casing, inverted siphon, encasement, or other related details.

Many of these details find repeated use in sewer projects. Developing standard details of these items, which may be reproduced for repeated use in multiple sewer contracts, is helpful.

10.2.4.13. *Special Details*

Details that do not pertain directly to the sewer piping and are not covered by standard details should be provided on miscellaneous detail sheets. The following would be included:

- *Special Structures.* Full details so that the finished work is structurally sound and hydraulically correct.
- *Special Castings.* Sufficient details for the manufacturers to prepare shop drawings. Standard casting items, such as manhole frames, covers, and manhole steps, will be identified by reference to a manufacturer's catalog number in the specifications.
- *Restoration Items.* Complete details for pavement, sidewalk, and curb repairs.

10.2.5. Record Drawings

During construction of the sewer project, the contractor or the engineer should measure and record the locations of all wyes, stubs for future connections, and other buried facilities which may have to be located in the future. All construction changes from the original drawings, rock profiles, and other special classes of excavated material also should be recorded by the contractor or the engineer.

Contract drawings should be revised to indicate this field information after the project is completed and a notation such as "Revised According to Field Construction Records" or "Record Drawing" should be made on each sheet. The term "As-Builts," once a common notation, is not recommended because it implies that all details illustrated in the drawings were constructed specifically as shown. Rarely is this true. Consequently, the engineer who certifies "As-Built" drawings may be exposed to potential liability. A qualifying statement should be placed on record drawings to the effect that the drawings are not warranted but are believed to represent conditions upon completion of construction within reasonable tolerances based on information furnished to, or obtained by, the engineer who certifies the drawings. It is recommended that record sets of such revised drawings should become a part of the owner's permanent sewer records.

10.3. PROJECT MANUAL

10.3.1. Introduction

The bound volume containing the bidding documents, the agreement forms, the conditions of the contract, and the specifications is preferably termed the Project Manual. The commonly used title "Specifications" is misleading in that the bound volume contains far more than the material and workmanship requirements for the project, which is the definition of the term "specifications."

The introduction to this chapter (Section 10.1) recommended that the standard documents of the Engineers' Joint Contract Documents Committee (EJCDC 2002a through 2002e) be used as a basis for developing contract documents. In addition, the "Commentary on Contract Documents" (Clark 1993) should also be used in the development of contract documents. Due to the increased legal exposure of the engineer and owner, the adoption of these standard documents (to benefit from mutual legal experience) should be considered.

These documents are closely related and coordinated. Changes in one may require changes in one or more of the others. Furthermore, the standard documents (as well as documents developed internally by the engineer or owner) should be reviewed by the attorneys for both the owner and the engineer for each project due to their legal consequences. It is especially important that coordination of documents be reviewed if non-standard documents are being used.

Standard documents generally provide acceptable documents and, through widespread usage, are better understood and less subject to misinterpretation. Furthermore, they offer the user language which has been tested by the courts.

Although there is a saving of time by utilizing standard documents in preparation of contract documents, one should guard against irrelevant or contradictory requirements within the entire contract document, particularly between standard and specific portions of the contract. Again, good legal advice is essential.

The most widely used standard for organizing specifications, MasterFormat™, is published by the Construction Specifications Institute (CSI). Recently, a new MasterFormat™ 2004 Edition (CSI 2004) was released, which replaced the previous 1995 edition. The newer version incorporates the most significant revisions in the product's 40-year history. The major differences between the 2004 and 1995 editions include the following: expansion of the number of divisions from 16 to 50; creation of additional separate divisions for specialty areas; reservation of divisions for future expansion; and expansion of the numbering system from a five- to a six-digit system.

10.3.2. Purpose

Documents contained in the Project Manual set forth the details of the contractual agreement between the contractor and owner. They describe the work to be done—complementing the information provided on the drawings—and establish the method of payment. They also set forth the details for the performance of the work, including necessary time schedules and requirements for insurance, permits, licenses, and other special procedures.

Documents contained in the Project Manual must be clear, concise, and complete (CSI 2005). All portions should be written to avoid ambiguity in interpretation. Specifications should be easily understood and should be devoid of unnecessary words and phrases, yet they must completely outline the requirements of the project. Reference to standards, such as those of ASTM International, can be used to reduce the bulk of the specifications without detracting from completeness. A high degree of writing skill and thorough knowledge of standards are needed to produce a quality set of documents.

10.3.3. Arrangement

The arrangement of the contents of the Project Manual varies, depending to an extent on the requirements of the owner and the practices of the engineer. Furthermore, arrangement and division of the contents are frequently subject to local legal requirements.

Many government agencies, private owners, and engineers have adopted the practices established by the CSI. As a means of standardizing the order of the documents and the location of contract subject matter within the Project Manual, and thereby improving communications among the construction team, it is recommended that the practices embodied in the Project Resource Manual—CSI Manual of Practice (CSI 2005) be considered in the preparation of the Project Manual. The EJCDC's Uniform Location of Subject Matter (EJCDC 1995) is also recommended as a guide in the preparation of Project Manuals to further improve communications.

Preferably, all parts of the Project Manual are bound in a single volume. However, for extensive programs of sewer construction, a Standard Project Manual may be developed, to be bound separately and incorporated by reference in manuals for individual but related projects.

The Project Manual is divided most logically into parts that each define a phase or function in the overall administration and performance of the contract. Details included in one section generally should not be repeated in others. Refer to Section 10.3.10 for a checklist with general notes on content of these sections.

The assembled Project Manual should be prefaced with a cover page, title page, and table of contents. A convenient arrangement which organizes the parts in their logical order of use is as follows:

- Addenda
- Procurement Requirements
- Contracting Requirements
- Specifications

General notes on these sections may be found on the pages following, with details covered in the included checklist.

10.3.4. Addenda

Addenda are issued during the bidding period to correct errors and omissions in the contract documents, to clarify questions raised by bidders, and to issue additions and deletions to the documents. Changes to both graphic and written documents may be accomplished by an addendum.

Procedures for issuing addenda are described in the Instructions to Bidders, and space for the bidder to acknowledge their receipt is provided in the Bid Form. Addenda must be issued sufficiently in advance of the bid opening to give bidders time to account for any modifications in the preparation of bids.

10.3.5. Procurement Requirements

Procurement requirements instruct the bidders or proposers about the established procedures for preparing and submitting their bids or proposals, and should be clearly set forth in the Project Manual. This section covers all requirements, instructions, and forms pertaining to the submission of pricing in the form of bids or proposals from prospective contractors.

With respect to the use of the standard documents, the EJCDC and CSI exclude procurement requirements from their definition of contract documents. This is primarily for two reasons: (1) much of their substance pertains to relationships before the agreement is signed and does not pertain to the performance of the work; and (2) the EJCDC Standard General Conditions of the contract apply to negotiated as well as to bid contracts. However, in special circumstances (such as a detailed Bid Form for a complex unit price contract), it may prove wise to attach the bid as an exhibit to the agreement.

The procurement requirements for obtaining bids differ from the requirements for obtaining proposals. The procurement requirements include the following items.

10.3.5.1. Solicitation

Bid solicitations fall into two categories: Invitations to Bid and Advertisements to Bid. The document should be brief and simple, containing only the information essential to permit a prospective bidder to determine whether the work is in his line, whether he has the capacity to perform, whether he satisfies the prequalification requirements, whether he will have time to prepare a bid, and how to obtain bid documents.

Soliciting a proposal requires a different type of process. This process is utilized to seek out unique solutions using delivery methods other than the traditional bidding process. In soliciting a proposal, a Request for Proposal (RFP) is usually prepared by the owner or engineer. The RFP describes what is desired by the owner.

10.3.5.2. Instructions for Procurement

The Instructions to Bidders furnish prospective bidders with detailed information and requirements for properly preparing and submitting bids. Included here are bidder's responsibilities and obligations; the method of preparation and submission of proposals; the manner in which bids will be canvassed, the successful bidder selected, and the contract executed; and other general information regarding the bid award procedure.

Unlike the bidding process, the proposal and negotiation process has not generated standard printed documents. The purpose of these instructions should be to establish proper methods of obtaining clarifications or interpretations and to define documents such as addenda.

10.3.5.3. Available Information

Information available to the bidder or proposer is listed in this section. The section for the bidding process could include such information as geotechnical reports, soil boring data, hazardous materials reports, descriptions of the site, resource drawings of existing buildings, and property survey information. The EJCDC's recommended practice is to include a reference to the information and describe the availability and location where the information may be reviewed. A disclaimer, however, may be advised to prevent the owner and the engineer from being held responsible for conclusions drawn from this information. This information, when made available, is for the bidders' use in preparing the bids but is not part of the contract documents.

The information made available for developing a proposal could include subsurface information and other existing conditions similar to a bid situation. It may also include documents the owner wishes to make available and drawings such as diagrams and schematics.

10.3.5.4. Procurement Forms and Supplements

The bid process requires a Bid Form be prepared and submitted to the owner. The purpose of a Bid Form is to ensure systematic submittal of pertinent data by all bidders in a form convenient for comparison. The Bid Form should not contain basic contractual provisions since it is only an offer to perform the work as required by and in accordance with the contract documents. It must be so worded and prepared that all bidders will be submitting prices on a uniform basis, allowing for equal consideration in awarding a contract.

The Bid Form is addressed to the owner of the proposed work and is to be signed by the bidder. The project for which the bid is submitted must be identified. The Bid Form contains spaces for insertion of unit or lump sum prices and may also contain spaces for each bidder's extensions for each item and a total bid price; however, it must be stated that bidder's extensions and total bid price are unofficial and subject to verification by the owner. Bid prices are commonly stated in both words and figures, with the word description governing in case of a discrepancy.

The Bid Form may provide for taking bids on alternative materials or methods of executing portions of the work. It may provide for combination bids on several contracts in the project. The basis for considering alternatives and combinations must be described in the Instructions to Bidders and set forth in the Bid Form. An informal comparison of bids first may be made, based on the totals given in the submitted Bid Forms. The formal bid comparison must be made after the extensions of unit prices and totals of contract items have been checked and determined by the owner to be correct.

The completion date or time is generally set by the owner so that all bidders are submitting prices on the same time basis and the only variable is price. However, on some projects, the time allowed for construction or completion may be set by the bidder as part of his bid. The latter practice is less common due to the complexities created in evaluating and comparing bids. This may be selected if time is more important than cost. If this course is chosen, criteria for evaluating the time for completion must be established in the Instructions to Bidders.

The Bid Form contains spaces for the bidders to acknowledge the receipt of addenda issued during the bidding period. Statements must also be included to the effect that the bidder has received or examined all documents pertaining to the project, that the requirements of addenda were taken into consideration in rendering the bid, and that all documents and the site have been examined.

Whether the process is formal or informal, the proposal process may utilize a similar form to the bid process. The Proposal Form provides assurances that all information requested by the owner is provided. The form can include acknowledgments similar to the Bid Form. Supplements to the Proposal Form can also be required to allow the owner to evaluate priorities other than price, such as schedule, value analysis, alternative products, or suggested modifications.

10.3.6. Contracting Requirements

The Contracting Requirements are the legal documents that describe the contractual requirements. Their purpose is to define the processes, rights, responsibilities, and relationships of the parties to the contract. They are comprised of the contracting forms, project forms, conditions of the contract, and special forms such as revisions, clarifications, and modifications.

10.3.6.1. Contracting Forms

The contracting forms section typically includes the Notice of Award, Construction Agreement, and attachments to the Agreement. The Construction Agreement, typically referred to as the Contract, is the document that legally obligates each party signing the Contract. The form of the Contract is regulated by the laws of the local jurisdiction and state in which it is executed. The Contract must, however, cover all items of work included in the bid except those that have been eliminated as alternative items. It also must bind the contracting parties to conformity with the provisions of all the contract documents.

10.3.6.2. Project Forms

The Project Forms section typically includes bond forms, certificates and other forms, clarification and modification forms, and closeout forms.

Signing of the Agreement is contingent on prior receipt of the executed bonds. Bonds must be executed by a financially responsible and acceptable surety company (U.S. Treasury Department, updated annually). General practice is to require performance and payment bonds and that each be in the amount of the contract bid price. Maintenance or guarantee bonds may also be required.

10.3.6.3. Conditions of the Contract

This portion of the contract documents is concerned with the administrative and legal relationships, rights and responsibilities between the owner, the owner's representatives, the contractor, subcontractors, the public, and other contractors. Conditions of the Contract should contain instructions on how to implement the provisions of the Contract. They should not include detailed specifications for materials, workmanship, or work-related administrative or nonlegal matters.

The conditions of the Contract consist of two parts: the General Conditions and the Supplementary Conditions.

Standard General Conditions have been developed by professional societies, as well as by government agencies and private owners. Consultants frequently use in-house standards for projects of their design.

Standard General Conditions usually require revisions, deletions, or additions to suit the needs of a particular project. These changes are incorporated into the Supplementary Conditions.

The carefully chosen language contained in the Standard General Conditions should be modified only when necessary for a specific project as may be required by the locale of the project, the requirements of the owner, or the complexity of the project. All modifications must be coordinated with the other documents to avoid the creation of contradictory requirements within the contract documents. The EJCDC Guide to the Preparation of Supplementary Conditions (EJCDC 2002d) gives examples of wording for modifications which occur frequently. However, modifications to the Standard General Conditions via the Supplementary Conditions should not occur without explicit approval and guidance from the owner, the owner's legal counsel, and the owner's insurance adviser.

Many currently published Standard General Conditions contain articles of a nonlegal nature, such as submittals. With the general acceptance of the MasterFormat™ 2004 Edition and its proper utilization, nonlegal matters more appropriately should be dealt with in "Division 01, General Requirements" of the specifications.

10.3.6.4. Revisions, Clarifications, and Modifications

Revisions consist of precontract revisions made prior to signing the Agreement. These are the written addenda or graphical documents issued to clarify, revise, add to, or delete information in the original bid documents or in previous addenda. An addendum is issued prior to the receipt of bids or proposals.

Clarifications consist of documents initiating changes or clarifications that have not been incorporated into the Contract by a formal contract modification.

Modifications include written amendments to the contract documents after the Construction Agreement has been signed. These modifications are typically accomplished by change orders, work change directives, and field orders.

Modifications serve to clarify, revise, delete from, or add to the existing contract documents. They serve to rectify errors, omissions, and discrepancies, and to institute design changes requested by the owner or made necessary by unanticipated conditions encountered in carrying out the work. The procedures for effecting modifications must be clearly spelled out in the contract documents.

10.3.7. Specifications

Specifications cover qualitative requirements for materials, equipment, and workmanship as well as administrative, work-related requirements. The two types of provisions should be kept separate.

The MasterFormat™ 2004 Edition provides a 50-division framework for the development of project specifications. Each division consists of a number of related sections. Division titles are fixed; sections under each division use a six-digit numbering system consisting of three pairs of two digits. Many divisions are not applicable to sewer work and would be deleted, with the remaining divisions utilized to define project requirements (see included checklist, Section 10.3.10).

10.3.7.1. General Requirements (CSI Division 01)

Individual characteristics regarding conditions of the work, procedures, access to the site, coordination with other contractors, scheduling, facilities available, and other nonlegal, work-related, and administrative details which are unique to the particular contract are placed properly in the General Requirements.

Division 01 sections expand on the administrative and procedural conditions of the Contract. These sections in Division 01 apply broadly to the execution of the work of all the other sections of the specifications. CSI practice is to state information only once and in the right place. Hence, Division 01 sections should be written in language broad enough to apply to the sections in all other divisions.

10.3.7.2. Material and Workmanship Specifications (CSI Divisions 02 through 49)

Specific details regarding materials or workmanship applicable to the project may be written especially for the Contract, or general specifications called Standard Specifications may be prepared which are intended to apply to many contracts. In addition, government and private organizations in many areas have developed standard specifications for sewer construction, which reflect local practices. These standards should be reviewed to obtain an understanding of common practices in a particular locale.

Commonly, materials are specified by reference to specifications of ASTM International, the American National Standards Institute (ANSI), the American Concrete Institute (ACI), the American Water Works Association (AWWA), and other similar organizations.

Standards of workmanship should be described in specific terms when feasible, but specification of construction means and methods and safety should always be avoided. Nonetheless, parameters and limits often must be specified to ensure that construction procedures will be consistent with design intent.

The MasterFormat™ 2004 Edition divides the specifications into subgroupings of divisions which contain work result sections applicable only to specific types of projects, such as building construction, heavy civil work, and process plant construction. The exception to this is for civil work, which is contained in the Site and Infrastructure subgroup located in the 30-series of the divisions.

For a major sewer project, the checklist (Section 10.3.10) covers the majority of material and workmanship specifications which may be required.

10.3.8. Supplementary Information

Frequently, special studies such as soil analyses and soil borings are made during the design phase of a project. The results of such soil investigations and samplings with soil classifications should be made available to the prospective bidders during the bidding period.

All available information pertinent to the work, especially with reference to subsurface conditions, should be made available to the bidders for examination. Bidders should be obliged to make their own interpretations of the subsoil information.

Arrangements should be made with the owner to permit bidders to inspect the soil samples and to make such additional soil borings as they may deem necessary prior to the bid date. Competent legal advice should be obtained in deciding whether or not such information should be made a part of the contract documents.

10.3.9. Standard Specifications

General specifications for workmanship and materials are intended to provide detailed descriptions of acceptable materials and performance standards which can be applied to all sewer contracts in a given jurisdiction. The description of acceptable construction procedures should be avoided in Standard Specifications. Any time that construction procedures are specified, care should be exercised to prevent the substitution of fixed concepts for the contractor's initiative. In every case, procedures must achieve, safely, the specified final results.

General specifications are usually aimed at more than any one specific contract. They may be used on a group of similar contracts or even for larger groups of dissimilar contracts.

A supplement to the Standard Specifications may be written to include special requirements modifying the Standard Specifications for a particular contract. In this regard, care must be taken in using Standard Specifications so that the work involved in writing supplements is not greater than the work that would be required to write completely new specifications for the contract. The use of computer-produced specifications facilitates converting Standard Specifications into project-specific ones.

It may be necessary to write a long supplementary specification due to the fact that the Standard Specifications are promulgated into law by a governing body. The engineer or owner cannot supersede this document except by supplementing the unwanted provisions in the governing body's standard specifications.

10.3.10. Project Manual Checklist

The Project Manual must cover all the legal, contractual, and specifications requirements for the contract. The MasterFormat™ 2004 Edition provides an excellent basis for a Project Manual checklist. The checklist delineates but does not classify these items in detail. Titles given in the listing refer to CSI Level 1 titles; reference to the MasterFormat™ 2004 Edition will assist in a more complete title breakdown.

As a guide for determining completeness of construction documents, the following checklist of subjects is offered:

10.3.10.1. Procurement Requirements

10.3.10.1.1. Solicitation

- Identification of owner or contracting agency.
- Name of project, contract number, or other positive means of identification.
- Time and place for receipt and opening of bids.
- Brief description of work to be performed.
- When and where contract documents may be examined.
- When and where contract documents may be obtained; deposits and refunds therefor.
- Amount and character of any required bid security.
- Reference to further instructions and legal requirements contained in the related documents.
- Statement of owner's right to reject any or all bids.
- Contractor's registration requirements.
- Bidder's prequalification, if required.
- Reference to special federal or state aid financing requirements.

10.3.10.1.2. Instructions for Procurement

- Instructions in regard to Bid Form, including method of preparing, signing, and submitting same; instructions on alternatives or options; data and formal documents to accompany bids; etc.

- Bid security requirements and conditions regarding return, retention, and forfeiture.
- Requirements for bidders to examine the documents and the site of the work.
- Use of stated quantities in unit price contracts.
- Withdrawals or modification of bid after submittal.
- Rejection of bids and disqualification of bidders.
- Evaluation of bids.
- Award and execution of contract.
- Failure of bidder to execute contract.
- Instructions pertaining to subcontractors.
- Instructions relative to resolution of ambiguities and discrepancies during the bid period.
- Contract bonding requirements.
- Governing laws and regulations.

10.3.10.1.3. Available Information

- Survey information
- Geotechnical data

10.3.10.1.4. Procurement Forms and Supplements

- Identification of contract
- Acknowledgment of receipt of addenda
- Bid prices (lump sum or unit prices)
- Construction time or completion date
- Amount of liquidated damages
- Financial statement
- Experience and equipment statements
- Subcontractor listing
- Contractor's statement of ownership
- Contractor's signature (and seal, if required)
- Bidder's qualifications
- Noncollusion affidavit
- Consent of surety

10.3.10.2. Contracting Requirements

10.3.10.2.1. Notice of Award

10.3.10.2.2. Agreement Form

- Identification of principal parties
- Date of execution

- Project description and identification
- Contract amount (sometimes with reference to and attachment of the contractor's bid)
- Contract time
- Liquidated damage clause, if any
- Progress payment provisions (may be covered in Conditions of the Contract or Division 01)
- List of documents comprising the contract (may be covered in Conditions of the Contract or Division 01)
- Authentication with signatures and seals

10.3.10.2.3. Special Forms

10.3.10.2.4. Bonds Forms

- Performance Bond
- Labor and Material Payment Bonds
- Maintenance and Guarantee Bonds (if required)

10.3.10.2.5. Notice to Proceed

10.3.10.3. *Conditions of the Contract*

- General Conditions
- Supplementary Conditions

10.3.10.4. *Revisions, Clarifications, and Modifications*

- Precontract Revisions
- Record Clarifications and Proposals
- Record Modifications

10.3.10.5. *Specifications*

10.3.10.5.1. General Requirements (CSI Division 01)

- Summary of work
- Price and payment procedures (many offices include this in the particular work item)
- Project management and coordination
- Submittal procedures
- Quality requirements
- Temporary facilities and controls (protection)
- Product requirements
- Execution and close-out requirements

10.3.10.5.2. Concrete (CSI Division 03)

- Concrete
  a. Concrete forming
  b. Concrete reinforcing
  c. Cast-in-place concrete
  d. Precast concrete
  e. Grouting

10.3.10.5.3. Finishes (CSI Division 09)

- Painting and coating
- Waterproofing

10.3.10.5.4. Earthwork (CSI Division 31)

- Site clearing (tree/shrub and pavement removal, soil stripping and stockpiling).
- Earth moving (grading, excavating, backfill and compaction, trenching, dewatering, erosion and sedimentation controls)
- Tunneling (excavating, drilling and blasting, construction)

10.3.10.5.5. Exterior Improvements (CSI Division 32)

- Paving and surfacing (streets, roadways, and sidewalks)
- Highways and railroad crossings

10.3.10.5.6. Utilities (CSI Division 33)

- Trenchless utility installation (directional drilling, pipe jacking, microtunneling, pipe ramming)
- Piping materials and jointing
- Manholes and appurtenances
- Pipe laying (control of alignment and control of grade)
- Service connections
- Connections to existing sewers
- Connections between different pipe materials
- Concrete encasement or cradle
- Sewer paralleling water main
- Sewer crossing water main
- Repair of damaged utility services
- Acceptance tests (infiltration, exfiltration, smoke, and air)

REFERENCES

Clark, J. R. (1993). "Commentary on agreements for engineering services and construction related documents," EJCDC Document No. 1910-9. Engineers' Joint Contract Documents Committee, available at <www.ejcdc.org>.

Construction Specifications Institute (CSI). (2005). "The project resource manual—CSI manual of practice," fifth ed., McGraw-Hill, New York.

CSI. (2004). "MasterFormat™ 2004 edition, master list of numbers and titles for the construction industry," CSI, Alexandria, Va.

Engineers' Joint Contract Documents Committee, (including as member organizations: American Society of Civil Engineers, National Society of Professional Engineers, and American Council of Engineering Companies (EJCDC) and The Construction Specifications Institute (CSI). (2002a). "Standard general conditions of the construction contract," Document No. C-700, available at <www.ejcdc.org>.

EJCDC/CSI. (2002b). "Standard form of agreement between owner and contractor." For construction contract (stipulated price), Document No. C-520; for construction contract (cost-plus), Document No. C-525, available at <www.ejcdc.org>.

EJCDC/CSI. (2002c). "Suggested instructions to bidders for construction contracts," Document No. C-200, available at <www.ejcdc.org>.

EJCDC/CSI. (2002d). "Guide to the preparation of supplementary conditions," Document No. C-800, available at <www.ejcdc.org>.

EJCDC/CSI (2002e). "Suggested bid form for construction contracts," Document No. C-410, available at <www.ejcdc.org>.

EJCDC/CSI. (1995). "Uniform location of subject matter," Document No. 1910-16, available at <www.ejcdc.org>.

U.S. Treasury Department. (Updated annually). "Companies holding certificates of authority as acceptable sureties on federal bonds and as acceptable reinsuring companies." Audit Staff Bureau of Accounts Circular 570, U.S. Treasury Department, Washington, D.C.

CHAPTER 11

CONSTRUCTION METHODS

11.1. INTRODUCTION

This chapter discusses construction methods in common use. Local conditions, of course, will dictate variations, and the ingenuities of the owner, engineer, and contractor must be applied continually if construction costs are to be minimized and acceptable standards are to be achieved.

Commencement of the construction phase normally introduces the third party—the contractor—to the sanitary sewer project, and the division of responsibility and liability must be understood by all. The role of the engineer changes from active direction and performance during design (a relationship to the owner) to that of professional and technical observation during construction. The engineer's duty during construction (during the contractual relationship between owner and contractor) is to determine for the owner that the work is substantially in accordance with the contract documents—in other words, acceptable. It is important to note, however, that the engineer's representative on the construction site is not expected to duplicate the detailed inspection of material and workmanship properly delegated to the manufacturer, supplier, and contractor.

Preconstruction conferences are helpful in deciding whether the contractor's proposed operations are compatible with contract requirements and whether they will result in finished construction acceptable to the owner and the engineer. These joint meetings of the owner, engineer, and contractor should culminate in definite construction protocol and administrative procedures to be followed throughout the life of the construction contract in accordance with that contract, and should include items such as progress schedules, payment details, method of making submittals for approval, and channels of communications. All of these aspects of construction should be mutually understood before construction begins.

11.2. PROJECT COSTS

Part of the evaluation of the feasibility of a sanitary sewer system is the total project cost. This consists of design costs, both direct and indirect construction costs, and administrative costs such as financial and legal expertise. Although the total project cost may not be the dominant factor in how a project is completed, it is important enough to review here.

11.2.1. Design Costs

The items included in design costs include planning and evaluating alternatives, surveying, engineering design, environmental assessments, permitting and review fees, land acquisition (if required), and geotechnical investigations. In most instances, design costs are a small percentage of the total project costs. Typical design fees are less than 10% of construction costs for new construction and less than 20% of construction costs for reconstruction projects.

11.2.2. Construction Costs

Construction costs include all items necessary to install the sanitary sewer system. In addition to the actual installation of the sanitary sewer (direct construction costs), there are indirect construction costs. These are the effects or consequences of the sanitary sewer installation. A brief description of direct and indirect costs follows.

11.2.3. Direct Construction Costs

By far the largest percentage of total project costs is the direct construction costs. This includes the materials, equipment, and labor required to construct the sanitary sewer system.

11.2.4. Indirect Construction Costs

11.2.4.1. Vehicular Traffic Disruption

Traffic disruptions may take two forms during construction projects: detours and delays. Detours force motorists to take longer routes than they would normally use. When constructing a sanitary sewer on an existing street and keeping the area open to traffic, delays will likely occur during delivery of materials or equipment, installing laterals, or connecting into existing systems. The additional time spent in negotiating these detours or construction zones results in additional fuel use by the affected motorists.

11.2.4.2. Road and Pavement Damage

The road and pavement damage referred to here is the portion in the construction zone. Depending upon soil conditions, either a portion of the roadway or the entire roadway may be removed during the trenching operation. In some cases where only a portion of the roadway is removed for sanitary sewer installation, the remainder of the roadway is effectively destroyed due to the heavy construction traffic that must use it to complete the project.

11.2.4.3. Damage to Adjacent Utilities

The construction portion of the project is when adjacent utilities, such as electrical, gas, and telephone facilities, are actually discovered. Even though records of these installations have been more accurate in recent years and the equipment used to locate them is more accurate, there still seem to be many surprises, such as unmarked utilities, on construction projects. The costs involved in repairing inadvertently damaged facilities consist of both the repair and the loss of production (down time) while making these repairs.

11.2.4.4. Damage to Adjacent Structures

The indirect cost attributable to damage to adjacent structures takes two forms. The first is the slower rate of progress in confined areas; the other is the time and expense required to repair the damaged structures. As with damaged utilities, the costs involved include both the cost of the repair and the lost production time while the repair is being completed.

11.2.4.5. Heavy Construction and Air Pollution

The construction process is, by nature, noisy and dirty. Although the contractor is used to this process and takes precautions to minimize impacts to the contractor's work force, the general public views this as an inconvenience. The noise, vibration, and exhaust fumes from the construction equipment can have an adverse effect on everyday life in the neighborhood where the work is occurring. This can range from noise inconvenience to possible health hazards from concentrated exhaust fumes. It is difficult to quantify these effects of the construction on the neighborhood, but steps can be taken to minimize these effects.

11.2.4.6. Pedestrian Safety

Providing pedestrian access through the construction site can be a very expensive proposition. In large projects in confined city streets, temporary pedestrian walkways may need to be established. In addition, walkways may need to be moved during construction as the sanitary sewer installation process moves along the street.

11.2.4.7. *Business and Trade Loss*

It is common for businesses along construction projects to experience a loss in business during construction projects in front of the business. Even if access is provided at all times, some customers will opt to not go to the business if there is any inconvenience at all.

11.2.4.8. *Damage to Detour Roads*

If a detour route is established for a project, the increased traffic on the detour route may cause damage to the existing pavement. Repairs to this pavement would be attributable to the construction project.

11.2.4.9. *Site Safety*

The site safety referred to here is public safety. This includes motorists, pedestrians, and emergency vehicles. Sufficient precautions must be established to permit crossing the work site, and access for emergency vehicles to reach homes and businesses along the line of work must be preserved at all times. Site safety for the contractor and subcontractors is considered a direct construction cost.

11.2.4.10. *Citizen Complaints*

When they occur, citizen complaints should be addressed immediately. Public relations are always a part of the construction process. The complaint should be heard and a response defining the corrective action, if any, should be given to the complaining party. Even a response indicating that nothing additional will be done for a complaint is better than no response.

11.2.4.11. *Environmental Impact*

In some instances, sanitary sewers are constructed in environmentally sensitive areas. Examples of this are construction through or under rivers and streams and through or near wetlands. During this type of installation, additional care should be taken to minimize disruption and environmental damage to these areas. Additional costs are incurred in the installation and maintenance of the best management practices used to keep these areas in their natural state.

11.3. CONSTRUCTION SURVEYS

11.3.1. General

The engineer should arrange through a licensed surveyor to establish baselines and benchmarks for sanitary sewer line and grade control along the route of the proposed construction. All control points should be referenced adequately to permanent objects located outside normal construction limits.

11.3.2. Preliminary Layouts

Prior to the start of any work, rights-of-way, work areas, clearing limits, and pavement cuts should be laid out to give proper recognition to, and protection for, adjacent properties. Access roads, detours, bypasses, and protective fences or barricades also should be laid out and constructed as required in advance of sanitary sewer construction. All layout work, if done by the contractor, should be reviewed by the engineer before any demolition or construction begins.

11.3.3. Setting Line and Grade

The transfer of line and grade from control points established by the engineer to the construction work should be the responsibility of the contractor, with spot checks by the engineer as work progresses. The preservation of stakes or other line and grade references provided by the engineer is similarly the responsibility of the contractor. (Generally, a charge is made for re-establishing stakes carelessly destroyed, and the charge is stated as part of the contract agreement.)

In general, the line and grade for the sanitary sewer may be set by one or a combination of the following methods:

1. Stakes, spikes, or crosses set on the surface on an offset from the sanitary sewer centerline.
2. Stakes set in the trench bottom on the sanitary sewer line as the rough grade for the sanitary sewer is completed.
3. Elevations given for the finished trench grade and sanitary sewer invert while sanitary sewer construction progresses.
4. A laser beam of light set in the manhole or a specified height above the sanitary sewer flowline.

Method 1 generally is used for small-diameter sanitary sewers. Methods 2 and 3 are used for large sanitary sewers or where sloped trench walls result in top-of-trench widths too great for practical use of short offsets. Method 4 is independent of the size of sanitary sewer.

In Method 1, stakes, spikes, or crosses are set at a uniform offset (insofar as practicable) from the sanitary sewer centerline on the opposite side of the trench from where excavated materials are to be cast. A cut sheet, as shown in Table 11-1, is prepared; this is a tabulation of the reference points giving sanitary sewer station, offset, and the vertical distance from each reference point to the proposed sanitary sewer invert.

TABLE 11-1. Typical Cut Sheet

"A" street ____________ sewer
4th Street to 5th Street
Stakes 5 ft left

Sheet 1 of 1 Sheet
Notes book ____________ Page ____________
Prepared by ____________ Date ____________

Station	Size, inches	Grade	Elevation		Cut		Remarks
			Invert	Stake	Feet	Hundredths	
0+00	12	0.0025	105.50	110.75	5	25	Existing manhole
0+25	—	0.0025	105.56	112.30	6	74	—
0+50	—	0.0025	105.63	109.70	4	07	Y-branch right
0+75	—	0.0025	105.69	110.35	4	66	—
1+00	12	0.0025	105.75	111.99	6	24	Etc.

Ft × 0.3 = m; inch × 2.54 = cm.

The line and grade may be transferred to the bottom of the sanitary sewer trench by the use of tape and level, or a patented bar tape and plumb bob unit.

Another method of setting grade is from offset crosses or stakes and the use of a grade rod with a target near the top. When the sanitary sewer invert is on grade, a sighting between grade rod and two or more consecutive offset bars or the double string line will show correct alignment.

Method 2, involving transfer of surface references to stakes along the trench bottom, is in some instances permitted. If stakes are established along the trench bottom, a string line should be drawn between no fewer than three points and checked in the manner used for batter boards.

When trench walls are not sheeted but are sloped to prevent caving, line-and-grade stakes are set in the trench bottom as the excavation proceeds. This procedure requires a field party to be at the work site almost constantly.

Method 3, which is applicable to large-diameter sanitary sewers or monolithic sections of sanitary sewers on flat grades, requires the line and grade for each pipe length or form section to be set by means of a transit and level from either on top or inside of the completed conduit.

In the construction of large sanitary sewer sections in an open trench, both line and grade may be set at or near the trench bottom. Line points and benchmarks may be established on cross-bracing where such bracing is in place and rigidly set. Later, alignment and grade must be determined by checking the setting of the forms.

Method 4, which is quite widely used, is the laser beam control. A laser is a device that projects a narrow beam of light down the centerline of the sewer pipe. It is usually set up in the invert of a manhole and then aligned horizontally. The proper slope is established by adjusting a dial on the machine and aiming the laser. A check elevation should be set about 100 ft (30 m) from the manhole to ensure that the proper slope is being maintained by the beam of light. A target set in the pipe centerline is then used to align the end of each pipe section. Care should be exercised in the use of the laser since temperature affects the aiming of the unit.

11.4. SITE PREPARATION

The amount of site preparation required varies from none to the extreme where the major portion of project costs is expended on items other than excavating for and construction of the sanitary sewer.

Operations that may properly be classified as site preparation are clearing and grubbing; removal of unsuitable soils; construction of access roads, detours, and bypasses; improvements to and modification of existing drainage; location, protection, or relocation of existing utilities; and pavement cutting. The extent and diversity of these operations make further discussion thereof impractical here. Note, however, that the contractor's success in keeping the project on schedule depends to a great degree on the thoroughness of the planning and execution of the site preparation work. Several engineers and contractors have adopted a practice of assembling extensive photographic or videotape evidence of the preconstruction condition of sidewalks, driveways, street surfaces, etc., to minimize post-construction claims by residents for construction-related damages.

11.5. OPEN-TRENCH CONSTRUCTION

11.5.1. Trench Dimensions

With plans and specifications competently prepared, it can be assumed that the location of the proposed sanitary sewers have been determined with proper regard for the known locations of existing underground utilities, surface improvements, and adjacent buildings. Barring unforeseen conditions, it becomes the contractor's objective to complete the work as shown on the plans at minimum cost and with minimum disturbance of adjacent facilities.

Because of load considerations, the width of trench at and below the top of the sanitary sewer should be only as wide as necessary for proper installation and backfilling, and consistent with safety. The Contract must provide for alternative methods or require corrective measures to be employed by the contractor if allowable trench widths are exceeded through overshooting of rock, caving of earth trenches, or overexcavation. The width of trench from a plane 1 ft (30 cm) above the top of the sanitary sewer to the ground surface is related primarily to its effect on the safety of the workers who must enter the trench, and on adjoining facilities such as other utilities, surface improvements, and nearby structures.

In undeveloped subdivisions and in open country, economic considerations often justify sloping the sides of the trench for earth stability from a plane 1 ft (30 cm) above the top of the finished sanitary sewer to the ground surface. This eliminates placing, maintaining, and removing substantial amounts of temporary sheeting and bracing unless safety regulations make some type of sheeting or bracing mandatory. Steel trench shields, or trench boxes, are often used to protect the workers where sheeting is deemed unnecessary.

In improved streets, on the other hand, it may be desirable to restrict the trench width so as to protect existing facilities and reduce the cost of surface restoration. Available working space, traffic conditions, and economics will all influence this decision.

11.5.2. Excavation

With favorable ground conditions, excavation can be accomplished in one simple operation. Under more adverse conditions it may require several steps. In general, stripping, drilling, blasting, and trenching will cover all phases of the excavation operation.

In all excavations, extreme care should be taken to properly locate, support, and protect existing utilities. The owners of the utilities should be contacted before the start of excavations.

11.5.2.1. Stripping

Stripping may be advantageous or required as a first step in trench excavation for a variety of reasons, the most common of which are:

- Removal of topsoil or other materials to be saved and used for site restoration.
- Removal of material unsatisfactory for backfill to ensure its separation from usable excavated soils.
- Removal of material having a low bearing value to a depth where there is material capable of supporting heavy construction equipment.
- To reduce cuts to depths down to which a backhoe can dig.
- To make it easier to charge drill holes.

11.5.2.2. Rock Removal

11.5.2.2.1. Drilling and Blasting

In some areas, sanitary sewer and house connections must be installed in hard rock. In addition, some shales and softer rocks that may be ripped in open excavation will require blasting before they can be removed in confined areas as required in trenching or excavating for structures.

Normally, the most economic method will involve preshooting (i.e., drilling and shooting rock before removal of overburden). In some instances the occurrence of wet, granular materials above the rock ledge will necessitate stripping before drilling, since holes cannot be maintained open through the overburden to permit placing of explosive charges.

For narrow trenches in soft rock, a single row of drill holes may be sufficient. One or more additional rows will be required in harder rock and for wider trenches. To reduce overbreak and improve bottom fragmentation, time delays should be used in blasting for trenches. In tight quarters, trench walls can be presplit and the center shot in successive short rounds to an open face to produce minimum vibration.

It must be recognized that there will be a minimum feasible trench width varying with the rock formation, and in the case of small sanitary sewers it may be necessary to design the conduit for the positive-projecting condition or extreme loads.

All ground and air pressures that result from blasting should be recorded on a sealed cassette seismograph. Surveys of adjacent structures for the instance of cracks before the blasting are advisable. Blasting should be done only by persons experienced in such operations.

11.5.2.2.2. Rock Trenching

In some instances, it may be advantageous to perform rock removal operations using a rock trencher. Rock trenchers remove the rock in the same way an earth trenching machine does. The rock removed from the trench is typically placed in a windrow set back from the trench. This material can then be used as backfill, if suitable, or removed from the site.

The advantages of rock trenching are:

- It leaves a vertical trench wall, minimizing the rock removal and restoration to be completed.
- It breaks down the rock into smaller sizes normally suitable to be utilized as backfill.

Since no blasting is done, there is no chance for rock or overburden to become airborne and possibly damage property or injure someone.

The disadvantages of rock trenching are:

- It can be very expensive in comparison to drilling and blasting.
- The size of the rock trenching machine makes it impractical to use in tight construction conditions.

11.5.2.3. Trenching

The method and equipment used for excavating the trench will depend on the type of material to be removed, the depth, the amount of space available for operation of equipment and storage of excavated material, and prevailing practice in the area.

Ordinarily the choice of method and equipment rests with the contractor. However, various types of equipment have practical and real limitations regarding minimum trench widths and depths. The contractor is therefore obligated to utilize only that equipment capable of meeting trench width limitations imposed by sewer pipe strength requirements or for other reasons as set forth in the technical specifications.

Spoil should be placed sufficiently back from the edge of the excavation to prevent caving of the trench wall and to permit safe access along the trench. With sheeted trenches, a minimum distance of 3 ft (1 m) from the edge of sheeting to the toe of spoil bank will normally provide safe and adequate access. Under such conditions the supports must be designed for the added surcharge. In unsupported trenches the minimum distance from the vertical projection of the trench wall to the toe of the spoil bank normally should be less than one-half the total depth of excavation. In most soils, this distance will be greater in order to provide safe access beyond the sloped trench walls.

11.5.2.3.1. Trenching Machines

This type of machine is generally used for shallow trenches less than 5 ft (1.5 m) deep. For installation of small sanitary sewers and for sewers in cohesive soils, the trenching machine can make rapid progress at low cost.

11.5.2.3.2. Backhoes

Backhoes like that shown on Fig. 11-1 are available with bucket capacities varying from ⅜ to 3 cu yd (0.3 to 2.3 m^3) and more. They are convenient for the excavation of trenches with widths exceeding 2 ft (0.7 m) and to depths down to 25 ft (8 m). They are the most satisfactory equipment for excavation in loosened rock. Minimum trench widths are compared with some common backhoe sizes in Table 11-2.

The backhoe is also used with a cable sling for lowering sewer pipe into the trench. By this means, a single piece of equipment can maintain the sewer pipe-laying close to the point of excavation. Where the soil does not require sheeting and bracing, this method becomes a very economical one. When sheeting and bracing must follow the excavation closely, the combination of a backhoe for excavation and a crane for placement of sewer pipe is a common practice.

FIGURE 11-1. Typical backhoe.
Courtesy of Chastain-Skillman, Inc.

11.5.2.3.3. Clamshells

When the protection of underground structures or soil conditions requires close sheeting and the use of vertical lift equipment, the clamshell bucket is used. In very deep trenches where two-stage excavation is required, the backhoe is sometimes used in combination with the clamshell, with the backhoe advancing the upper part of the excavation

TABLE 11-2. Trench Widths Associated with Various Backhoe Bucket Capacities

	Minimum Trench Width, inches	
Bucket Capacity, cu yd	Without Side Cutters	With Side Cutters
3/8	22	24–28
1/2	27	28–32
3/4	28	28–38
1	34	34–44
1 1/4	37	37–46
1 1/2	38	38–46
2	50	50–58

Inch × 2.54 = cm; cu yd × 0.76 = m^3.

and the clamshell following for the lower. Sheeting and bracing of the upper part are installed and driven as required prior to the clamming of the lower part and the installation of the lower stage of sheeting.

11.5.2.3.4. Draglines

In open country, for stream crossings, or in a wide right-of-way, it may be feasible to do a large part of the excavation by means of a dragline, allowing the sides of the trenches to acquire their natural slope. In cases of very deep trench excavation, say 30 to 50 ft (9 to 15 m), the dragline has been used for the upper part of the excavation, with a backhoe operating at an intermediate level. By rotating the backhoe, the material thus excavated can be relayed to the dragline, which then lifts it to the spoil bank or to trucks at the surface.

11.5.2.3.5. Front-End Loaders

The principal use of loaders is in bringing sewer pipe, manholes, and granular bedding material to the trenches. In wide, deep trenches, the front-end loader has sometimes been used as an auxiliary to a backhoe or clamshell. In this arrangement the backhoe excavates the upper part of the trench and, perhaps, the center section of the lower part, leaving the bottom bench or benches for the front-end loader or dozer which completes the excavation, placing the spoil within the reach of the backhoe or clamshell.

11.5.3. Sheeting and Bracing

Trench sheeting and bracing should be adequate to prevent cave-in of the trench walls, subsidence of areas adjacent to the trench, and sloughing of the base of the excavation from water seepage. Responsibility for the adequacy of any required sheeting and bracing usually is stipulated to be with the contractor. The strength design of the system of supports should be based on the principles of soil mechanics and structural engineering as they apply to the materials encountered. Sheeting and bracing always must comply with applicable safety requirements.

For wider and deeper trenches, a system of wales and cross-struts of heavy timber (or steel sections), as shown in Fig. 11-2, is often used. Sheeting is installed outside the horizontal wales as required to maintain the stability of the trench walls. Jacks mounted on one end of the cross-struts maintain pressure against the wales and sheeting.

In some soil conditions, it has been found economical and practical to use steel trench shields, or trench boxes, which are pulled forward as sewer pipe-laying progresses. Care must be exercised in pulling shields forward so as not to drag or otherwise disturb the previously laid pipe and bedding or to create conditions not assumed in calculating trench loads.

FIGURE 11-2. Sheeting and bracing system.
Courtesy of Bruce Corwin, CDM.

In noncohesive soils containing considerable groundwater, it may be necessary to use continuous steel sheet piling to prevent excessive soil movements. Such steel piling usually extends several feet (meters) below the bottom of the trench unless the lower part of the trench is in firm material. In the latter case, the width of the trench in the upper granular material may be widened so that the steel sheet piling can toe-in to the lower strata.

In some soils, steel sheet piling, such as shown in Fig. 11-3, can be used with a backhoe operation for the upper part of excavation, but the piling usually needs to be braced before the excavation has reached its full depth. The remaining excavation then is performed by vertical lift equipment, such as a clamshell.

Another means of trench sheeting occasionally adopted involves the use of vertical H-beams as "soldier beams" with horizontal wood lagging.

FIGURE 11-3. Steel sheeting.
Courtesy of David Howell, Midwest Mole.

This is sometimes advantageous for trenches under existing overhead viaducts where overhead clearances are low and spread footings lie alongside the trench walls. The holes for the vertical beams can be partially excavated and then the beams can be tilted into these holes and driven. As excavation progresses downward, the lagging is installed between adjacent pairs of soldier beams. For deep trenches with limited overhead clearances, the soldier beams can be delivered to the site in shorter lengths and their ends field-welded as driving progresses.

The removal of sheeting following pipe laying may affect the earth load on the sanitary sewer or adjacent structures. This possibility must be considered during the design phase. If removal is to be permitted, appropriate requirements must be placed in the technical specifications. Whenever sheeting is removed, it must be done properly, taking care to backfill thoroughly the voids thus created.

11.5.4. Dewatering

Trenches should be dewatered for concrete placement and sewer pipe laying, and they should be kept continuously dewatered for as long as necessary. Unfortunately, the disposal of large quantities of water from this operation, in the absence of storm drains or adjacent water courses, may present problems. Other means of disposal being unavailable, the possibility of draining the water through the complete sanitary sewer to a permissible point of discharge should be considered, provided sufficient precautions are taken to prevent scour of freshly placed concrete or mortar and to prevent the transport of sediment. In all cases, care must be exercised to ensure that property damage, including silt deposits in sanitary sewer and on streets, does not result from the disposal of trench drainage.

Crushed stone or gravel may be used as a subdrain to facilitate drainage to trench or sump pumps. It is good practice to provide clay dams in the subdrain to minimize the possibility of undercutting the sewer foundation from excessive groundwater flows.

An excessive quantity of water, particularly when it creates an unstable soil condition, may require the use of a well-point system. A system of this type consists of a series of perforated pipes driven or jetted into the water-bearing strata on either side of the sanitary sewer trench and connected with a header pipe leading to a pump. The equipment for a well-point system is expensive and specialized. General contractors often seek the help of special dewatering contractors for such work. Well-point systems must be run continuously to avoid disturbing the excavated trench bottom by uplift pressure.

Where excavation is in coarse water-bearing material, turbine well pumps may be used to lower the water table during construction. Chemical or cement grouting and freezing of the soil adjacent to the excavation have been used in extremely unstable water-bearing strata. Water from all types of dewatering systems should be checked to ensure that fine-grained material is not being removed from beneath the pipe, which might cause future settlement.

11.5.5. Foundations

Firm, cohesive soils provide adequate sewer pipe foundations when properly prepared. Occasionally, the trench bottom may be shaped to fit the sewer pipe barrel and holes may be dug to receive projecting joint elements. A frequent practice is to overexcavate and backfill with granular material, such as crushed stone, crushed slag, or gravel to provide uniform bedding of the sewer pipe. Such granular bedding is used because it is both practical and economical.

In very soft bottoms, it is necessary first, as a minimum, to overexcavate to greater depths and stabilize the trench bottom by the addition of gravel or crushed slag or rock compacted to receive the load. The stabilizing material must be graded to prevent movement of subgrade up into the stabilized base and the base into the bedding material. Increasingly, specialized filter fabrics are used to prevent this movement. The required depth of stabilization should be determined by tests and observations on the job.

In those instances where the trench bottom cannot be stabilized satisfactorily with a crushed rock or gravel bed, and where limited and intermittent areas of unequal settlement are anticipated, a timber cribbing, piling, or reinforced concrete cradle may be necessary.

Where the bottom of the trench is rock, it must be overexcavated to make room for an adequate bedding of granular material which will uniformly support the conduit. The trench bottom must be cleaned of shattered and decomposed rock or shale prior to placement of bedding.

In some instances, sanitary sewer pipe (i.e., not cast-in-place) must be constructed for considerable distances in areas generally subject to subsidence, and consideration should be given to constructing them on a timber platform or reinforced concrete cradle supported by piling. The sewer's support should be adequate to sustain the weight of the full sewer and backfill. In this case, piling is sometimes driven to grade with a follower prior to making the excavation. This practice avoids subsidence of trench walls resulting from pipe-driving vibrations. Extreme care must be taken to locate all underground structures.

11.5.6. Pipe Sanitary Sewers

Proper sanitary sewer construction requires that quality materials and acceptable laying methods are to be used. Diligence in ensuring both is required of all project personnel.

11.5.6.1. Sewer Pipe Quality

Sewer pipe inspection should be properly conducted by the manufacturer and by independent testing and inspection laboratories. Moreover, inspection of sanitary sewer pipe at the pipe plant is usually desirable due to the transportation charges, which may constitute a substantial portion of material costs. Inspection may consist of visual inspection of workmanship, surface finish, and markings; physical check of length, thickness, diameter, and joint tolerances; proof of crushing strength (rigid pipe) or pipe stiffness (flexible pipe) design materials tests; and tests of representative specimens. If three-edge-bearing (T.E.B.) tests are not used on precast concrete pipe, core or cylinder tests should be required. However, because standard cylinder tests are not practical with the mixes used in some manufacturing methods, core tests are generally used. Cores also permit checking tolerances on placement of reinforcing cages.

Sewer pipe suppliers should furnish certificates of compliance with specifications that can be easily checked as the loads of sewer pipe arrive at the site. Sewer pipe also should be checked visually at time of delivery for possible damage in transit, and again as it is laid for damage in storage or handling.

11.5.6.2. *Sewer Pipe Handling*

Care must be exercised in handling and bedding all precast sewer pipe, regardless of cross-sectional shape. All phases of construction should be undertaken to ensure that, insofar as practical, pipe is installed as designed. Sewer pipe should be handled during delivery in a manner that eliminates any possibility of high-impact or point loading, with care taken to always protect joint elements.

11.5.6.3. *Sewer Pipe Placement*

Sewer pipe should be laid on a firm but slightly yielding bedding true to line and grade, with uniform bearing under the full length of the barrel of the sewer pipe, without break from structure to structure, and with the socket ends of bell-and-spigot or tongue-and-groove sewer pipe joints facing upgrade. Sewer pipe should be supported free of the bedding during the jointing process to avoid disturbance of the subgrade. A suitable excavation should be made to receive sewer pipe bells and joint collars, where applicable, so the bottom reaction and support are confined only to the sewer pipe barrel. Adjustment to line and grade should be made by scraping away or adding adequately compacted foundation material under the sewer pipe, and not by using wedges and blocks or by beating on the sewer pipe.

Extreme care should be taken in jointing to ensure that the bell and spigot are clean and free of any foreign materials. Joint materials vary with the type of sewer pipe used. All pipe joints should be made properly, using the jointing materials and methods specified. All pipe joints should be sufficiently tight to meet infiltration or exfiltration tests.

In large-diameter sewers with compression-type joints, considerable force will be required to insert the spigot fully into the bell. Come-alongs and winches (or the crane itself) may be rigged to provide the necessary force. Inserts should be used to prevent the sewer pipe from being thrust completely home prior to checking gasket location. After the gasket is checked, the inserts can be removed and the joint completed.

The operation of equipment over small-diameter sewer pipe, or other actions that would otherwise disturb any conduit after pipe jointing, must not be permitted.

At the close of each day's work or when sewer pipe is not being laid, the end of the sewer pipe should be protected by a close-fitting stopper to keep the pipe clean, with adequate precautions taken to overcome possible uplift. The elevation of the last sewer pipe placed should be checked the next morning before work resumes.

If the sewer pipe load factor is increased with either arch or total encasement, contraction joints should be provided at regular intervals in the encasement coincident with the pipe joints to increase flexibility of the encased conduit.

11.5.7. Backfilling

11.5.7.1. General Considerations

Backfilling of the sanitary sewer trench is a very important consideration which seldom receives the attention and inspection it deserves. The methods and equipment used in placing fill must be selected to prevent dislocation or damage to the sewer pipe. The method of backfilling varies with the width of the trench, the character of the materials excavated, the method of excavation, and the degree of compaction required.

11.5.7.2. Degree of Compaction

In improved streets or streets programmed for immediate paving, a high degree of compaction should be required. In less important streets or in sparsely inhabited subdivisions where flexible pavement (asphaltic concrete) roadways are used, a more moderate specification for backfilling may be justified. Along outfall sewers in open country, it may be sufficient to mound the trench and, after natural settlement, return to regrade the area. Compaction results should be determined in accordance with current American Association of State Highway and Transportation Officials (AASHTO) or ASTM test procedures. Laboratory tests to establish optimum moisture content are commonly done according to the modified Proctor Method, AASHTO T-180, or ASTM D1557. Field tests to determine actual compaction may be done by any of several mechanical methods or by the use of nuclear density meters.

11.5.7.3. Methods of Compaction

11.5.7.3.1. Cohesive Materials

Cohesive materials with high clay content are characterized by small particle size and low internal friction. They have small ranges of moisture content over which they may be compacted satisfactorily, and are very impervious in a dense state. Because of the strong adhesive forces of the soil particles, strong pressures must be exerted in order to shear the adhesive forces and remold the particles in a dense soil mass. These characteristics dictate the use of impact-type equipment for most satisfactory results in compaction. In confined areas, pneumatic tampers and engine-driven rammers may give good results. The upper portion of the trench can be consolidated by self-propelled rammers where trench widths are relatively narrow. In wide trenches, sheepsfoot rollers may be used. If the degree of compaction required is not high, dozers and loaders may be used to compact the fill.

Regardless of equipment used, the soil must be near optimum moisture content and compacted in multiple lifts if satisfactory results are to be obtained. The trench bottom must be free of excessive water before the first lift of backfill is placed.

If the material had a high moisture content at the time of excavation, some preparation of the material probably will be required before spreading in the trench. This may include pulverizing, drying, or blending with dry or granular materials to improve placement and consolidation.

11.5.7.3.2. Noncohesive Materials

Noncohesive materials are granular with little adhesion but with high internal friction. Moisture content at the time of compaction is not so critical; consolidation is effected by reducing the surface friction between particles, thus allowing them to rearrange in a more compact mass. Considering the characteristics of noncohesive material, the most satisfactory compaction is achieved through the use of vibratory equipment.

In confined areas, vibratory plates give the best results. For wider trenches, vibratory rollers are most satisfactory. Again, if the degree of compaction required is not high and if the layers are thin, the vibration impacted by dozer or loader tracks may result in satisfactory consolidation.

In some areas, water is used to consolidate granular materials. However, unless the fill is saturated and immersion vibrations are used, the degree and uniformity of compaction cannot be controlled closely. With some materials, adequate compaction may be obtained by draining off, through drains constructed in manhole walls, water used to saturate or puddle fill. These drains are capped after the backfill has drained.

11.5.7.3.3. Borrow Materials

Sometimes the material removed from the trench may be entirely unsatisfactory for backfill. In this case, selected materials must be hauled in from other sources.

Both cohesive and noncohesive materials are used, but an assessment must be made of the possible change in groundwater movement that the use of outside materials may cause. For example, the use of cohesive materials to backfill a trench in rock could result in a dam impervious to groundwater traveling in rock faults, seams, and crevices. On the other hand, granular materials placed in a clay trench could result in a very effective subdrain.

11.5.7.4. Backfilling Sequence

Backfilling should proceed immediately upon curing of trench-made joints and after the concrete cradle, arch, or other structures gain sufficient strength to withstand loads without damage. Backfill is generally specified as consisting of three zones, with different criteria for each. The first zone (embedment) extends from the foundation material to 12 inches (30 cm) above top of sewer pipe or structure; an intermediate zone generally contains the major volume of the fill; and the upper zone consists of pavement subgrade or finish grading materials.

The first zone should consist of selected materials placed by hand or by suitable equipment in such manner as not to disturb the sewer pipe, and compacted to a density consistent with design assumptions. In some instances, the material used for granular bedding is brought above the sewer pipe to ensure high-density backfill with minimum compactive effort. When installing flexible pipes, attention must be given to proper placement and compaction of the haunching material from the base of the pipe to the springline. When high water tables are anticipated, embedment materials without substantial voids are required to prevent soil migration.

Compaction of the intermediate zone is usually controlled by the location. Under traffic areas or other improved existing surfaces, 95% of modified Proctor density should be required. In other general urban areas, 90% may be adequate. In undeveloped areas, little compaction may be required. In general, the degree of compaction required will often affect the choice of material. The use of excavated material, if suitable, is usually best in areas that are subject to frost heave so that excavated areas will not move more or move less than undisturbed areas.

Depth and compaction of the upper zone are dependent on the type of finish grade to be provided. If the construction area is to be seeded or sodded, the upper 18 inches (450 mm) may consist of 14 inches (350 mm) of select material slightly mounded over the trench and lightly rolled, covered by 4 inches (100 mm) of topsoil. If the area is to be paved, the upper zone must be constructed to the proper elevation for receiving base and paving courses under conditions matching design assumptions for the subgrade. If the trench backfill is completed in advance of paving, the top 16 inches (150 mm) of the upper zone should be scarified and recompacted prior to paving. In such instances, it may be necessary to install a temporary surface to be replaced at a later date with permanent pavement.

Before and during the backfilling of a trench, precautions should be taken against the flotation of pipe lines due to the entry of large quantities of water into the trench, which would cause uplift on the empty or partly filled pipe line.

11.5.8. Surface Restoration

On completion of backfill, the surface should be restored fully to a condition at least equal to that which existed prior to the sanitary sewer construction. Portland cement or asphaltic concrete pavements should be saw-cut and removed to a point beyond any caving or disturbance of the base materials prior to patching. If this results in narrow, unstable panels, pavement should be removed to the next existing contraction or construction joint. Before replacing permanent pavement, the subgrade must be restored and compacted until smooth and unyielding.

The final grade in unpaved areas should match existing grades at construction limits without producing drainage problems. Restoration of grass, shrubs, and other plantings should be done in conformance with construction contract documents. Tree damage should be repaired in accordance with good arboricultural practice.

11.6. SPECIAL CONSTRUCTION

11.6.1. Railroad Crossings

Sanitary sewers at times must be constructed under railroad tracks, which may be at street level, on a raised embankment above street grade, or on an existing railroad viaduct. Crossing of tracks at grade or on an embankment usually can be accomplished most economically by jacking, boring, tunneling, or a combination thereof. Usually a casing pipe is inserted and the sewer pipe is placed inside.

When the distance from the base of the rail to the top of the sanitary sewer is insufficient to allow jacking or tunneling (less than one diameter clearance), it is necessary either to remove the tracks and interrupt service during an open-cut operation, or to build a temporary structure for support of the railroad tracks, after which the sanitary sewer may be constructed in an open trench below the structure.

Construction of sanitary sewers under existing railroad viaducts involves a wide variation in methods, depending on the size of the sanitary sewer, its location in plan and elevation with respect to viaduct footing, type of footings, and the nature of the soil. Where the soil is stable and the sanitary sewer is of sufficient size to allow the use of tunnel methods and is located satisfactorily with respect to viaduct footings, tunneling may be both safe and economical.

When the proposed sanitary sewer does not meet these criteria, special methods of sheeting and bracing must be devised. To prevent subsequent movement of soil beneath the footings, all sheeting and bracing should be left in place.

Early planning with the railroad authorities is essential, since most companies have extensive design, inspection, and permit requirements.

11.6.2. Crossing of Principal Traffic Arteries

Residential and secondary traffic arteries usually can be closed to traffic during the construction of sanitary sewer crossings. However, on heavily traveled streets and highways where public convenience is a major factor, it may be desirable to use tunneling or jacking methods for the crossing.

When required, limited traffic movements across open trenches can be accommodated by temporary decking. Trenches of narrow or medium width can be spanned with prefabricated decks placed on timber mudsills at the edges of the trench. Where the top of the trench is wider than 16 to

20 ft (5.3 to 6.7 m), temporary piling for end support (and in some cases, center support) may be required. Where center supports pass through the sanitary sewer section, provisions must be made for such piling to remain until the sanitary sewer is completed. On a project in Chicago (described in the 1982 edition of this Manual), a center piling of steel was set on the centerline of a proposed twin-barrel sanitary sewer and later encased in the sanitary sewer section.

11.6.3. Stream and River Crossings

11.6.3.1. Sanitary Sewer Crossing under Waterway

Stream and river crossings may be constructed either in the dry through use of cofferdams and diversion channels, or subaqueously. Open trenches may be excavated from barges with sewer pipe laying and pipe jointing done by divers. Ball-joint sewer pipe may be effectively used, especially for force mains. For shallow stream crossings, it may be possible to install an earth embankment and construct half of the crossings at a time. If constructed in the dry, planning and scheduling of construction should be such that completed portions of the line are not subject to damage in the event of cofferdam overtopping.

Concrete encasement, if required, should be placed with construction joints at 30- to 40-ft (9- to 12-m) intervals coincident with pipe joints. Sewer pipe may be set conveniently to line and grade by supporting it on burlap bags filled with a dry-batched concrete mix. These bags also may be placed for construction of bulkheads in subaqueous concrete placements.

After placing the crossings beneath the bottom of the stream, it is usually advisable to place a layer of large rip-rap to form an armor course to protect the sewer pipe from erosion or hanging boat anchors.

11.6.3.2. Sanitary Sewer Crossing Spanning Waterway

In Chapter 7, various methods of spanning obstacles are described, including hanging and fastening sanitary sewers to structural supports and the construction of sanitary sewer pipe beams. The latter type of construction consists of a manhole or other supporting structure on either side of the waterway and the spanning member itself. Where the crossing is of considerable width, intermediate piers or supports may be necessary. Figure 11-4 shows a typical aerial span crossing with concrete pipe.

11.6.4. Outfall Structures

11.6.4.1. Riverbank Structures

Sanitary outfall sewers and head walls may be located above or below surface water levels. When they are partly submerged, it is necessary to provide some form of cofferdam during construction.

FIGURE 11-4. Installing concrete pipe aerial crossing.
Courtesy of American Concrete Pressure Pipe Association, Reston, Va.

In shallow water, an earth dike or timber piling may be sufficient to maintain a dry pit. In deep water, steel sheet piling cofferdams are desirable. Usually a single-wall cofferdam with adequate bracing is sufficient, but in excessive depths at the banks of main navigation channels a double wall may be required. Standard practice of cofferdam design and construction should govern.

11.6.4.2. Ocean Outfalls

For long ocean outfalls there are two distinct phases of construction: the inshore section through the surf zone, and the offshore section. The surf zone usually extends to a depth of 50 ft (15 m) but may be shallower or deeper depending on local ocean conditions. The inshore or surf zone section requires positive support and lateral restraint for the outfall sewer pipe. The inshore section usually requires a temporary pier for driving sheet piling to maintain the trench through the breakers and for pipe installation. If the shore is all sand, suitable piles must be driven for support and anchorage of the outfall pipe.

The offshore section of outfall sewer pipe is usually laid from floating equipment, such as shown in Fig. 11-5, and is often placed directly on the ocean floor, provided the grade is satisfactory. Gravel or rock side fill to

FIGURE 11-5. Construction of ocean outfall risers—3.5 m diameter, Mumbai, India.
Courtesy of John Trypus, Black & Veatch.

the springline of the pipe is frequently added to prevent lateral currents from scouring local potholes, which might cause pipe movement.

Ocean outfall pipes have been made of cast iron, reinforced concrete, protected steel pipe, plastic pipe, and a combination of these materials. Small pipelines may be assembled on shore and then pulled or floated into position. Large lines must be laid in sections, although it is obviously advantageous to make the lengths as long as possible to minimize the number of underwater pipe joints which must be assembled by divers. It is also desirable to select a pipe joint type and construction procedure that will facilitate underwater connections. This is especially important if the outfall sewer is in water more than 150 ft (50 m) deep because of the limited time divers can work at this depth. Flexible pipe joints or ball-and-socket cast iron pipe (laid with a cradle) permit the jointing of sewer pipe above water, thus eliminating the use of divers and underwater operations.

11.7. SEWER APPURTENANCES

Recent improvements in both sewer pipe and pipe joints have made it possible to construct a very watertight sanitary sewer system. It is important to be sure that large quantities of extraneous flows are not admitted at poorly constructed sanitary sewer appurtenances and service connections. Increased attention to this phase of sanitary sewer construction is essential if flow of surface and groundwaters in sanitary sewers is to be reduced.

The sound principles of construction that apply to reinforced concrete and masonry structures must be applied also to sanitary sewer manholes. The use of watertight covers, pipe-to-manhole connection seals, and proper waterproofing at joints between the frame and the top of the structure are also very important.

Sanitary sewer connections should be permitted only at wye or tee branches or at machine-cut, watertight-jointed taps. The fitting should be supported adequately during and after the pipe joint is made. Bell-and-spigot, compression-type flexible pipe joints should be used at the junction of the house sewer and service tap. Caps and plugs for any dead-end branches or house service connections should be made as watertight as any other pipe joint and be anchored to hold against internal pressure or external force.

11.8. PROJECT ACCEPTANCE

Upon completion of the sanitary sewer installation, the system should be tested to verify that there will not be groundwater infiltration into the system. The following test methods are common ways to verify system integrity:

- Infiltration/exfiltration testing
- Low-pressure air testing
- Mandrel testing
- Televising

Leakage limits may be stated in terms of water leakage quantities and should include both a maximum allowable test section rate and a maximum allowable system average rate. Current information indicates that a maximum allowable infiltration rate of 50 to 100 gal/inch-diameter/mi (5 to 10 L/m-diameter/km) of sewer pipe per day can be achieved without additional construction costs.

Manholes need to be tested separately from the sanitary sewer. A leakage allowance of 0.1 gal/hr/ft-diameter/ft head (4 L/hr/m-diameter/m head) is deemed appropriate.

Further discussion on system testing is presented in Chapter 6, Design of Sanitary Sewer Systems.

11.8.1. Infiltration/Exfiltration Testing

When groundwater is observed to be at least 4 ft (1.2 m) above the top of the sewer pipe, the infiltration test can be used to determine the integrity of the sewer line. Any leakage can be measured with a V-notch weir or similar flow measuring device. If no leakage is observed, it can be assumed that the line passes the test.

If the groundwater level is not at least 4 ft (1.2 m) above the top of the sewer pipe, then an exfiltration test is required. This is performed by plugging the manhole at the lower end of the test section and filling the line with water to at least 4 ft (1.2 m) above either the top of the sewer pipe or the measured groundwater level, whichever is greater. If leakage does not exceed the limits specified, then the section tested is accepted. If leakage exceeds the limits specified, the leak must be located and repaired.

11.8.2. Low-Pressure Air Testing

A low-pressure air test may be used to detect leaks in sewer pipe in lieu of using an exfiltration or infiltration test. Two air test methods used are the constant pressure method and the time pressure method, with the latter the most commonly used. The constant pressure method utilizes an air flow measuring device operated at 3 psi (20 kPa) greater pressure than the average back-pressure of any groundwater. In the time pressure drop method, the air supply is disconnected and the time required for the pressure to drop from 3.5 psi (24 kPa) to 2.5 psi (17 kPa) gauge is determined. Test procedures and calculations are available from ASTM International.

In applying the low-pressure air test, the following factors should be understood and precautions followed during the test. The air test is intended to detect defects in the sewer line and establish the integrity of the line under sewer conditions. Since the pipe will be in a moist environment when in service, testing the pipe in wet conditions is appropriate. Plugs should be securely braced and should not be removed until all air pressure in the test section has been reduced to ambient pressure.

For safety reasons, no one should be allowed in the trench or manhole while the test is being conducted. The testing apparatus should be equipped with a pressure relief device to prevent the possibility of loading the test section with full compressor capacity.

11.8.3. Mandrel Testing

Sanitary sewer constructed with flexible pipe should also be required to pass a mandrel test. The mandrel must pass through the entire section being tested when pulled by hand without the use of excessive force. The mandrel should be a rigidly constructed cylinder or other approved shape which will not change shape or size when subjected to forces exerted on it by the pipe wall.

The dimensions of the testing device should be based on the specified maximum allowable deflection. Mandrel testing should not be permitted until a minimum of 30 days after placement of the backfill. Dimensions should be stated as a mandrel diameter equal to the nominal pipe diameter minus the allowable deflection percentage and a minimum mandrel length of not less than the nominal diameter of the pipe. This test is also commonly referred to as a go/no-go test.

11.8.4. Televising

Sanitary sewer may also be televised as part of the acceptance procedure. If televising is to be used, it is recommended a pan and tilt camera be used. This type of camera can be stopped at the lateral connections and turned to provide a view for a distance up the lateral.

The records should include starting manhole, ending manhole, distance along the line from the starting manhole, location of all laterals, and notation of any areas showing leaks or standing water.

11.9. CONSTRUCTION RECORDS

In addition to the acceptance testing previously discussed, it may be the responsibility of the engineer to record details of construction as accomplished in the field. This data should be incorporated into a final revision of the contract drawings to represent the most reliable record for future use. The contractor should maintain a drawing set during construction that is regularly (at least weekly) marked to record as-constructed changes from the design drawings, and is delivered to the engineer at completion of the work.

Records should be sufficient to allow future recovery of the sewer itself, underground structures, connections, and services. Invert elevations should be recorded for each manhole, structure connection, and house service. In some instances, it may be advisable to set concrete reference markers flush with finished grade to facilitate future recovery. Locations and sizes of all sanitary laterals and crossings of other underground utilities should also be included in the record drawings.

BIBLIOGRAPHY

American Concrete Pipe Association (ACPA). (2001). "Concrete pipe design manual," ACPA, Irving, Tex.

Associated General Contractors of America (AGCA). (2003). "Manual of accident prevention for construction," AGCA, Alexandria, Va.

American Society of Civil Engineers (ASCE). (2003). "How to work effectively with consulting engineers: Getting the best project at the best price," Manual and Rep. No. 45, ASCE, Reston, Va.

American Water Works Association (AWWA). (2005). "Installation of ductile iron water mains and their appurtenances," Standard C-600, AWWA, Denver, Colo.

AWWA. (1995). "Concrete pressure pipe," Manual M9, AWWA, Denver, Colo.

AWWA. (1979). "PVC Pipe—Design and installation," Manual M23, AWWA, Denver, Colo.

ASTM International (ASTM). (Active as of January, 2007). "Standard practice for installing vitrified clay pipe lines," Standard C12-03, ASTM, West Conshohocken, Penn.

International Society of Explosives Engineers (ISEE). (2003). "Blasters handbook," ISEE, Cleveland, Ohio.

National Clay Pipe Institute (NCPI). (1995). "Clay pipe engineering manual," NCPI, Lake Geneva, Wisc.

National Fire Protection Association (NFPA). (2001). "Explosive materials code," NFPA 495, NFPA, Quincy, Mass.

Uni-Bell PVC Pipe Association (Uni-Bell). (1993). "Handbook of PVC pipe—Design and construction," Uni-Bell, Dallas, Tex.

CHAPTER 12

TRENCHLESS DESIGN AND CONSTRUCTION

12.1. INTRODUCTION AND COMPARISON OF TRENCHLESS TECHNOLOGY METHODS

12.1.1. Introduction

12.1.1.1. Summary of Trenchless Technology Benefits

The term trenchless technology (or No-Dig) is relatively new to the utility industry in the United States. Although trenchless methods (albeit not called trenchless) have been used in one form or another for a number of years throughout the world, trenchless methods as a family of methods have only been formally recognized in the United States since June, 1990, when the first technical society dedicated to trenchless technology methods was formed [the North American Society for Trenchless Technology (NASTT)].[1] Trenchless methods encompass a family of new installation methods as well as rehabilitation methods. This chapter will focus primarily on new installation methods. Other ASCE and Water Environment Federation (WEF) manuals address trenchless rehabilitation (also known as renewal) methods [specifically, ASCE Manual of Practice No. 62/WEF MOP FD-6 (WEF/ASCE, 1994)].

Public and private utility agencies throughout the United States and the world have realized the benefits of trenchless technology methods due to their inherent advantages of minimizing project area construction impacts to the public and the environment. Trenchless methods are less intrusive construction methods for installing new pipes and conduits or rehabilitating existing pipes and conduits while reducing:

- Indirect construction costs (known as social costs to third parties).
- Adverse environmental impacts and permitting concerns.
- Problems with handling and disposing of contaminated soils and groundwater .

- Cost of utility conflicts and the resulting relocations.
- Costs for surface restoration, including pavement reconstruction.

12.1.1.2. Trenchless Technology Market Share

The primary uses of trenchless methods are for construction of new water and wastewater pipes and rehabilitation of existing ones. These methods are also used in other utility industries, such as the:

- Natural gas distribution market to rehabilitate existing gas pipes and install new ones.
- Telecommunications market to install fiber-optic lines.
- Electrical distribution market to install high-voltage cables.

For new installations, the trenchless methods are used to cost-effectively install new pipes:

- In urban environments.
- In environmentally sensitive areas, such as wetlands.
- In areas of contaminated soils and groundwater.
- Under existing high-traffic-volume roadways.
- Under waterways, such as rivers.
- When deep pipes are used to eliminate pump stations.

12.1.2. Two Main Divisions of Trenchless Technology Methods

There are two main divisions of trenchless methods: those for rehabilitation of existing underground infrastructure (including sewer pipes) and those for new installations. ASCE Manual of Practice No. 62/WEF MOP FD-6 identifies and reviews the various trenchless rehabilitation methods, and is discussed herein only to distinguish the difference between trenchless rehabilitation methods and new installation trenchless methods. The second division of trenchless methods is for installing new pipes and conduits. Since ASCE Manual of Practice No. 60/WEF MOP FD-5 addresses the design and construction of new sanitary sewers, these methods are identified and reviewed herein.

12.1.2.1. Trenchless Installation Methods—New Pipes and Conduits

Trenchless installation methods to install new pipes and conduits are listed and summarized in Table 12-1. A number of these methods have manuals of practice or guidelines published by ASCE, the Gas Industry Research Institute, the National Association of Sewer Service Companies (NASSCO), and others. In 2005, WEF published a complete reference guideline book (Najafi, 2005) for pipeline and utility design, construction,

TABLE 12-1. Trenchless Installation Methods—New Pipes and Conduits

Method	Purpose
Microtunneling	To install new pipes at precise line and grade or below groundwater
Pipe Ramming	To install new pipes under railroads and highways where precise line and grade is not an issue, and to install new pipes for relatively short distances
Auger Boring	To install new pipes under railroads and highways where precise line and grade is not an issue and for relatively short distances when groundwater is not present
Pipe Jacking	To install new large-diameter pipes [>36 inches (900 mm)] and when groundwater is not present (control of line and grade is difficult but not impossible)
Pipe Bursting	To replace and possibly upsize an existing pipe to increase carrying capacity
Horizontal Directional Drilling	To install new pressure pipes, including siphons (can be used to install new pipe several thousand feet (meters), under ideal conditions)

and renewal using trenchless technology methods. Pipe bursting can be considered trenchless construction or trenchless renewal, depending on whether the existing pipe being burst is being upsized to increase capacity, or being burst to install equivalent size pipe for rehabilitating the existing pipe. This chapter focuses primarily on trenchless construction methods relative to new water and wastewater piping.

12.1.2.2. Trenchless Rehabilitation Methods—Rehabilitation of Existing Pipes and Conduits

Trenchless methods used to rehabilitate existing pipes and conduits are listed in Table 12-2. Since this chapter is primarily about new trenchless installation methods, Table 12-2 provides only a brief description of trenchless rehabilitation methods. For detailed information on these methods, see ASCE Manual of Practice No. 62/WEF MOP FD-6 (WEF/ASCE, 1994).

ASTM International has compiled standards relative to pipe materials, rehabilitation materials, and rehabilitation methods relative to trenchless technology methods. The compiled standards were published by ASTM in 1999 and updated in 2006 (ASTM, 2006).

TABLE 12-2. Trenchless Rehabilitation Methods—To Rehabilitate Existing Pipes and Conduits

Method	Purpose
Cured-in-Place Pipe Lining	To rehabilitate existing deteriorated pipe (including noncircular, asymmetrical pipes with limitations and under specific conditions) for improving structural integrity or increasing capacity (flows generally need to be bypassed)
Plastic Pipe Deform and Reform (Fold & Form)	To rehabilitate existing deteriorated circular pipe [generally <18 inches (450 mm)] for improving structural integrity (flows generally need to be bypassed)
Sliplining—Segmental Method	To rehabilitate existing large-diameter, circular pipe [generally >24 inches (600 mm)] when flow diversion is an issue (generally, loss of carrying capacity occurs)
Sliplining—Continuous Method	To rehabilitate existing circular pipe with continuous pipe string, such as fused plastic pipe when flow diversion is possible (generally, loss of carrying capacity occurs)
Pipe Bursting	To rehabilitate existing relatively small-diameter [<36 inches (900 mm)] pipe for improving structural integrity, where increased capacity is not an issue
Cementitous Lining	To cement a line of existing water mains and improve structural integrity of large-diameter sewer pipes [>48 inches (1,220 mm)] when carrying capacity is not an issue (bypass pumping of flows is generally required)
Manhole Rehabilitation	To rehabilitate existing manholes to improve structural integrity and to reduce infiltration and inflow
Internal Spot Repairs	To repair specific sections of an existing pipe from within the pipe to repair localized structural defects or reduce infiltration and inflow

12.2. COSTS OF UTILITY CONSTRUCTION USING TRENCHLESS INSTALLATION METHODS

Construction costs associated with installing utilities using trenchless methods will vary depending on a number of factors. Generally, *direct* construction costs are higher for trenchless methods than nontrenchless methods (conventional open-trench methods) when the new utility is installed at relatively shallow depths. However, the cost for installing new utilities by trenchless methods at deeper depths remains relatively constant, whereas the costs for installing new utilities by nontrenchless methods (open-trench excavation) increase with depth. In addition, when a number of other factors (e.g., environmentally sensitive areas, high traffic volumes, contaminated soil and groundwater, and relocation of other utilities is required), the construction costs associated with trenchless methods may be more cost-effective than nontrenchless methods (conventional open-trench methods) at shallower depths. See Chapter 1 of this Manual for cost estimating and planning.

There are also a number *indirect* construction costs imposed on the public in general; they are known as social costs and are discussed in more detail in Section 12.2.2.2. Reduced social costs are a major benefit to trenchless methods but also apply to conventional open-cut sanitary sewer design and construction. Social costs or indirect construction costs are associated with all construction activities on public lands and right-of-ways. These costs are other than the costs for labor, material, and equipment associated with conventional construction methods.

12.2.1. Preconstruction Design Costs

Preconstruction design costs are the costs for planning, field investigations, report preparation, final design cost, bidding costs, and administrative costs associated with sewer design. In general, the preconstruction design costs for conventional sewer installation are between 5% and 10% of the total construction cost. However, for trenchless installation projects, the preconstruction design costs may be as high as 10% to 15% of the total cost. Although there may be higher preconstruction design costs, these may be generally offset by savings in construction costs associated with handling and disposal of contaminated soils, surface and pavement restoration, and environmental mitigation measures. A detailed investigation of site-specific conditions should be conducted prior to developing cost estimates, to identify items that may affect the overall cost. Engineers experienced in trenchless installation methods may prepare generalized feasibility-level cost estimates prior to the detailed field investigation work in order to determine the order of magnitude of the various alternatives for the project funding authorities to consider.

12.2.2. Trenchless Installation Construction Costs

12.2.2.1. Direct Construction Costs

The direct construction costs for new trenchless construction methods generally consist of the same components as other types of sewer pipe construction methods, including costs of materials, general labor, and equipment. However, due to the specialized nature of trenchless construction methods, the costs for mobilization, specialized labor, and some materials may be increased. There are also costs associated with pavement reconstruction, traffic control, and environmental controls, albeit not as high as for traditional open-trench construction methods. In addition, due to the cost of excavation and earth support requirements for conventional open-trench excavation, the deeper a proposed pipeline is, the more cost-effective it becomes to install it by trenchless methods.

12.2.2.2. Indirect Construction (Social) Costs

Indirect construction cost or social costs are costs borne by third parties due to the impacts of conventional construction methods. With conventional construction methods, social costs are generally not taken into consideration. Trenchless construction methods generally reduce social costs (sometimes significantly) as compared to conventional methods. These social costs can also have political as well as economic considerations. Social costs include:

- Vehicular traffic disruption.
- Road and pavement damage.
- Damage to adjacent utilities.
- Damage to adjacent structures.
- Noise, vibration, and air pollution.
- Pedestrian safety impacts.
- Business and trade loss.
- Damage to detour roads.
- Site safety.
- Citizen complaints.
- Environmental impacts.

12.3. DESIGN CONSIDERATIONS FOR TRENCHLESS PIPELINE CONSTRUCTION METHODS

Chapter 2 of this Manual addresses surveys and investigations associated with the installation of new sewers. The methods addressed in Chapter 2 are generally for conventional methods but are also required for trenchless methods. Trenchless methods require *extensive* subsurface investigations to minimize risk associated with underground construction. This section addresses the additional investigation requirements for installing new utilities by trenchless methods.

12.3.1. Surface Survey

Chapter 2 of this Manual identifies many of the required surface survey requirements that are also required in trenchless construction methods. See Chapter 2 for a list of the surface survey requirements.

12.3.2. Subsurface Investigations

12.3.2.1. Existing Utilities

Location of underground existing utilities is critical for trenchless installation methods in order to avoid conflict between the equipment used for the proposed trenchless installation and for installing working excavations while minimizing required utility relocations. Appropriate true scales showing actual widths of underground utilities should be considered. For example, at a 1 inch = 40 ft scale, all pipes larger than 24 inches (610 mm) should be shown with two lines representing the true width of the utility.

12.3.2.2. Geotechnical Investigations

12.3.2.2.1. General Geotechnical Review

During the planning phase of a project proposing to use trenchless methods, a review should be performed to determine the appropriate trenchless method. Available information regarding geological subsurface conditions should be reviewed to determine whether the proposed trenchless installation method is appropriate for the geological conditions. In addition, areas of contaminated soils should be identified so that appropriate handling and mitigation methods can be developed during the final design phase. Useful existing information includes current and historical aerial photographs, geology maps, previous project investigations in the vicinity, and land use records.

12.3.2.2.2. Geotechnical Survey

Once the review of the subsurface conditions is completed using available information and the appropriate trenchless installation method is selected, subsurface borings should be taken at appropriate depths and intervals, using judgment and experience to confirm or revise the preliminary model developed from existing information. Laboratory and field tests will also be needed to understand soil or rock properties and infer their behavior. In addition, any contaminated soils and/or groundwater should be tested to determine handling and disposal requirements, as well as to identify which pipe material and other installation materials are appropriate.

12.3.3. Alignment Considerations

The proposed horizontal and vertical alignments of sewers that will be installed by trenchless methods must take into consideration the location of and potential conflicts with existing utilities, changes in subsurface conditions, potential obstructions (such as boulders, fill debris, structures, old piles or piers), areas of contaminated soils and groundwater, surface traffic, entrances and exits to parking areas and loading docks, and environmentally sensitive areas such as wetlands.

12.4. PIPE MATERIALS

This section should be reviewed in conjunction with Chapter 8, Materials for Sewer Construction. Chapter 8 reviews various pipe materials such as concrete, vitrified clay pipe (VCP), ductile iron pipe (DIP), steel pipe, and plastic pipe. However, these pipe materials are intended for conventional pipe associated with the conventional installation of new sewers. Although some of the pipe materials identified in Chapter 8 can be used in trenchless installation methods, Chapter 8 does not discuss details specific to pipe materials used in trenchless designs today. For example, pipe materials such as centrifugally cast reinforced fiber-glass pipe and polymer concrete pipe are much more common in trenchless construction than in conventional construction. In addition, "conventional" materials, such as DIP, are often modified for use in pipe jacking, pipe bursting, and horizontal directional drilling (HDD). It should be noted that trenchless installation methods require special pipe joints capable of handling installation stresses. These modifications may include additional reinforcing or changes in the jointing techniques to improve their performance in trenchless installations. See the Bibliography at the end of this chapter for additional information on trenchless pipe materials. Engineers and owners should consult special manuals, such as ASCE microtunneling construction guidelines and HDD good practices guidelines, as described in the Bibliography. Chapter 8 of this Manual should be consulted when considering pipe material and pipe hydraulics with trenchless installation of sewers.

12.5. HORIZONTAL AUGER BORING

12.5.1. Introduction and Method Description

Horizontal auger boring (HAB), also known as jack and bore, is used to install steel casing pipes under railroads and highways where precise line and grade is not an issue. It is generally used to install small-diameter [less than 60 inches (1,524 mm)] casing pipes for relatively short distances [less than 600 ft (200 m)] when groundwater is not present. However, HAB has been used to install casing pipes as large as 66 inches (1,676 mm) in diameter, and bores as long as 800 ft (250 m) have been completed.

Chapter 11 of this Manual describes HAB and pipe jacking as a special construction method which may be used for installing casing and pipelines under railroads, busy highways, and environmentally sensitive areas. These areas requiring special installation methods fall under the umbrella of trenchless installation.

Refer to ASCE Manual and Report on Engineering Practice No. 106, Horizontal Auger Boring Projects, for additional information on this method. Figure 12-1 is a schematic of an HAB operation.

12.6. PIPE RAMMING

12.6.1. Introduction and Method Description

Pipe ramming is a nonsteerable, trenchless construction method for installing steel casing, which uses pneumatic force to drive the casing through the ground. Pipe ramming is a technique for installing casing pipe under rail and road crossings where open excavation methods are not feasible. It is generally limited to pipe diameters up to 60 inches (1,500 mm) and lengths less than 200 linear ft (60 m), although a casing as large as 144 inches (3,600 mm) in diameter has been successfully installed using pipe ramming. Although the process is not depth-limited, the majority of work is performed in shallow ground conditions less than 20 ft (6 m) deep. The process is generally not suitable for use in rock or soils with large boulders or where groundwater is present. However, pipe ramming can be successful in areas that have boulders and rock if a pneumatic hammer is attached to leading edge of the pipe-ramming device to break

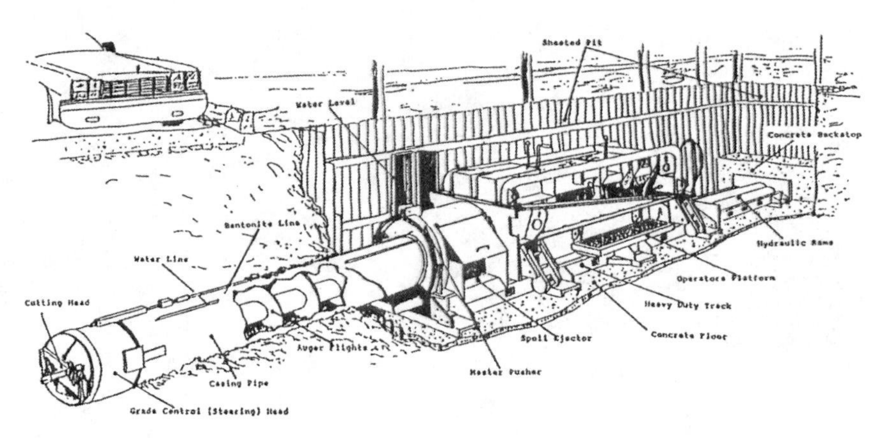

FIGURE 12-1. Schematic of a horizontal auger boring operation.
Courtesy of National Utility Contractors Association, Arlington, Va.

up such obstacles. In this case, use of pipe jacks in a jacking pit should be used to push the casing forward as the hammer excavates the rock.

This process requires entrance and receiving pits unless the pipe grades are above the surrounding ground (e.g., a railroad embankment crossing), and is advantageous where setup space is limited. The entrance and receiving pits, when used, must be relatively long to accommodate the pipe and hammer. Spoil material, captured as the pipe or casing is advanced, is removed using screw augers, a bucket with chain, or manual methods. Because the alignment is controlled by the drive equipment guide rail system and that method is not steerable, installation accuracy can vary widely. Steel casing is used for pipe ramming, and welded or proprietary interlocking systems are used to join the steel pipe during the installation process. Figure 12-2 illustrates a pipe ramming operation.

12.7. PIPE JACKING

12.7.1. Introduction and Method Description

Pipe jacking is a method for installing casing and/or carrier pipe by means of jacking and advancing the pipe through the ground while earth at the leading edge of the lead pipe (the face) is excavated by personnel or

FIGURE 12-2. Typical pipe ramming operation.
Courtesy of TT Technologies, Aurora, Ill., available at <www.tttechnologies.com/products/gram/index.html>.

personnel-controlled excavating tools at the face. Pipe jacking requires man-entry-sized pipe to allow personnel access to the face, and sufficient space for removal of the excavated material; this is typically accomplished with track-mounted muck skips and winches. Figure 12-3 illustrates a pipe jacking operation.

FIGURE 12-3. A pipe jacking operation.
Courtesy of Akkerman, Brownsdale, Minn., available at <www.akkerman.com/nav.php?link=products>.

12.7.2. Main Features and Application Range

12.7.2.1. Pipe Diameter Range

The pipe diameters used in pipe jacking generally range from 48 inches to 144 inches (1,200 mm to 3,600 mm). With pipe diameters larger than 144 inches (3,600 mm), the method is considered to be utility tunneling. Pipe jacking is different from tunneling in that tunneling uses temporary supports and jacks are not required in shaft. If a pipe is being pushed through the soil formation from a common starting point (a jacking pit) and excavation at the face and removal of the spoil is not remotely controlled, it is pipe jacking, regardless of pipe diameter.

12.7.2.2. Drive Length

Maximum drive lengths are influenced by ground conditions, available jacking thrust, pipe type, and pipe joint configuration, as well as available shaft locations and required manhole spacing. Intermediate jacking stations (IJSs) and lubricants may be used to increase drive lengths. Single-span drives of more than 3,000 ft (900 m) have been achieved using these methods, but drives of less than 700 ft (210 m) are more typical.

12.7.2.3. Type of Casing

Casing pipes can consist of steel pipe, Portland cement concrete pipe, polymer mortar concrete pipe, or centrifugally cast fiberglass reinforced polymer mortar pipe; however, steel casing is typical. These pipe materials can also be used as carrier pipes and can be directly jacked into place if the medium being transported in the carrier pipe is compatible with the pipe material.

12.7.2.4. Required Working Space

Jacking operations will generally occur from one confined location. Jacking shaft sizes are dependent on the length and diameter of the individual pipe segments and the size of the pipe jacking system. In addition, support equipment on the surface occupies additional space within the work zone. The surface support equipment generally consists of pipe laydown areas, hydraulic power packs for the pipe jacking system, a crane for lowering pipe into the jacking shaft and removal of excavated earth, and a spoils handling area.

12.7.2.5. Productivity

Productivity rates will depend on availability of skilled labor and the ground conditions. With mechanized earth excavating equipment at the face and efficient spoils removal and disposal, productivity rates of up to

150 linear ft (45 m) per work shift can be achieved. Even in more difficult conditions, productivity rates of 45 to 75 linear ft per day (15 to 25 m per day) are common. However, with diameters less than about 7 ft (2 m), productivity can be significantly limited (especially on relatively long drives) because of the small-capacity mucking carts that must be used. Shift production rates as low as 20 linear ft (6 m) per day are not unusual for pipe diameters less than 60 inches (1,500 mm) and drive lengths longer than 700 linear ft (210 m).

12.7.2.6. Major Advantages

The major advantage of installing pipe by pipe jacking is the ability to install sewer pipe under active railroad lines or very busy roadways with limited impacts to the railroad lines or roadways.

12.7.2.7. Major Limitations

Because of the open-face nature of the earth excavating method, groundwater control can be a significant issue. Therefore, pipe jacking generally is appropriate only above groundwater levels. In addition, achieving precise line and grade may be difficult but can be accomplished with skilled contractors experienced in the site-specific ground conditions.

12.8. HORIZONTAL DIRECTIONAL DRILLING

12.8.1. Introduction and Background

Horizontal directional drilling (HDD) is a trenchless method originally developed in the 1970s for the oil and gas industry. HDD was further developed in the 1980s as a result of research and development by the natural gas and electric power industries. HDD is generally used for installing cables and pressure pipe, such as water mains and force mains, under roadways, railroads, waterways, and environmentally sensitive areas. In recent years, HDD has been used on a limited basis for installing gravity sewers if the grade is steep enough to overcome this method's inherent lack of precise line and grade control.

12.8.2. Method Description

The HDD process uses a hydraulic drilling machine and a series of threaded drill pipes with a steerable head to establish a small pilot borehole along the desired pipeline alignment. Aboveground tracking systems and monitoring transmitting devices in the steering head can be used to control the alignment. The pilot borehole is then enlarged to accommodate the host pipe by back-reaming with increasingly larger reaming tools. Finally, the host pipe is pulled back through the finished borehole using the HDD machine. Because of its installation length capabilities, HDD has become a common method for river crossings. Pipe diameters up to 60 inches (1,500 mm) and installed lengths of more than a mile (1.5 km) have been successfully completed. HDD can be used in a wide variety of soil and rock conditions.

Small HDD equipment requires very little space for setup and the process generally requires only small entry or exit pits, but layout space for the total length of host pipe to be pulled back can be a constraint. High-density polyethylene (HDPE) and steel pipe are the most common pipe materials used. Fusible and segmented polyvinyl chloride (PVC) pipe and restrained-joint DIP have also been used in some applications. In addition, fiberglass pipe has also been used recently. Drilling fluid is pumped into the borehole to stabilize the bore, cool the drilling tools and transmitter, transport cuttings, and reduce friction forces during pullback. Grout may be injected after completion of the bore to attempt to fill the annular space between the host pipe and borehole. The potential for "frac-outs" (uncontrolled leakage of the drilling muds) should be addressed as a part of a comprehensive surface spill and frac-out contingency plan. Figure 12-4 presents an HDD rig.

FIGURE 12-4. A horizontal directional drilling rig.
Courtesy of Vermeer Manufacturing, Pella, Iowa, available at <www.vermeer.com/vcom/trenchlessequipment/trenchless-equipment.htm>.

12.9. MICROTUNNELING

12.9.1. Introduction

ASCE published its Standard Construction Guidelines for Microtunneling (CI/ASCE 36-01) in 2001. This manual defines the method, pipe materials, and other pertinent information necessary for installing pipe by the microtunneling method. This microtunneling guideline should be consulted when planning, designing, and constructing sewer pipe by microtunnelling. Figure 12-5 illustrates a microtunnel boring machine (MTBM).

12.9.2. Method Description

Microtunneling is defined as a trenchless method for installing pipelines. It uses a remote-controlled, guided self-excavating boring machine that provides continuous support at the face while the casing or carrier pipe is jacked in behind the boring machine. Figure 12-6 presents a schematic of a microtunneling operation. The difference between microtunneling and the pipe jacking method is that microtunneling is remote-controlled excavation at the face of the machine and remote-controlled removal of the excavated material from the face to the surface, whereas pipe jacking is controlled from the jacking pit, with no directional control of the actual cutting head.

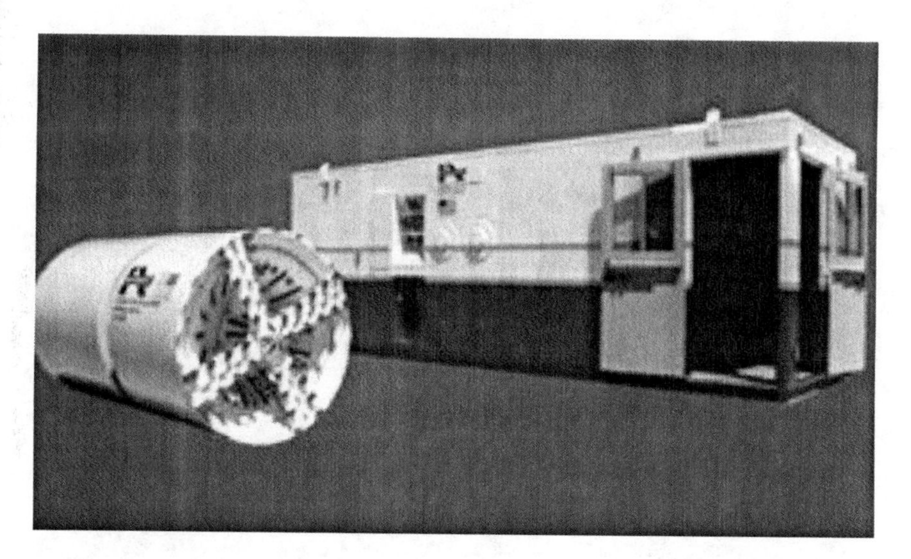

FIGURE 12-5. A microtunnel boring machine.
Courtesy of Akkerman, Brownsdale, Minn., available at <www.akkerman.com/nav.php?link=products>.

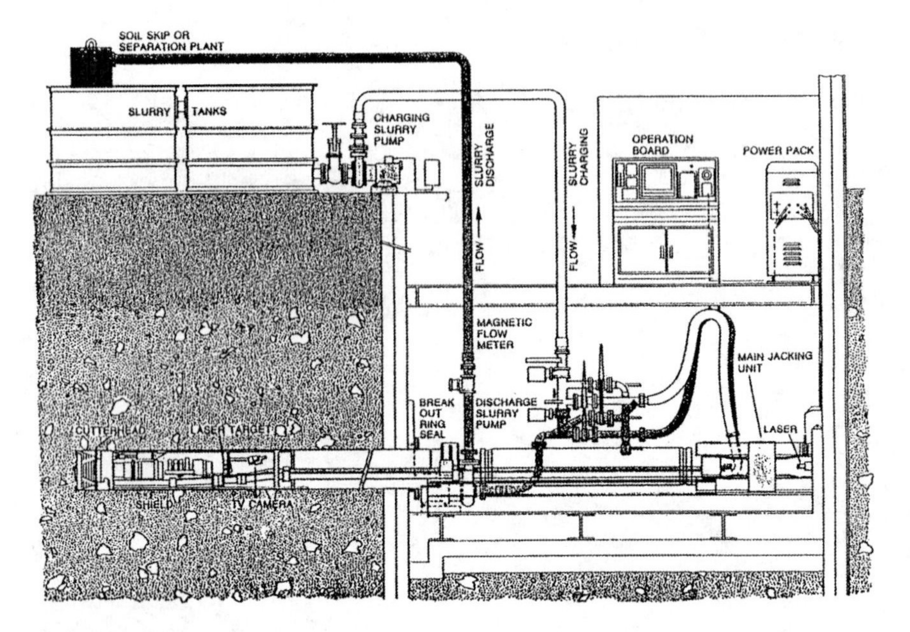

FIGURE 12-6. Schematic of a microtunneling operation.
Courtesy of National Utility Contractors Association, Arlington, Va.

12.9.3. Main Features and Application Range

12.9.3.1. Diameter Range

Pipes installed by microtunneling range in size from 10 inches to 144 inches (300 mm to 3,600 mm) in diameter. It should be noted that the European definition of microtunneling is generally 36 inches (914 mm) or less in diameter, or non-man entry.

12.9.3.2. Drive Length

The drive lengths are dependent on the ground conditions, available jacking thrust, pipe type, pipe joint configuration, and the excavated material removal system. The use of IJSs and/or lubricants to reduce pipe friction will increase pipe drive lengths.

12.9.3.3. Required Working Space

Microtunneling operations will generally occur from one confined location. Jacking shaft sizes are dependent on the length of the individual pipe segments and the size of the pipe jacking system. In addition, support equipment on the surface occupies additional space within the work zone. The surface support equipment generally consists of pipe lay-down areas, hydraulic power packs for the pipe jacking system, a crane for lowering pipe into the jacking shaft, a spoils handling area, and a microtunneling machine control unit. An additional shaft at the opposite end of the microtunnel drive is required for retrieval and removal of the boring machine when microtunnel operations are completed. The work zone for this area needs only to be of sufficient size for a crane and shaft to remove the machine.

12.9.3.4. *Major Advantage*

Microtunneling can be accomplished to accurate line and grade, and below groundwater.

12.9.3.5. *Major Limitations*

Microtunneling, because of its sophistication, is relatively expensive compared to other trenchless installation methods. In addition, mixed ground conditions can be problematic and can increase risk of failure during the installation of the pipe.

12.10. PILOT-TUBE MICROTUNNELING

Pilot-tube microtunneling (PTMT) is an alternative to conventional microtunneling. PTMT combines the accuracy of microtunneling, the steering mechanism of a directional drill, and the spoil removal system of an auger boring machine. PTMT employs an auger and a guidance system, using a camera-mounted theodolite and target with electric light-emitting diodes (LEDs) to secure high accuracy in line and grade. Typically, accuracy will be on the order of ±1 inch (±2.5 cm). When conditions are favorable, PTMT can be a cost-effective tool for the installation of small-diameter pipes of sewer lines or water lines. This technique can also be used for house connections, direct from the main-line sewers. Typically, pilot-tube machines can be used in soft soils and at relatively shallow depths. PTMT is typically used for 6- to 10-inch (150- to 250-mm) pipe, typical of small-diameter gravity sewers. Jacking distances are typically limited to 300 ft (90 m) or less, although this distance has been increasing. Pipe for sanitary sewer microtunneling is typically fiberglass, vitrified clay, or ductile iron. Figure 12-7 illustrates the three phases of PTMT.

12.11. PIPE BURSTING METHOD

Pipe bursting is a method whereby an existing pipe is replaced in-place by means of inserting a tool of slightly larger diameter than the inside diameter of the existing pipe and forcing the tool through the pipe, thus fracturing the existing pipe and pushing the fractured pieces out into the

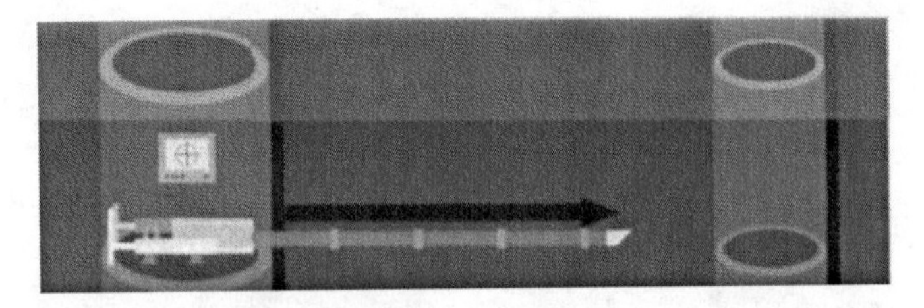

In the first phase, a steerable pilot tube is jacked into the ground, displacing the soil.

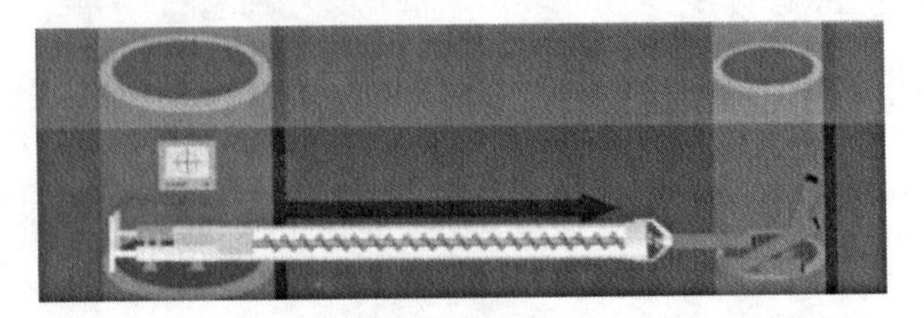

The second phase installs a temporary steel casing.

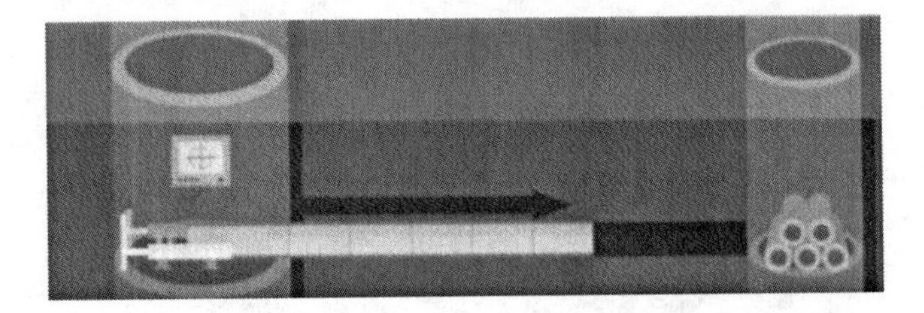

In the third phase, the product pipe is installed.

FIGURE 12-7. The three phases of pilot-tube microtunneling. Courtesy of Wirth Soltau, Mount Pleasant, S.C., available at <www.microtunneling.com/soltau/rvs80.htm>.

surrounding soil. As the tool is advanced through the existing pipe, a replacement pipe is pulled or pushed behind the tool, filling the void left by the former existing pipe with a new pipe. As of this writing, ASCE has a forthcoming Manual of Practice on pipe bursting. This new manual should be referred to when planning, designing, and installing replacement pipe by the pipe bursting method (ASCE 2007). Generally speaking, when the existing pipe is upsized due to capacity requirements, pipe bursting is considered to be a new installation. However, when pipe bursting is used to replace an existing pipe due to rehabilitation requirements, it is considered rehabilitation. Figure 12-8 presents some pipe bursting equipment.

FIGURE 12-8. Pipe bursting equipment.
Courtesy of TT Technologies, Aurora, Ill., available at <www.tttechnologies.com/products/pipeburst/index.html>.

NOTE

1. North American Society for Trenchless Technology (NASTT). Information available at <www.nastt.org>, accessed March 20, 2007.

BIBLIOGRAPHY

American Society of Civil Engineers (ASCE). (2007). "Pipe bursting projects," ASCE Manuals and Reports on Engineering Practice No. 112, ASCE, Reston, Va.

ASCE. (2004). "Horizontal auger boring projects," ASCE Manuals and Reports on Engineering Practice No. 106, ASCE, Reston, Va.

ASCE. (2001). "Standard construction guidelines for microtunneling," CI/ASCE Standard 36-01, ASCE, Reston, Va.

ASTM, International. (2006). "ASTM Standards Related to Trenchless Technology," ASTM International, West Conshohocken, PA.

Kramer, S. R., W. J. McDonald, and J. C. Thomson. (1992). "An introduction to trenchless technology," Van Nostrand Reinhold, New York, N.Y.

Najafi, M. (2005). "Trenchless technology: Pipeline and utility design, construction and renewal," McGraw-Hill, New York, N.Y.

National Association of Sewer Service Companies (NASSCO). (1996). "Manual of practices—Wastewater collection systems," second ed., NAASCO, Maitland, Fla.
North American Society for Trenchless Technology (NASTT). (2005). "Pipe bursting guidelines," Tech. Rep. No. 2001.02, Trenchless Technology Center, Louisiana Technological University, Ruston, La.
NASTT. (2004). "Guidelines for directional drilling projects: HDD good practices guidelines," North American Society for Trenchless Technology, Ruston, La.
Trenchless Technology Center. (2001). "Guidelines for pipe ramming," Tech. Rep. No. 2001.04, Trenchless Technology Center, Louisiana Technological University, Ruston, La. Available at www.ttc.latech.edu/publications/guidelines_pb_im_pr/ramming.pdf, accessed March 20, 2007.
Joint Task Force of the Water Environment Federation and the American Society of Civil Engineers (WEF/ASCE). (1994). "Existing sewer evaluation and rehabilitation," second ed., ASCE Manuals and Reports on Engineering Practice No. 62, ASCE, Reston, Va.

INDEX

Pages followed by *f* indicate figures.
Pages followed by *t* indicate tables.

O

P

R

S

T

U